Graduate Texts in Mathematics **190**

Graduate Texts in Mathematics

1 TAKEUTI/ZARING. Introduction to Axiomatic Set Theory. 2nd ed.
2 OXTOBY. Measure and Category. 2nd ed.
3 SCHAEFER. Topological Vector Spaces. 2nd ed.
4 HILTON/STAMMBACH. A Course in Homological Algebra. 2nd ed.
5 MAC LANE. Categories for the Working Mathematician. 2nd ed.
6 HUGHES/PIPER. Projective Planes.
7 J.-P. SERRE. A Course in Arithmetic.
8 TAKEUTI/ZARING. Axiomatic Set Theory.
9 HUMPHREYS. Introduction to Lie Algebras and Representation Theory.
10 COHEN. A Course in Simple Homotopy Theory.
11 CONWAY. Functions of One Complex Variable I. 2nd ed.
12 BEALS. Advanced Mathematical Analysis.
13 ANDERSON/FULLER. Rings and Categories of Modules. 2nd ed.
14 GOLUBITSKY/GUILLEMIN. Stable Mappings and Their Singularities.
15 BERBERIAN. Lectures in Functional Analysis and Operator Theory.
16 WINTER. The Structure of Fields.
17 ROSENBLATT. Random Processes. 2nd ed.
18 HALMOS. Measure Theory.
19 HALMOS. A Hilbert Space Problem Book. 2nd ed.
20 HUSEMOLLER. Fibre Bundles. 3rd ed.
21 HUMPHREYS. Linear Algebraic Groups.
22 BARNES/MACK. An Algebraic Introduction to Mathematical Logic.
23 GREUB. Linear Algebra. 4th ed.
24 HOLMES. Geometric Functional Analysis and Its Applications.
25 HEWITT/STROMBERG. Real and Abstract Analysis.
26 MANES. Algebraic Theories.
27 KELLEY. General Topology.
28 ZARISKI/SAMUEL. Commutative Algebra. Vol.I.
29 ZARISKI/SAMUEL. Commutative Algebra. Vol.II.
30 JACOBSON. Lectures in Abstract Algebra I. Basic Concepts.
31 JACOBSON. Lectures in Abstract Algebra II. Linear Algebra.
32 JACOBSON. Lectures in Abstract Algebra III. Theory of Fields and Galois Theory.
33 HIRSCH. Differential Topology.
34 SPITZER. Principles of Random Walk. 2nd ed.
35 ALEXANDER/WERMER. Several Complex Variables and Banach Algebras. 3rd ed.
36 KELLEY/NAMIOKA et al. Linear Topological Spaces.
37 MONK. Mathematical Logic.
38 GRAUERT/FRITZSCHE. Several Complex Variables.
39 ARVESON. An Invitation to C^*-Algebras.
40 KEMENY/SNELL/KNAPP. Denumerable Markov Chains. 2nd ed.
41 APOSTOL. Modular Functions and Dirichlet Series in Number Theory. 2nd ed.
42 J.-P. SERRE. Linear Representations of Finite Groups.
43 GILLMAN/JERISON. Rings of Continuous Functions.
44 KENDIG. Elementary Algebraic Geometry.
45 LOÈVE. Probability Theory I. 4th ed.
46 LOÈVE. Probability Theory II. 4th ed.
47 MOISE. Geometric Topology in Dimensions 2 and 3.
48 SACHS/WU. General Relativity for Mathematicians.
49 GRUENBERG/WEIR. Linear Geometry. 2nd ed.
50 EDWARDS. Fermat's Last Theorem.
51 KLINGENBERG. A Course in Differential Geometry.
52 HARTSHORNE. Algebraic Geometry.
53 MANIN. A Course in Mathematical Logic.
54 GRAVER/WATKINS. Combinatorics with Emphasis on the Theory of Graphs.
55 BROWN/PEARCY. Introduction to Operator Theory I: Elements of Functional Analysis.
56 MASSEY. Algebraic Topology: An Introduction.
57 CROWELL/FOX. Introduction to Knot Theory.
58 KOBLITZ. p-adic Numbers, p-adic Analysis, and Zeta-Functions. 2nd ed.
59 LANG. Cyclotomic Fields.
60 ARNOLD. Mathematical Methods in Classical Mechanics. 2nd ed.
61 WHITEHEAD. Elements of Homotopy Theory.
62 KARGAPOLOV/MERLZJAKOV. Fundamentals of the Theory of Groups.
63 BOLLOBAS. Graph Theory.

(continued after index)

M. Ram Murty
Jody Esmonde

Problems in Algebraic Number Theory

Second Edition

M. Ram Murty
Department of Mathematics and Statistics
Queen's University
Kingston, Ontario K7L 3N6
Canada
murty@mast.queensu.ca

Jody Esmonde
Graduate School of Education
University of California at Berkeley
Berkeley, CA 94720
USA
esmonde@uclink.berkeley.edu

Mathematics Subject Classification (2000): 11Rxx 11-01

Library of Congress Cataloging-in-Publication Data
Esmonde, Jody
Problems in algebraic number theory / Jody Esmonde, M. Ram Murty.—2nd ed.
p. cm. — (Graduate texts in mathematics ; 190)
Includes bibliographical references and index.
ISBN 0-387-22182-4 (alk. paper)
1. Algebraic number theory—Problems, exercises, etc. I. Murty, Maruti Ram. II. Title.
III. Series.
QA247.E76 2004
512.7′4—dc22 2004052213

ISBN 0-387-22182-4 Printed on acid-free paper.

Printed in the United States of America. (MP)

9 8 7 6 5 4 3 2 1 SPIN 10950647

springeronline.com

It is practice first and knowledge afterwards.
Vivekananda

Preface to the Second Edition

Since arts are more easily learnt by examples than precepts, I have thought fit to adjoin the solutions of the following problems.

Isaac Newton, in *Universal Arithmetick*

Learning is a mysterious process. No one can say what the precise rules of learning are. However, it is an agreed upon fact that the study of good examples plays a fundamental role in learning. With respect to mathematics, it is well-known that problem-solving helps one acquire routine skills in how and when to apply a theorem. It also allows one to discover nuances of the theory and leads one to ask further questions that suggest new avenues of research. This principle resonates with the famous aphorism of Lichtenberg, "What you have been obliged to discover by yourself leaves a path in your mind which you can use again when the need arises."

This book grew out of various courses given at Queen's University between 1996 and 2004. In the short span of a semester, it is difficult to cover enough material to give students the confidence that they have mastered some portion of the subject. Consequently, I have found that a problem-solving format is the best way to deal with this challenge. The salient features of the theory are presented in class along with a few examples, and then the students are expected to teach themselves the finer aspects of the theory through worked examples.

This is a revised and expanded version of "Problems in Algebraic Number Theory" originally published by Springer-Verlag as GTM 190. The new edition has an extra chapter on density theorems. It introduces the reader to the magnificent interplay between algebraic methods and analytic methods that has come to be a dominant theme of number theory.

I would like to thank Alina Cojocaru, Wentang Kuo, Yu-Ru Liu, Stephen Miller, Kumar Murty, Yiannis Petridis and Mike Roth for their corrections and comments on the first edition as well as their feedback on the new material.

Kingston, Ontario
March 2004

Ram Murty

Preface to the First Edition

It is said that Ramanujan taught himself mathematics by systematically working through 6000 problems[1] of Carr's *Synopsis of Elementary Results in Pure and Applied Mathematics.* Freeman Dyson in his Disturbing the Universe describes the mathematical days of his youth when he spent his summer months working through hundreds of problems in differential equations. If we look back at our own mathematical development, we can certify that problem solving plays an important role in the training of the research mind. In fact, it would not be an exaggeration to say that the ability to do research is essentially the art of asking the "right" questions. I suppose Pólya summarized this in his famous dictum: if you can't solve a problem, then there is an easier problem you can't solve – find it!

This book is a collection of about 500 problems in algebraic number theory. They are systematically arranged to reveal the evolution of concepts and ideas of the subject. All of the problems are completely solved and no doubt, the solutions may not all be the "optimal" ones. However, we feel that the exposition facilitates independent study. Indeed, any student with the usual background of undergraduate algebra should be able to work through these problems on his/her own. It is our conviction that the knowledge gained by such a journey is more valuable than an abstract "Bourbaki-style" treatment of the subject.

How does one do research? This is a question that is on the mind of every graduate student. It is best answered by quoting Pólya and Szegö: "General rules which could prescribe in detail the most useful discipline of thought are not known to us. Even if such rules could be formulated, they would not be very useful. Rather than knowing the correct rules of thought theoretically, one must have them assimilated into one's flesh and blood ready for instant and instinctive use. Therefore, for the schooling of one's powers of thought only the practice of thinking is really useful. The

[1] Actually, Carr's *Synopsis* is not a problem book. It is a collection of theorems used by students to prepare themselves for the Cambridge Tripos. Ramanujan made it famous by using it as a problem book.

independent solving of challenging problems will aid the reader far more than aphorisms."

Asking how one does mathematical research is like asking how a composer creates a masterpiece. No one really knows. However, it is clear that some preparation, some form of training, is essential for it. Jacques Hadamard, in his book *The Mathematician's Mind*, proposes four stages in the process of creation: preparation, incubation, illumination, and verification. The preparation is the conscious attention and hard work on a problem. This conscious attention sets in motion an unconscious mechanism that searches for a solution. Henri Poincaré compared ideas to atoms that are set in motion by continued thought. The dance of these ideas in the crucible of the mind leads to certain "stable combinations" that give rise to flashes of illumination, which is the third stage. Finally, one must verify the flash of insight, formulate it precisely, and subject it to the standards of mathematical rigor.

This book arose when a student approached me for a reading course on algebraic number theory. I had been thinking of writing a problem book on algebraic number theory and I took the occasion to carry out an experiment. I told the student to round up more students who may be interested and so she recruited eight more. Each student would be responsible for one chapter of the book. I lectured for about an hour a week stating and sketching the solution of each problem. The student was then to fill in the details, add ten more problems and solutions, and then typeset it into TEX. Chapters 1 to 8 arose in this fashion. Chapters 9 and 10 as well as the supplementary problems were added afterward by the instructor.

Some of these problems are easy and straightforward. Some of them are difficult. However, they have been arranged with a didactic purpose. It is hoped that the book is suitable for independent study. From this perspective, the book can be described as a first course in algebraic number theory and can be completed in one semester.

Our approach in this book is based on the principle that questions focus the mind. Indeed, quest and question are cognates. In our quest for truth, for understanding, we seem to have only one method. That is the Socratic method of asking questions and then refining them. Grappling with such problems and questions, the mind is strengthened. It is this exercise of the mind that is the goal of this book, its raison d'être. If even one individual benefits from our endeavor, we will feel amply rewarded.

Kingston, Ontario
August 1998

Ram Murty

Acknowledgments

We would like to thank the students who helped us in the writing of this book: Kayo Shichino, Ian Stewart, Bridget Gilbride, Albert Chau, Sindi Sabourin, Tai Huy Ha, Adam Van Tuyl and Satya Mohit.

We would also like to thank NSERC for financial support of this project as well as Queen's University for providing a congenial atmosphere for this task.

J.E.
M.R.M.
August 1998

Contents

II Solutions

Part I

Problems

Chapter 1

Elementary Number Theory

1.1 Integers

The nineteenth century mathematician Leopold Kronecker wrote that "all results of the profoundest mathematical investigation must ultimately be expressible in the simple form of properties of integers." It is perhaps this feeling that made him say "God made the integers, all the rest is the work of humanity" [B, pp. 466 and 477].

In this section, we will state some properties of integers. Primes, which are integers with exactly two positive divisors, are very important in number theory. Let $\mathbb{Z}$ represent the set of integers.

Theorem 1.1.1 *If a, b are relatively prime, then we can find integers x, y such that $ax + by = 1$.*

Proof. We write $a = bq + r$ by the Euclidean algorithm, and since a, b are relatively prime we know $r \neq 0$ so $0 < r < |b|$. We see that b, r are relatively prime, or their common factor would have to divide a as well. So, $b = rq_1 + r_1$ with $0 < r_1 < |r|$. We can then write $r = r_1 q_2 + r_2$, and continuing in this fashion, we will eventually arrive at $r_k = 1$ for some k. Working backward, we see that $1 = ax + by$ for some $x, y \in \mathbb{Z}$. □

Remark. It is convenient to observe that

$$\begin{aligned}\begin{pmatrix} a \\ b \end{pmatrix} &= \begin{pmatrix} q & 1 \\ 1 & 0 \end{pmatrix}\begin{pmatrix} b \\ r \end{pmatrix} = \begin{pmatrix} q & 1 \\ 1 & 0 \end{pmatrix}\begin{pmatrix} q_1 & 1 \\ 1 & 0 \end{pmatrix}\begin{pmatrix} r \\ r_1 \end{pmatrix} \\ &= \begin{pmatrix} q & 1 \\ 1 & 0 \end{pmatrix}\begin{pmatrix} q_1 & 1 \\ 1 & 0 \end{pmatrix}\cdots\begin{pmatrix} q_k & 1 \\ 1 & 0 \end{pmatrix}\begin{pmatrix} r_{k-1} \\ r_k \end{pmatrix} \\ &= A\begin{pmatrix} r_{k-1} \\ r_k \end{pmatrix},\end{aligned}$$

(say). Notice $\det A = \pm 1$ and A^{-1} has integer entries whose bottom row gives $x, y \in \mathbb{Z}$ such that $ax + by = 1$.

Theorem 1.1.2 *Every positive integer greater than* 1 *has a prime divisor.*

Proof. Suppose that there is a positive integer having no prime divisors. Since the set of positive integers with no prime divisors is nonempty, there is a least positive integer n with no prime divisors. Since n divides itself, n is not prime. Hence we can write $n = ab$ with $1 < a < n$ and $1 < b < n$. Since $a < n$, a must have a prime divisor. But any divisor of a is also a divisor of n, so n must have a prime divisor. This is a contradiction. Therefore every positive integer has at least one prime divisor. □

Theorem 1.1.3 *There are infinitely many primes.*

Proof. Suppose that there are only finitely many primes, that is, suppose that there is no prime greater than n where n is an integer. Consider the integer $a = n! + 1$ where $n \geq 1$. By Theorem 1.1.2, a has at least one prime divisor, which we denote by p. If $p \leq n$, then $p \mid n!$ and $p \mid (a - n!) = 1$. This is impossible. Hence $p > n$. Therefore we can see that there is a prime greater than n for every positive integer n. Hence there are infinitely many primes. □

Theorem 1.1.4 *If p is prime and $p \mid ab$, then $p \mid a$ or $p \mid b$.*

Proof. Suppose that p is prime and $p \mid ab$ where a and b are integers. If p does not divide a, then a and p are coprime. Then $\exists x, y \in \mathbb{Z}$ such that $ax + py = 1$. Then we have $abx + pby = b$ and $pby = b - abx$. Hence $p \mid b - abx$. Thus $p \mid b$. Similarly, if p does not divide b, we see that $p \mid a$. □

Theorem 1.1.5 *$\mathbb{Z}$ has unique factorization.*

Proof.

Existence. Suppose that there is an integer greater than 1 which cannot be written as a product of primes. Then there exists a least integer m with such a property. Since m is not a prime, m has a positive divisor d such that $m = de$ where e is an integer and $1 < d < m$, $1 < e < m$. Since m is the least integer which cannot be written as a product of primes, we can write d and e as products of primes such that $d = p_1 p_2 \cdots p_r$ and $e = q_1 q_2 \cdots q_s$. Hence $m = de = p_1 p_2 \cdots p_r \cdot q_1 q_2 \cdots q_s$. This contradicts our assumption about m. Hence all integers can be written as products of primes.

Uniqueness. Suppose that an integer a is written as

$$a = p_1 \cdots p_r = q_1 \cdots q_s,$$

where p_i and q_j are primes for $1 \leq i \leq r$, $1 \leq j \leq s$. Then $p_1 \mid q_1 \cdots q_s$, so there exists q_j such that $p_1 \mid q_j$ for some j. Without loss of generality, we can let q_j be q_1. Since p_1 is a prime, we see that $p_1 = q_1$. Dividing $p_1 \cdots p_r = q_1 \cdots q_s$ by $p_1 = q_1$, we have $p_2 \cdots p_r = q_2 \cdots q_s$. Similarly there exists q_j such that $p_2 \mid q_j$ for some j. Let q_j be q_2. Then $q_2 = p_2$. Hence there exist $q_1, \ldots, q_r$ such that $p_i = q_j$ for $1 \le i \le r$ and $r \le s$. If $r < s$, then we see that $1 = q_{r+1} \cdots q_s$. This is impossible. Hence $r = s$. Therefore the factorization is unique. □

Example 1.1.6 Show that

$$S = 1 + \frac{1}{2} + \cdots + \frac{1}{n}$$

is not an integer for $n > 1$.

Solution. Let $k \in \mathbb{Z}$ be the highest power of 2 less than n, so that $2^k \le n < 2^{k+1}$. Let m be the least common multiple of $1, 2, \ldots, n$ excepting 2^k. Then

$$mS = m + \frac{m}{2} + \cdots + \frac{m}{n}.$$

Each of the numbers on the right-hand side of this equation are integers, except for $m/2^k$. If $m/2^k$ were an integer, then 2^k would have to divide the least common multiple of the number $1, 2, \ldots, 2^k - 1, 2^k + 1, \ldots, n$, which it does not. So mS is not an integer, which implies that S cannot be an integer.

Exercise 1.1.7 Show that

$$1 + \frac{1}{3} + \frac{1}{5} + \cdots + \frac{1}{2n-1}$$

is not an integer for $n > 1$.

We can use the same method to prove the following more general result.

Exercise 1.1.8 Let $a_1, \ldots, a_n$ for $n \ge 2$ be nonzero integers. Suppose there is a prime p and positive integer h such that $p^h \mid a_i$ for some i and p^h does not divide a_j for all $j \ne i$.

Then show that

$$S = \frac{1}{a_1} + \cdots + \frac{1}{a_n}$$

is not an integer.

Exercise 1.1.9 Prove that if n is a composite integer, then n has a prime factor not exceeding $\sqrt{n}$.

Exercise 1.1.10 Show that if the smallest prime factor p of the positive integer n exceeds $\sqrt[3]{n}$, then n/p must be prime or 1.

Exercise 1.1.11 Let p be prime. Show that each of the binomial coefficients $\binom{p}{k}$, $1 \leq k \leq p-1$, is divisible by p.

Exercise 1.1.12 Prove that if p is an odd prime, then $2^{p-1} \equiv 1 \pmod{p}$.

Exercise 1.1.13 Prove Fermat's little Theorem: If $a, p \in \mathbb{Z}$ with p a prime, and $p \nmid a$, prove that $a^{p-1} \equiv 1 \pmod{p}$.

For any integer n we define $\phi(n)$ to be the number of positive integers less than n which are coprime to n. This is known as the Euler ϕ-function.

Theorem 1.1.14 *Given $a, n \in \mathbb{Z}$, $a^{\phi(n)} \equiv 1 \pmod{n}$ when $\gcd(a, n) = 1$. This is a theorem due to Euler.*

Proof. The case where n is prime is clearly a special case of Fermat's little Theorem. The argument is basically the same as that of the alternate solution to Exercise 1.1.13.

Consider the ring $\mathbb{Z}/n\mathbb{Z}$. If a, n are coprime, then $\overline{a}$ is a unit in this ring. The units form a multiplicative group of order $\phi(n)$, and so clearly $\overline{a}^{\phi(n)} = \overline{1}$. Thus, $a^{\phi(n)} \equiv 1 \pmod{n}$. □

Exercise 1.1.15 Show that $n \mid \phi(a^n - 1)$ for any $a > n$.

Exercise 1.1.16 Show that $n \nmid 2^n - 1$ for any natural number $n > 1$.

Exercise 1.1.17 Show that

$$\frac{\phi(n)}{n} = \prod_{p|n} \left(1 - \frac{1}{p}\right)$$

by interpreting the left-hand side as the probability that a random number chosen from $1 \leq a \leq n$ is coprime to n.

Exercise 1.1.18 Show that ϕ is multiplicative (i.e., $\phi(mn) = \phi(m)\phi(n)$ when $\gcd(m, n) = 1$) and $\phi(p^\alpha) = p^{\alpha-1}(p-1)$ for p prime.

Exercise 1.1.19 Find the last two digits of 3^{1000}.

Exercise 1.1.20 Find the last two digits of 2^{1000}.

Let $\pi(x)$ be the number of primes less than or equal to x. The prime number theorem asserts that

$$\pi(x) \sim \frac{x}{\log x}$$

as $x \to \infty$. This was first proved in 1896, independently by J. Hadamard and Ch. de la Vallée Poussin.

We will not prove the prime number theorem here, but derive various estimates for $\pi(x)$ by elementary methods.

Exercise 1.1.21 Let p_k denote the kth prime. Prove that

$$p_{k+1} \leq p_1 p_2 \cdots p_k + 1.$$

Exercise 1.1.22 Show that

$$p_k < 2^{2^k},$$

where p_k denotes the kth prime.

Exercise 1.1.23 Prove that $\pi(x) \geq \log(\log x)$.

Exercise 1.1.24 By observing that any natural number can be written as sr^2 with s squarefree, show that

$$\sqrt{x} \leq 2^{\pi(x)}.$$

Deduce that

$$\pi(x) \geq \frac{\log x}{2 \log 2}.$$

Exercise 1.1.25 Let $\psi(x) = \sum_{p^\alpha \leq x} \log p$ where the summation is over prime powers $p^\alpha \leq x$.

(i) For $0 \leq x \leq 1$, show that $x(1-x) \leq \frac{1}{4}$. Deduce that

$$\int_0^1 x^n (1-x)^n \, dx \leq \frac{1}{4^n}$$

for every natural number n.

(ii) Show that $e^{\psi(2n+1)} \int_0^1 x^n (1-x)^n \, dx$ is a positive integer. Deduce that

$$\psi(2n+1) \geq 2n \log 2.$$

(iii) Prove that $\psi(x) \geq \frac{1}{2} x \log 2$ for $x \geq 6$. Deduce that

$$\pi(x) \geq \frac{x \log 2}{2 \log x}$$

for $x \geq 6$.

Exercise 1.1.26 By observing that

$$\prod_{n < p \leq 2n} p \,\Bigg|\, \binom{2n}{n},$$

show that

$$\pi(x) \leq \frac{9x \log 2}{\log x}$$

for every integer $x \geq 2$.

1.2 Applications of Unique Factorization

We begin this section with a discussion of nontrivial solutions to Diophantine equations of the form $x^l + y^m = z^n$. Nontrivial solutions are those for which $xyz \neq 0$ and $(x,y) = (x,z) = (y,z) = 1$.

Exercise 1.2.1 Suppose that $a, b, c \in \mathbb{Z}$. If $ab = c^2$ and $(a,b) = 1$, then show that $a = d^2$ and $b = e^2$ for some $d, e \in \mathbb{Z}$. More generally, if $ab = c^g$ then $a = d^g$ and $b = e^g$ for some $d, e \in \mathbb{Z}$.

Exercise 1.2.2 Solve the equation $x^2 + y^2 = z^2$ where x, y, and z are integers and $(x,y) = (y,z) = (x,z) = 1$.

Exercise 1.2.3 Show that $x^4+y^4 = z^2$ has no nontrivial solution. Hence deduce, with Fermat, that $x^4 + y^4 = z^4$ has no nontrivial solution.

Exercise 1.2.4 Show that $x^4 - y^4 = z^2$ has no nontrivial solution.

Exercise 1.2.5 Prove that if $f(x) \in \mathbb{Z}[x]$, then $f(x) \equiv 0 \pmod p$ is solvable for infinitely many primes p.

Exercise 1.2.6 Let q be prime. Show that there are infinitely many primes p so that $p \equiv 1 \pmod q$.

We will next discuss integers of the form $F_n = 2^{2^n} + 1$, which are called the Fermat numbers. Fermat made the conjecture that these integers are all primes. Indeed, $F_0 = 3, F_1 = 5, F_2 = 17, F_3 = 257$, and $F_4 = 65537$ are primes but unfortunately, $F_5 = 2^{2^5} + 1$ is divisible by 641, and so F_5 is composite. It is unknown if F_n represents infinitely many primes. It is also unknown if F_n is infinitely often composite.

Exercise 1.2.7 Show that F_n divides $F_m - 2$ if n is less than m, and from this deduce that F_n and F_m are relatively prime if $m \neq n$.

Exercise 1.2.8 Consider the nth Fermat number $F_n = 2^{2^n} + 1$. Prove that every prime divisor of F_n is of the form $2^{n+1}k + 1$.

Exercise 1.2.9 Given a natural number n, let $n = p_1^{\alpha_1} \cdots p_k^{\alpha_k}$ be its unique factorization as a product of prime powers. We define the squarefree part of n, denoted $S(n)$, to be the product of the primes p_i for which $\alpha_i = 1$. Let $f(x) \in \mathbb{Z}[x]$ be nonconstant and monic. Show that $\liminf S(f(n))$ is unbounded as n ranges over the integers.

1.3 The *ABC* Conjecture

Given a natural number n, let $n = p_1^{\alpha_1} \cdots p_k^{\alpha_k}$ be its unique factorization as a product of prime powers. Define the radical of n, denoted $\mathrm{rad}(n)$, to be the product $p_1 \cdots p_k$.

In 1980, Masser and Oesterlé formulated the following conjecture. Suppose we have three mutually coprime integers A, B, C satisfying $A+B=C$. Given any $\varepsilon > 0$, it is conjectured that there is a constant $\kappa(\varepsilon)$ such that

$$\max\bigl(|A|,|B|,|C|\bigr) \le \kappa(\varepsilon)\bigl(\mathrm{rad}(ABC)\bigr)^{1+\varepsilon}.$$

This is called the *ABC* Conjecture.

Exercise 1.3.1 Assuming the *ABC* Conjecture, show that if $xyz \neq 0$ and $x^n + y^n = z^n$ for three mutually coprime integers x, y, and z, then n is bounded.

[The assertion $x^n + y^n = z^n$ for $n \geq 3$ implies $xyz = 0$ is the celebrated Fermat's Last Theorem conjectured in 1637 by the French mathematician Pierre de Fermat (1601–1665). After a succession of attacks beginning with Euler, Dirichlet, Legendre, Lamé, and Kummer, and culminating in the work of Frey, Serre, Ribet, and Wiles, the situation is now resolved, as of 1995. The *ABC* Conjecture is however still open.]

Exercise 1.3.2 Let p be an odd prime. Suppose that $2^n \equiv 1 \pmod{p}$ and $2^n \not\equiv 1 \pmod{p^2}$. Show that $2^d \not\equiv 1 \pmod{p^2}$ where d is the order of $2 \pmod{p}$.

Exercise 1.3.3 Assuming the *ABC* Conjecture, show that there are infinitely many primes p such that $2^{p-1} \not\equiv 1 \pmod{p^2}$.

Exercise 1.3.4 Show that the number of primes $p \leq x$ for which

$$2^{p-1} \not\equiv 1 \pmod{p^2}$$

is $\gg \log x / \log\log x$, assuming the *ABC* Conjecture.

In 1909, Wieferich proved that if p is a prime satisfying

$$2^{p-1} \not\equiv 1 \pmod{p^2},$$

then the equation $x^p + y^p = z^p$ has no nontrivial integral solutions satisfying $p \nmid xyz$. It is still unknown without assuming *ABC* if there are infinitely many primes p such that $2^{p-1} \not\equiv 1 \pmod{p^2}$. (See also Exercise 9.2.15.)

A natural number n is called squarefull (or powerfull) if for every prime $p \mid n$ we have $p^2 \mid n$. In 1976 Erdös [Er] conjectured that we cannot have three consecutive squarefull natural numbers.

Exercise 1.3.5 Show that if the Erdös conjecture above is true, then there are infinitely many primes p such that $2^{p-1} \not\equiv 1 \pmod{p^2}$.

Exercise 1.3.6 Assuming the ABC Conjecture, prove that there are only finitely many n such that $n-1, n, n+1$ are squarefull.

Exercise 1.3.7 Suppose that a and b are odd positive integers satisfying

$$\operatorname{rad}(a^n - 2) = \operatorname{rad}(b^n - 2)$$

for every natural number n. Assuming ABC, prove that $a = b$. (This problem is due to H. Kisilevsky.)

1.4 Supplementary Problems

Exercise 1.4.1 Show that every proper ideal of $\mathbb{Z}$ is of the form $n\mathbb{Z}$ for some integer n.

Exercise 1.4.2 An ideal I is called *prime* if $ab \in I$ implies $a \in I$ or $b \in I$. Prove that every prime ideal of $\mathbb{Z}$ is of the form $p\mathbb{Z}$ for some prime integer p.

Exercise 1.4.3 Prove that if the number of prime Fermat numbers is finite, then the number of primes of the form $2^n + 1$ is finite.

Exercise 1.4.4 If $n > 1$ and $a^n - 1$ is prime, prove that $a = 2$ and n is prime.

Exercise 1.4.5 An integer is called *perfect* if it is the sum of its divisors. Show that if $2^n - 1$ is prime, then $2^{n-1}(2^n - 1)$ is perfect.

Exercise 1.4.6 Prove that if p is an odd prime, any prime divisor of $2^p - 1$ is of the form $2kp + 1$, with k a positive integer.

Exercise 1.4.7 Show that there are no integer solutions to the equation $x^4 - y^4 = 2z^2$.

Exercise 1.4.8 Let p be an odd prime number. Show that the numerator of

$$1 + \frac{1}{2} + \frac{1}{3} + \cdots + \frac{1}{p-1}$$

is divisible by p.

Exercise 1.4.9 Let p be an odd prime number greater than 3. Show that the numerator of

$$1 + \frac{1}{2} + \frac{1}{3} + \cdots + \frac{1}{p-1}$$

is divisible by p^2.

Exercise 1.4.10 (Wilson's Theorem) Show that $n > 1$ is prime if and only if n divides $(n-1)! + 1$.

Exercise 1.4.11 For each $n > 1$, let Q be the product of all numbers $a < n$ which are coprime to n. Show that $Q \equiv \pm 1 \pmod{n}$.

Exercise 1.4.12 In the previous exercise, show that $Q \equiv 1 \pmod{n}$ whenever n is odd and has at least two prime factors.

Exercise 1.4.13 Use Exercises 1.2.7 and 1.2.8 to show that there are infinitely many primes $\equiv 1 \pmod{2^r}$ for any given r.

Exercise 1.4.14 Suppose p is an odd prime such that $2p + 1 = q$ is also prime. Show that the equation

$$x^p + 2y^p + 5z^p = 0$$

has no solutions in integers.

Exercise 1.4.15 If x and y are coprime integers, show that if

$$(x + y) \quad \text{and} \quad \frac{x^p + y^p}{x + y}$$

have a common prime factor, it must be p.

Exercise 1.4.16 (Sophie Germain's Trick) Let p be a prime such that $2p + 1 = q > 3$ is also prime. Show that

$$x^p + y^p + z^p = 0$$

has no integral solutions with $p \nmid xyz$.

Exercise 1.4.17 Assuming ABC, show that there are only finitely many consecutive cubefull numbers.

Exercise 1.4.18 Show that

$$\sum_p \frac{1}{p} = +\infty,$$

where the summation is over prime numbers.

Exercise 1.4.19 (Bertrand's Postulate) (a) If $a_0 \geq a_1 \geq a_2 \geq \cdots$ is a decreasing sequence of real numbers tending to 0, show that

$$\sum_{n=0}^{\infty} (-1)^n a_n \leq a_0 - a_1 + a_2.$$

(b) Let $T(x) = \sum_{n \leq x} \psi(x/n)$, where $\psi(x)$ is defined as in Exercise 1.1.25. Show that

$$T(x) = x \log x - x + O(\log x).$$

(c) Show that

$$T(x) - 2T\left(\frac{x}{2}\right) = \sum_{n \leq x} (-1)^{n-1} \psi\left(\frac{x}{n}\right) = (\log 2)x + O(\log x).$$

Deduce that

$$\psi(x) - \psi\left(\frac{x}{2}\right) \geq \tfrac{1}{3}(\log 2)x + O(\log x).$$

Chapter 2

Euclidean Rings

2.1 Preliminaries

We can discuss the concept of divisibility for any commutative ring R with identity. Indeed, if $a, b \in R$, we will write $a \mid b$ (a divides b) if there exists some $c \in R$ such that $ac = b$. Any divisor of 1 is called a *unit*. We will say that a and b are *associates* and write $a \sim b$ if there exists a unit $u \in R$ such that $a = bu$. It is easy to verify that $\sim$ is an equivalence relation.

Further, if R is an integral domain and we have $a, b \neq 0$ with $a \mid b$ and $b \mid a$, then a and b must be associates, for then $\exists c, d \in R$ such that $ac = b$ and $bd = a$, which implies that $bdc = b$. Since we are in an integral domain, $dc = 1$, and d, c are units.

We will say that $a \in R$ is *irreducible* if for any factorization $a = bc$, one of b or c is a unit.

Example 2.1.1 Let R be an integral domain. Suppose there is a map $n : R \to \mathbb{N}$ such that:

(i) $n(ab) = n(a)n(b)\ \forall a, b \in R$; and

(ii) $n(a) = 1$ if and only if a is a unit.

We call such a map a *norm map*, with $n(a)$ the norm of a. Show that every element of R can be written as a product of irreducible elements.

Solution. Suppose b is an element of R. We proceed by induction on the norm of b. If b is irreducible, then we have nothing to prove, so assume that b is an element of R which is not irreducible. Then we can write $b = ac$ where neither a nor c is a unit. By condition (i),

$$n(b) = n(ac) = n(a)n(c)$$

and since a, c are not units, then by condition (ii), $n(a) < n(b)$ and $n(c) < n(b)$.

If a, c are irreducible, then we are finished. If not, their norms are smaller than the norm of b, and so by induction we can write them as products of irreducibles, thus finding an irreducible decomposition of b.

Exercise 2.1.2 Let D be squarefree. Consider $R = \mathbb{Z}[\sqrt{D}]$. Show that every element of R can be written as a product of irreducible elements.

Exercise 2.1.3 Let $R = \mathbb{Z}[\sqrt{-5}]$. Show that $2, 3, 1+\sqrt{-5}$, and $1-\sqrt{-5}$ are irreducible in R, and that they are not associates.

We now observe that $6 = 2 \cdot 3 = (1+\sqrt{-5})(1-\sqrt{-5})$, so that R does not have unique factorization into irreducibles.

We will say that R, an integral domain, is a *unique factorization domain* if:

(i) every element of R can be written as a product of irreducibles; and

(ii) this factorization is essentially unique in the sense that if $a = \pi_1 \cdots \pi_r$ and $a = \tau_1 \cdots \tau_s$, then $r = s$ and after a suitable permutation, $\pi_i \sim \tau_i$.

Exercise 2.1.4 Let R be a domain satisfying (i) above. Show that (ii) is equivalent to (ii*): if π is irreducible and π divides ab, then $\pi \mid a$ or $\pi \mid b$.

An ideal $I \subseteq R$ is called *principal* if it can be generated by a single element of R. A domain R is then called a principal ideal domain if every ideal of R is principal.

Exercise 2.1.5 Show that if π is an irreducible element of a principal ideal domain, then (π) is a maximal ideal, (where (x) denotes the ideal generated by the element x).

Theorem 2.1.6 *If R is a principal ideal domain, then R is a unique factorization domain.*

Proof. Let S be the set of elements of R that cannot be written as a product of irreducibles. If S is nonempty, take $a_1 \in S$. Then a_1 is not irreducible, so we can write $a_1 = a_2 b_2$ where a_2, b_2 are not units. Then $(a_1) \subsetneqq (a_2)$ and $(a_1) \subsetneqq (b_2)$. If both $a_2, b_2 \notin S$, then we can write a_1 as a product of irreducibles, so we assume that $a_2 \in S$. We can inductively proceed until we arrive at an infinite chain of ideals,

$$(a_1) \subsetneqq (a_2) \subsetneqq (a_3) \subsetneqq \cdots \subsetneqq (a_n) \subsetneqq \cdots.$$

Now consider $I = \bigcup_{i=1}^{\infty} (a_i)$. This is an ideal of R, and because R is a principal ideal domain, $I = (\alpha)$ for some $\alpha \in R$. Since $\alpha \in I$, $\alpha \in (a_n)$ for some n, but then $(a_n) = (a_{n+1})$. From this contradiction, we conclude that the set S must be empty, so we know that if R is a principal ideal domain, every element of R satisfies the first condition for a unique factorization domain.

Next we would like to show that if we have an irreducible element π, and $\pi \mid ab$ for $a, b \in R$, then $\pi \mid a$ or $\pi \mid b$. If $\pi \nmid a$, then the ideal $(a, \pi) = R$, so $\exists x, y$ such that

$$\begin{aligned} ax + \pi y &= 1, \\ \Rightarrow \qquad abx + \pi by &= b. \end{aligned}$$

Since $\pi \mid abx$ and $\pi \mid \pi by$ then $\pi \mid b$, as desired. By Exercise 2.1.4, we have shown that R is a unique factorization domain. □

The following theorem describes an important class of principal ideal domains:

Theorem 2.1.7 *If R is a domain with a map $\phi : R \to \mathbb{N}$, and given $a, b \in R$, $\exists q, r \in R$ such that $a = bq + r$ with $r = 0$ or $\phi(r) < \phi(b)$, we call R a Euclidean domain. If a ring R is Euclidean, it is a principal ideal domain.*

Proof. Given an ideal $I \subseteq R$, take an element a of I such that $\phi(a)$ is minimal among elements of I. Then given $b \in I$, we can find $q, r \in R$ such that $b = qa + r$ where $r = 0$ or $\phi(r) < \phi(a)$. But then $r = b - qa$, and so $r \in I$, and $\phi(a)$ is minimal among the norms of elements of I. So $r = 0$, and given any element b of I, $b = qa$ for some $q \in R$. Therefore a is a generator for I, and R is a principal ideal domain. □

Exercise 2.1.8 If F is a field, prove that $F[x]$, the ring of polynomials in x with coefficients in F, is Euclidean.

The following result, called *Gauss' lemma*, allows us to relate factorization of polynomials in $\mathbb{Z}[x]$ with the factorization in $\mathbb{Q}[x]$. More generally, if R is a unique factorization domain and K is its field of fractions, we will relate factorization of polynomials in $R[x]$ with that in $K[x]$.

Theorem 2.1.9 *If R is a unique factorization domain, and $f(x) \in R[x]$, define the content of f to be the gcd of the coefficients of f, denoted by $\mathcal{C}(f)$. For $f(x), g(x) \in R[x]$, $\mathcal{C}(fg) = \mathcal{C}(f)\mathcal{C}(g)$.*

Proof. Consider two polynomials $f, g \in R[x]$, with $\mathcal{C}(f) = c$ and $\mathcal{C}(g) = d$. Then we can write

$$f(x) = ca_0 + ca_1x + \cdots + ca_nx^n$$

and

$$g(x) = db_0 + db_1x + \cdots + db_mx^m,$$

where $c, d, a_i, b_j \in R$, $a_n, b_m \neq 0$. We define a primitive polynomial to be a polynomial f such that $\mathcal{C}(f) = 1$. Then $f = cf^*$ where $f^* = a_0 + a_1 x + \cdots + a_n x^n$, a primitive polynomial, and $g = dg^*$, with g^* a primitive polynomial. Since $fg = cf^* dg^* = cd(f^* g^*)$, it will suffice to prove that the product of two primitive polynomials is again primitive.

Let

$$f^* g^* = k_0 + k_1 x + \cdots + k_{m+n} x^{m+n},$$

and assume that this polynomial is not primitive. Then all the coefficients k_i are divisible by some $\pi \in R$, with π irreducible. Since f^* and g^* are primitive, we know that there is at least one coefficient in each of f^* and g^* that is not divisible by π. We let a_i and b_j be the first such coefficients in f^* and g^*, respectively.

Now,

$$k_{i+j} = (a_0 b_{i+j} + \cdots + a_{i-1} b_{j+1}) + a_i b_j + (a_{i+1} b_{j-1} + \cdots + a_{i+j} b_0).$$

We know that $k_{i+j}, a_0, a_1, \ldots, a_{i-1}, b_0, b_1, \ldots, b_{j-1}$ are all divisible by π, so $a_i b_j$ must also be divisible by π. Since π is irreducible, then $\pi \mid a_i$ or $\pi \mid b_j$, but we chose these elements specifically because they were not divisible by π. This contradiction proves that our polynomial $f^* g^*$ must be primitive.

Then $fg = cdf^* g^*$ where $f^* g^*$ is a primitive polynomial, thus proving that $\mathcal{C}(fg) = cd = \mathcal{C}(f)\mathcal{C}(g)$. □

Theorem 2.1.10 *If R is a unique factorization domain, then $R[x]$ is a unique factorization domain.*

Proof. Let k be the set of all elements a/b, where $a, b \in R$, and $b \neq 0$, such that $a/b = c/d$ if $ad - bc = 0$. It is easily verified that k is a field; we call k the fraction field of R. Let us examine the polynomial ring $k[x]$. We showed in Exercise 2.1.8 that $k[x]$ is a Euclidean domain, and we showed in Theorem 2.1.7 that all Euclidean domains are unique factorization domains. We shall use these facts later.

First notice that given any nonzero polynomial $f(x) \in k[x]$, we can write this polynomial uniquely (up to multiplication by a unit) as $f(x) = cf^*(x)$, where $f^*(x) \in R[x]$ and $f^*(x)$ is primitive. We do this by first finding a common denominator for all the coefficients of f and factoring this out. If we denote this constant by β, then we can write $f = f'/\beta$, where $f' \in R[x]$. We then find the *content* of f' (which we will denote by α), and factor this out as well. We let $\alpha/\beta = c$ and write $f = cf^*$, noting that f^* is primitive.

We must prove the uniqueness of this expression of f. If

$$f(x) = cf^*(x) = df'(x),$$

with both $f^*(x)$ and $f'(x)$ primitive, then we can write

$$f'(x) = (c/d) f^*(x) = (a/b) f^*(x),$$

where $\gcd(a, b) = 1$. Since the coefficients of $f'(x)$ are elements of R, then $b \mid a\gamma_i$ for all i, where γ_i are the coefficients of f^*. But since $\gcd(a, b) = 1$, then $b \mid \gamma_i$ for all i. Since f^* is a primitive polynomial, then b must be a unit of R. Similarly, we can write $f^*(x) = (b/a)f'(x)$, and by the same argument as above, a must be a unit as well. This shows that $f^*(x) \sim f'(x)$.

Let us suppose that we have a polynomial $f(x) \in R[x]$. Then we can factor this polynomial as $f(x) = g(x)h(x)$, with $g(x), h(x) \in k[x]$ (because $k[x]$ is a unique factorization domain). We can also write $cf^*(x) = d_1g^*(x)d_2h^*(x)$, where $g^*(x), h^*(x) \in R[x]$, and $g^*(x), h^*(x)$ are primitive. We showed above that the polynomial $g^*(x)h^*(x)$ is primitive, and we know that this decomposition $f(x) = cf^*(x)$ is unique. Therefore we can write $f^*(x) = g^*(x)h^*(x)$ and thus $f(x) = cg^*(x)h^*(x)$. But both $f(x)$ and $f^*(x) = g^*(x)h^*(x)$ have coefficients in R, and $f^*(x)$ is primitive. So c must be an element of R.

Thus, when we factored $f(x) \in k[x]$, the two factors were also in $R[x]$. By induction, if we decompose f into all its irreducible components in $k[x]$, each of the factors will be in $R[x]$, and we know that this decomposition will be essentially unique because $k[x]$ is a unique factorization domain. This shows that $R[x]$ is a unique factorization domain. □

2.2 Gaussian Integers

Let $\mathbb{Z}[i] = \{a + bi \mid a, b \in \mathbb{Z}, i = \sqrt{-1}\}$. This ring is often called the ring of Gaussian integers.

Exercise 2.2.1 Show that $\mathbb{Z}[i]$ is Euclidean.

Exercise 2.2.2 Prove that if p is a positive prime, then there is an element $x \in \mathbb{F}_p := \mathbb{Z}/p\mathbb{Z}$ such that $x^2 \equiv -1 \pmod{p}$ if and only if either $p = 2$ or $p \equiv 1 \pmod{4}$. (Hint: Use Wilson's theorem, Exercise 1.4.10.)

Exercise 2.2.3 Find all integer solutions to $y^2 + 1 = x^3$ with $x, y \neq 0$.

Exercise 2.2.4 If π is an element of R such that when $\pi \mid ab$ with $a, b \in R$, then $\pi \mid a$ or $\pi \mid b$, then we say that π is prime. What are the primes of $\mathbb{Z}[i]$?

Exercise 2.2.5 A positive integer a is the sum of two squares if and only if $a = b^2c$ where c is not divisible by any positive prime $p \equiv 3 \pmod{4}$.

2.3 Eisenstein Integers

Let $\rho = (-1 + \sqrt{-3})/2$. Notice that $\rho^2 + \rho + 1 = 0$, and $\rho^3 = 1$. Notice also that $\rho^2 = \overline{\rho}$. Define the *Eisenstein integers* as the set $\mathbb{Z}[\rho] = \{a + b\rho : a, b \in \mathbb{Z}\}$. Notice that $\mathbb{Z}[\rho]$ is closed under complex conjugation.

Exercise 2.3.1 Show that $\mathbb{Z}[\rho]$ is a ring.

Exercise 2.3.2 (a) Show that $\mathbb{Z}[\rho]$ is Euclidean.

(b) Show that the only units in $\mathbb{Z}[\rho]$ are ± 1, $\pm\rho$, and $\pm\rho^2$.

Notice that $(x^2+x+1)(x-1) = x^3-1$ and that we have

$$(x-\rho)(x-\overline{\rho}) = (x-\rho)(x-\rho^2) = x^2+x+1$$

so that

$$(1-\rho)(1-\rho^2) = 3 = (1+\rho)(1-\rho)^2 = -\rho^2(1-\rho)^2.$$

Exercise 2.3.3 Let $\lambda = 1-\rho$. Show that λ is irreducible, so we have a factorization of 3 (unique up to unit).

Exercise 2.3.4 Show that $\mathbb{Z}[\rho]/(\lambda)$ has order 3.

We can apply the arithmetic of $\mathbb{Z}[\rho]$ to solve $x^3+y^3+z^3=0$ for integers x, y, z. In fact we can show that $\alpha^3+\beta^3+\gamma^3=0$ for $\alpha,\beta,\gamma \in \mathbb{Z}[\rho]$ has no nontrivial solutions (i.e., where none of the variables is zero).

Example 2.3.5 Let $\lambda = 1-\rho$, $\theta \in \mathbb{Z}[\rho]$. Show that if λ does not divide θ, then $\theta^3 \equiv \pm 1 \pmod{\lambda^4}$. Deduce that if α,β,γ are coprime to λ, then the equation $\alpha^3+\beta^3+\gamma^3=0$ has no nontrivial solutions.

Solution. From the previous problem, we know that if λ does not divide θ then $\theta \equiv \pm 1 \pmod{\lambda}$. Set $\xi = \theta$ or $-\theta$ so that $\xi \equiv 1 \pmod{\lambda}$. We write ξ as $1+d\lambda$. Then

$$\begin{aligned}
\pm(\theta^3 \mp 1) &= \xi^3 - 1 \\
&= (\xi-1)(\xi-\rho)(\xi-\rho^2) \\
&= (d\lambda)(d\lambda+1-\rho)(1+d\lambda-\rho^2) \\
&= d\lambda(d\lambda+\lambda)(d\lambda-\lambda\rho^2) \\
&= \lambda^3 d(d+1)(d-\rho^2).
\end{aligned}$$

Since $\rho^2 \equiv 1 \pmod{\lambda}$, then $(d-\rho^2) \equiv (d-1) \pmod{\lambda}$. We know from the preceding problem that λ divides one of d, $d-1$, and $d+1$, so we may conclude that $\xi^3 - 1 \equiv 0 \pmod{\lambda^4}$, so $\xi^3 \equiv 1 \pmod{\lambda^4}$ and $\theta \equiv \pm 1 \pmod{\lambda^4}$. We can now deduce that no solution to $\alpha^3+\beta^3+\gamma^3 = 0$ is possible with α, β, and γ coprime to λ, by considering this equation mod λ^4. Indeed, if such a solution were possible, then somehow the equation

$$\pm 1 \pm 1 \pm 1 \equiv 0 \pmod{\lambda^4}$$

could be satisfied. The left side of this congruence gives ± 1 or ± 3; certainly ± 1 is not congruent to 0 $\pmod{\lambda^4}$ since λ^4 is not a unit. Also, ± 3 is not congruent to 0 (mod λ^4) since λ^2 is an associate of 3, and thus λ^4 is not. Thus, there is no solution to $\alpha^3+\beta^3+\gamma^3=0$ if α,β,γ are coprime to λ.

Hence if there is a solution to the equation of the previous example, one of α,β,γ is divisible by λ. Say $\gamma=\lambda^n\delta$, $(\delta,\lambda)=1$. We get $\alpha^3+\beta^3+\delta^3\lambda^{3n}=0$, δ,α,β coprime to λ.

Theorem 2.3.6 *Consider the more general*

$$\alpha^3+\beta^3+\varepsilon\lambda^{3n}\delta^3=0 \tag{2.1}$$

for a unit ε. Any solution for δ,α,β coprime to λ must have $n\ge 2$, but if (2.1) can be solved with $n=m$, it can be solved for $n=m-1$. Thus, there are no solutions to the above equation with δ,α,β coprime to λ.

Proof. We know that $n\ge 1$ from Example 2.3.5. Considering the equation mod λ^4, we get that $\pm1\pm1\pm\varepsilon\lambda^{3n}\equiv 0 \pmod{\lambda^4}$. There are two possibilities: if $\lambda^{3n}\equiv\pm2 \pmod{\lambda^4}$, then certainly n cannot exceed 1; but if $n=1$, then our congruence implies that $\lambda\mid 2$ which is not true. The other possibility is that $\lambda^{3n}\equiv 0 \pmod{\lambda^4}$, from which it follows that $n\ge 2$.

We may rewrite (2.1) as

$$\begin{aligned} -\varepsilon\lambda^{3n}\delta^3 &= \alpha^3+\beta^3 \\ &= (\alpha+\beta)(\alpha+\rho\beta)(\alpha+\rho^2\beta). \end{aligned}$$

We will write these last three factors as A_1, A_2, and A_3 for convenience. We can see that λ^6 divides the left side of this equation, since $n\ge 2$. Thus $\lambda^6\mid A_1A_2A_3$, and $\lambda^2\mid A_i$ for some i. Notice that

$$\begin{aligned} A_1-A_2 &= \lambda\beta, \\ A_1-A_3 &= \lambda\beta\rho^2, \end{aligned}$$

and

$$A_2-A_3=\lambda\beta\rho.$$

Since λ divides one of the A_i, it divides them all, since it divides their differences. Notice, though, that λ^2 does not divide any of these differences, since λ does not divide β by assumption. Thus, the A_i are inequivalent mod λ^2, and only one of the A_i is divisible by λ^2. Since our equation is unchanged if we replace β with $\rho\beta$ or $\rho^2\beta$, then without loss of generality we may assume that $\lambda^2\mid A_1$. In fact, we know that

$$\lambda^{3n-2}\mid A_1.$$

Now we write

$$\begin{aligned} B_1 &= A_1/\lambda, \\ B_2 &= A_2/\lambda, \\ B_3 &= A_3/\lambda. \end{aligned}$$

We notice that these B_i are pairwise coprime, since if for some prime p, we had $p \mid B_1$ and $p \mid B_2$, then necessarily we would have

$$p \mid B_1 - B_2 = \beta$$

and

$$p \mid \lambda B_1 + B_2 - B_1 = \alpha.$$

This is only possible for a unit p since $\gcd(\alpha, \beta) = 1$. Similarly, we can verify that the remaining pairs of B_i are coprime. Since $\lambda^{3n-2} \mid A_1$, we have $\lambda^{3n-3} \mid B_1$. So we may rewrite (2.1) as

$$-\varepsilon\lambda^{3n-3}\delta^3 = B_1 B_2 B_3.$$

From this equation we can see that each of the B_i is an associate of a cube, since they are relatively prime, and we write

$$\begin{aligned} B_1 &= e_1\lambda^{3n-3}C_1^3, \\ B_2 &= e_2 C_2^3, \\ B_3 &= e_3 C_3^3, \end{aligned}$$

for units e_i, and pairwise coprime C_i. Now recall that

$$\begin{aligned} A_1 &= \alpha + \beta, \\ A_2 &= \alpha + \rho\beta, \\ A_3 &= \alpha + \rho^2\beta. \end{aligned}$$

From these equations we have that

$$\begin{aligned} \rho^2 A_3 + \rho A_2 + A_1 &= \alpha(\rho^2 + \rho + 1) + \beta(\overline{\rho}^2 + \overline{\rho} + 1) \\ &= 0 \end{aligned}$$

so we have that

$$0 = \rho^2\lambda B_3 + \rho\lambda B_2 + \lambda B_1$$

and

$$0 = \rho^2 B_3 + \rho B_2 + B_1.$$

We can then deduce that

$$\rho^2 e_3 C_3^3 + \rho e_2 C_2^3 + e_1\lambda^{3n-3}C_1^3 = 0$$

so we can find units e_4, e_5 so that

$$C_3^3 + e_4 C_2^3 + e_5\lambda^{3n-3}C_1^3 = 0.$$

Considering this equation mod λ^3, and recalling that $n \geq 2$, we get that $\pm 1 \pm e_4 \equiv 0 \pmod{\lambda^3}$ so $e_4 = \mp 1$, and we rewrite our equation as

$$C_3^3 + (\mp C_2)^3 + e_5\lambda^{3(n-1)}C_1^3 = 0.$$

This is an equation of the same type as (2.1), so we can conclude that if there exists a solution for (2.1) with $n = m$, then there exists a solution with $n = m - 1$.

This establishes by descent that no nontrivial solution to (2.1) is possible in $\mathbb{Z}[\rho]$. □

2.4 Some Further Examples

Example 2.4.1 Solve the equation $y^2 + 4 = x^3$ for integers x, y.

Solution. We first consider the case where y is even. It follows that x must also be even, which implies that $x^3 \equiv 0 \pmod 8$. Now, y is congruent to 0 or 2 (mod 4). If $y \equiv 0 \pmod 4$, then $y^2 + 4 \equiv 4 \pmod 8$, so we can rule out this case. However, if $y \equiv 2 \pmod 4$, then $y^2 + 4 \equiv 0 \pmod 8$. Writing $y = 2Y$ with Y odd, and $x = 2X$, we have $4Y^2 + 4 = 8X^3$, so that

$$Y^2 + 1 = 2X^3$$

and

$$(Y + i)(Y - i) = 2X^3 = (1 + i)(1 - i)X^3.$$

We note that $Y^2 + 1 \equiv 2 \pmod 4$ and so X^3 is odd. Now,

$$\begin{aligned} X^3 &= \frac{(Y+i)(Y-i)}{(1+i)(1-i)} \\ &= \left(\frac{1+Y}{2} + \frac{1-Y}{2}i\right)\left(\frac{1+Y}{2} - \frac{1-Y}{2}i\right) \\ &= \left(\frac{1+Y}{2}\right)^2 + \left(\frac{1-Y}{2}\right)^2. \end{aligned}$$

We shall write this last sum as $a^2 + b^2$. Since Y is odd, a and b are integers. Notice also that $a + b = 1$ so that $\gcd(a, b) = 1$. We now have that

$$X^3 = (a + bi)(a - bi).$$

We would like to establish that $(a+bi)$ and $(a-bi)$ are relatively prime. We assume there exists some nonunit d such that $d \mid (a + bi)$ and $d \mid (a - bi)$. But then $d \mid [(a + bi) + (a - bi)] = 2a$ and $d \mid (a + bi) - (a - bi) = 2bi$. Since $\gcd(a, b) = 1$, then $d \mid 2$, and thus d must have even norm. But then it is impossible that $d \mid (a + bi)$ since the norm of $(a + bi)$ is $a^2 + b^2 = X^3$ which is odd. Thus $(a + bi)$ and $(a - bi)$ are relatively prime, and each is therefore a cube, since $\mathbb{Z}[i]$ is a unique factorization domain. We write

$$a + bi = (s + ti)^3 = s^3 - 3st^2 + (3s^2t - t^3)i.$$

Comparing real and imaginary parts yields

$$\begin{aligned} a &= s^3 - 3st^2, \\ b &= 3s^2t - t^3. \end{aligned}$$

Adding these two equations yields $a+b = s^3-3st^2+3s^2t-t^3$. But $a+b=1$, so we have

$$\begin{aligned} 1 &= s^3 - 3st^2 + 3s^2t - t^3 \\ &= (s-t)(s^2+4st+t^2). \end{aligned}$$

Now, $s,t \in \mathbb{Z}$ so $(s-t) = \pm 1$ and $(s^2+4st+t^2) = \pm 1$. Subtracting the second equation from the square of the first we find that $-6st = 0$ or 2. Since s and t are integers, we rule out the case $-6st = 2$ and deduce that either $s = 0$ or $t = 0$. Thus either $a = 1$, $b = 0$ or $a = 0$, $b = 1$. It follows that $Y = \pm 1$, so the only solutions in $\mathbb{Z}$ to the given equation with y even are $x = 2$, $y = \pm 2$.

Next, we consider the case where y is odd. We write $x^3 = (y+2i)(y-2i)$. We can deduce that $(y+2i)$ and $(y-2i)$ are relatively prime since if d divided both, d would divide both their sum and their difference, i.e., we would have $d \mid 2y$ and $d \mid 4i$. But then d would have even norm, and since y is odd, $(y+2i)$ has odd norm; thus d does not divide $(y+2i)$. Hence, $(y+2i)$ is a cube; we write

$$y + 2i = (q+ri)^3 = q^3 - 3qr^2 + (3q^2r - r^3)i.$$

Comparing real and imaginary parts we have that $2 = 3q^2r - r^3$ so that $r \mid 2$, and the only values r could thus take are ± 1 and ± 2. We get that the only possible pairs (q,r) we can have are $(1,1)$, $(-1,1)$, $(1,-2)$, and $(-1,-2)$. Solving for y, and excluding the cases where y is even, we find that $x = 5, y = \pm 11$ is the only possible solution when y is odd.

Exercise 2.4.2 Show that $\mathbb{Z}[\sqrt{-2}]$ is Euclidean.

Exercise 2.4.3 Solve $y^2 + 2 = x^3$ for $x, y \in \mathbb{Z}$.

Example 2.4.4 Solve $y^2 + 1 = x^p$ for an odd prime p, and $x, y \in \mathbb{Z}$.

Solution. Notice that the equation $y^2 + 1 = x^3$ from an earlier problem is a special case of the equation given here. To analyze the solutions of this equation, we first observe that for odd y, $y^2 \equiv 1 \pmod 4$. Thus x would need to be even, but then if we consider the equation mod 4 we find that it cannot be satisfied; $y^2 + 1 \equiv 2 \pmod 4$, while $x^p \equiv 0 \pmod 4$. Thus y is even; it is easy to see that x must be odd. If $y = 0$, then $x = 1$ is a solution for all p. We call this solution a trivial solution; we proceed to investigate solutions other than the trivial one. Now we write our equation as

$$(y+i)(y-i) = x^p.$$

If $y \neq 0$, then we note that if some divisor δ were to divide both $(y+i)$ and $(y-i)$, then it would divide $2i$; if δ is not a unit, then δ will thus divide 2, and also y, since y is even. But then it is impossible that δ also divide $y+i$ since i is a unit. We conclude that $(y+i)$ and $(y-i)$ are relatively prime when $y \neq 0$. Thus $(y+i)$ and $(y-i)$ are both pth powers, and we may write

$$(y+i) = e(a+bi)^p$$

for some unit e and integers a and b. We have analyzed the units of $\mathbb{Z}[i]$; they are all powers of i, so we write

$$(y+i) = i^k(a+bi)^p.$$

Now,

$$(y-i) = \overline{(y+i)} = (-i)^k(a-bi)^p.$$

Thus

$$\begin{aligned}(y+i)(y-i) &= i^k(a+bi)^p(-i)^k(a-bi)^p \\ &= (a^2+b^2)^p \\ &= x^p,\end{aligned}$$

and it follows that $x = (a^2+b^2)$. We know that x is odd, so one of a and b is even (but not both). We now have that

$$\begin{aligned}(y+i)-(y-i) &= 2i \\ &= i^k(a+bi)^p - (-i)^k(a-bi)^p.\end{aligned}$$

We consider two cases separately:

Case 1. k is odd.

In this case we use the binomial theorem to determine the imaginary parts of both sides of the above equation. We get

$$\begin{aligned}2 &= \operatorname{Im}[(i)^k((a+bi)^p + (a-bi)^p)] \\ &= \operatorname{Im}\left[(i)^k\left(\sum_{j=0}^{p} a^{p-j}(bi)^j\binom{p}{j} + \sum_{j=0}^{p} a^{p-j}(-bi)^j\binom{p}{j}\right)\right] \\ &= 2(-1)^{(k-1)/2}\sum_{\substack{\text{even } j,\\ 0\leq j<p}} a^{p-j}(b)^j(-1)^{j/2}\binom{p}{j}.\end{aligned}$$

Thus

$$1 = (-1)^{(k-1)/2}\sum_{\substack{\text{even } j,\\ 0\leq j<p}} a^{p-j}(b)^j(-1)^{j/2}\binom{p}{j}.$$

Since a divides every term on the right-hand side of this equation, then $a \mid 1$ and $a = \pm 1$. We observed previously that only one of a, b is odd; thus b is even. We now substitute $a = \pm 1$ into the last equation above to get

$$\begin{aligned} \pm 1 &= \sum_{\substack{\text{even } j, \\ 0 \le j < p}} (b)^j (-1)^{j/2} \binom{p}{j} \\ &= 1 - b^2 \binom{p}{2} + b^4 \binom{p}{4} - \cdots \pm b^{p-1} \binom{p}{p-1}. \end{aligned}$$

If the sign of 1 on the left-hand side of this equality were negative, we would have that $b^2 \mid 2$; b is even and in particular $b \neq \pm 1$, so this is impossible. Thus

$$\begin{aligned} 0 &= -b^2 \binom{p}{2} + b^4 \binom{p}{4} - \cdots \pm b^{p-1} \binom{p}{p-1} \\ &= -\binom{p}{2} + b^2 \binom{p}{4} - \cdots \pm b^{p-3} \binom{p}{p-1}. \end{aligned}$$

Now we notice that $2 \mid b$, so $2 \mid \binom{p}{2}$. If $p \equiv 3 \pmod 4$, then we are finished because $2 \nmid \binom{p}{2}$. Suppose in fact that 2^q is the largest power of 2 dividing $\binom{p}{2}$. We shall show that 2^{q+1} will then divide every term in $b^2 \binom{p}{4} - \cdots \pm b^{p-3} \binom{p}{p-1}$, and this will establish that no b will satisfy our equation. We consider one of these terms given by $(b)^{j-2} \binom{p}{j}$, for an even value of j; we rewrite this as $b^{2m-2} \binom{p}{2m}$ (we are not concerned with the sign of the term). We see that

$$\begin{aligned} \binom{p}{2m} &= \binom{p-2}{2m-2} \frac{(p)(p-1)}{2m(2m-1)} \\ &= \binom{p-2}{2m-2} \binom{p}{2} \frac{1}{m(2m-1)}, \end{aligned}$$

so we are considering a term

$$\binom{p-2}{2m-2} \binom{p}{2} \frac{b^{2m-2}}{m(2m-1)}.$$

Now, $2^q \mid \binom{p}{2}$ by assumption. Recall that b is even; thus $2^{2m-2} \mid b^{2m-2}$. Now $m \geq 2$; it is easy to see then that $2m - 2 \geq m$, so 2^{2m-2} does not divide m. Thus when we reduce the fraction

$$\frac{b^{2m-2}}{m(2m-1)}$$

to lowest terms, the numerator is even and the denominator is odd. Therefore,

$$2(2^q) \mid \binom{p-2}{2m-2} \binom{p}{2} \frac{b^{2m-2}}{m(2m-1)}.$$

Thus 2^{q+1} divides every term in $b^2\binom{p}{4} - \cdots \pm \binom{p}{p-1}b^{p-3}$ and we deduce that no value of b can satisfy our equation.

Case 2. k is even.

This case is almost identical to the first case; we mention only the relevant differences. When we expand

$$(y+i)-(y-i) = 2i = i^k(a+bi)^p - (-i)^k(a-bi)^p$$

and consider imaginary parts, we get

$$1 = (-1)^{k/2} \sum_{\substack{\text{odd } j,\\ 0<j\leq p}} a^{p-j}(b)^j(-1)^{(j-1)/2}\binom{p}{j}.$$

We are able to deduce that $b = \pm 1$; substituting we get the equation

$$\begin{aligned} \pm 1 &= \sum_{\substack{\text{odd } j,\\ 0<j\leq p}} a^{p-j}(b)^j(-1)^{(j-1)/2}\binom{p}{j} \\ &= 1 - a^2\binom{p}{2} + a^4\binom{p}{4} - \cdots \pm \binom{p}{p-1}a^{p-1}, \end{aligned}$$

which we can see is identical to the equation we arrived at in Case 1, with b replaced by a. Thus we can reproduce the proof of Case 1, with b replaced by a, to establish that there are no nontrivial solutions with k even. We conclude that the equation $y^2 + 1 = x^p$ has no nontrivial solution with $x, y \in \mathbb{Z}$.

Exercise 2.4.5 Show that $\mathbb{Z}[\sqrt{2}]$ is Euclidean.

Exercise 2.4.6 Let $\varepsilon = 1+\sqrt{2}$. Write $\varepsilon^n = u_n + v_n\sqrt{2}$. Show that $u_n^2 - 2v_n^2 = \pm 1$.

Exercise 2.4.7 Show that there is no unit η in $\mathbb{Z}[\sqrt{2}]$ such that $1 < \eta < 1+\sqrt{2}$. Deduce that every unit (greater than zero) of $\mathbb{Z}[\sqrt{2}]$ is a power of $\varepsilon = 1+\sqrt{2}$.

2.5 Supplementary Problems

Exercise 2.5.1 Show that $R = \mathbb{Z}[(1+\sqrt{-7})/2]$ is Euclidean.

Exercise 2.5.2 Show that $\mathbb{Z}[(1+\sqrt{-11})/2]$ is Euclidean.

Exercise 2.5.3 Find all integer solutions to the equation $x^2 + 11 = y^3$.

Exercise 2.5.4 Prove that $\mathbb{Z}[\sqrt{3}]$ is Euclidean.

Exercise 2.5.5 Prove that $\mathbb{Z}[\sqrt{6}]$ is Euclidean.

Exercise 2.5.6 Show that $\mathbb{Z}[(1+\sqrt{-19})/2]$ is not Euclidean for the norm map.

Exercise 2.5.7 Prove that $\mathbb{Z}[\sqrt{-10}]$ is not a unique factorization domain.

Exercise 2.5.8 Show that there are only finitely many rings $\mathbb{Z}[\sqrt{d}]$ with $d \equiv 2$ or 3 (mod 4) which are norm Euclidean.

Exercise 2.5.9 Find all integer solutions of $y^2 = x^3 + 1$.

Exercise 2.5.10 Let $x_1, ..., x_n$ be indeterminates. Evaluate the determinant of the $n \times n$ matrix whose (i, j)-th entry is x_i^{j-1}. (This is called the *Vandermonde determinant.*)

Chapter 3

Algebraic Numbers and Integers

3.1 Basic Concepts

A number $\alpha \in \mathbb{C}$ is called an *algebraic number* if there exists a polynomial $f(x) = a_n x^n + \cdots + a_0$ such that $a_0, \ldots, a_n$, not all zero, are in $\mathbb{Q}$ and $f(\alpha) = 0$. If α is the root of a monic polynomial with coefficients in $\mathbb{Z}$, we say that α is an *algebraic integer*. Clearly all algebraic integers are algebraic numbers. However, the converse is false.

Example 3.1.1 Show that $\sqrt{2}/3$ is an algebraic number but not an algebraic integer.

Solution. Consider the polynomial $f(x) = 9x^2 - 2$, which is in $\mathbb{Q}[x]$. Since $f(\sqrt{2}/3) = 0$, we know that $\sqrt{2}/3$ is an algebraic number.

Assume $\sqrt{2}/3$ is an algebraic integer. Then there exists a monic polynomial in $\mathbb{Z}[x]$, say $g(x) = x^n + b_{n-1}x^{n-1} + \cdots + b_0$, which has $\alpha = \sqrt{2}/3$ as a root. So

$$\begin{aligned} g(\alpha) = \left(\frac{\sqrt{2}}{3}\right)^n + b_{n-1}\left(\frac{\sqrt{2}}{3}\right)^{n-1} + \cdots + b_0 &= 0, \\ \Rightarrow \qquad (\sqrt{2})^n + b_{n-1}(\sqrt{2})^{n-1}(3) + \cdots + b_0(3)^n &= 0. \end{aligned}$$

If i is odd, $(\sqrt{2})^i$ is not an integer. So we can separate our equation into two smaller equations:

$$\sum_{i \text{ odd}} b_i \sqrt{2}^i 3^{n-i} = 0 \quad \Rightarrow \quad \sqrt{2} \sum_{i \text{ odd}} b_i 2^{(i-1)/2} 3^{n-i} = 0$$

and

$$\sum_{i \text{ even}} b_i \sqrt{2}^i 3^{n-i} = 0$$

for $i = 0, \dots, n$. Since $3 \mid 0$, each sum above must be divisible by 3. In particular, because each summand containing $b_i, i \neq n$, has a factor of 3, 3 divides the summand containing $b_n = 1$. This tells us that $3 \mid 2^{(n-1)/2}$ if n is odd, and $3 \mid 2^{n/2}$ if n is even. In either case, this is false and hence we can conclude that $\sqrt{2}/3$ is not an algebraic integer.

Exercise 3.1.2 Show that if $r \in \mathbb{Q}$ is an algebraic integer, then $r \in \mathbb{Z}$.

Exercise 3.1.3 Show that if $4 \mid (d+1)$, then

$$\frac{-1 \pm \sqrt{-d}}{2}$$

is an algebraic integer.

Theorem 3.1.4 *Let α be an algebraic number. There exists a unique polynomial $p(x)$ in $\mathbb{Q}[x]$ which is monic, irreducible and of smallest degree, such that $p(\alpha) = 0$. Furthermore, if $f(x) \in \mathbb{Q}[x]$ and $f(\alpha) = 0$, then $p(x) \mid f(x)$.*

Proof. Consider the set of all polynomials in $\mathbb{Q}[x]$ for which α is a root and pick one of smallest degree, say $p(x)$. If $p(x)$ is not irreducible, it can be written as a product of two lower degree polynomials in $\mathbb{Q}[x]$: $p(x) = a(x)b(x)$. However, $p(\alpha) = a(\alpha)b(\alpha) = 0$ and since $\mathbb{C}$ is an integral domain, either $a(\alpha) = 0$ or $b(\alpha) = 0$. But this contradicts the minimality of $p(x)$, so $p(x)$ must be irreducible.

Suppose there were two such polynomials, $p(x)$ and $q(x)$. By the division algorithm,

$$p(x) = a(x)q(x) + r(x),$$

where $a(x)$, $r(x) \in \mathbb{Q}[x]$, and either $\deg(r) = 0$ or $\deg(r) < \deg(q)$. But $p(\alpha) = a(\alpha)q(\alpha) + r(\alpha) = 0$ and $q(\alpha) = 0$ together imply that $r(\alpha) = 0$. Because $p(x)$ and $q(x)$ are the smallest degree polynomials with α as a root, $r = 0$. So $p(x) = a(x)q(x)$ and $a(x) \in \mathbb{Q}^*$ (the set of all nonzero elements of $\mathbb{Q}$), since $\deg(p) = \deg(q)$. Thus $p(x)$ is unique up to a constant and so we may suppose its leading coefficient is 1.

Now suppose $f(x)$ is a polynomial in $\mathbb{Q}[x]$ such that $f(\alpha) = 0$. If $p(x)$ does not divide $f(x)$ then, since $p(x)$ is irreducible, $\gcd(p(x), f(x)) = 1$. So we can find $a(x), b(x) \in \mathbb{Q}[x]$ such that $a(x)p(x) + b(x)f(x) = 1$. However, putting $x = \alpha$ yields a contradiction. Thus, $p(x) \mid f(x)$. $\square$

The degree of $p(x)$ is called the *degree* of α and is denoted $\deg(\alpha)$; $p(x)$ is called the *minimal polynomial* of α.

Complex numbers which are not algebraic are called *transcendental*. Well before an example of a transcendental number was known, mathematicians were assured of their existence.

Example 3.1.5 Show that the set of algebraic numbers is countable (and hence the set of transcendental numbers is uncountable).

Solution. All polynomials in $\mathbb{Q}[x]$ have a finite number of roots. The set of rational numbers, $\mathbb{Q}$, is countable and so the set $\mathbb{Q}[x]$ is also countable. The set of algebraic numbers is the set of all roots of a countable number of polynomials, each with a finite number of roots. Hence the set of algebraic numbers is countable.

Since algebraic numbers and transcendental numbers partition the set of complex numbers, $\mathbb{C}$, which is uncountable, it follows that the set of transcendental numbers is uncountable.

Exercise 3.1.6 Find the minimal polynomial of $\sqrt{n}$ where n is a squarefree integer.

Exercise 3.1.7 Find the minimal polynomial of $\sqrt{2}/3$.

3.2 Liouville's Theorem and Generalizations

In 1853, Liouville showed that algebraic numbers cannot be too well approximated by rationals.

Theorem 3.2.1 (Liouville) *Given α, a real algebraic number of degree $n \neq 1$, there is a positive constant $c = c(\alpha)$ such that for all rational numbers p/q, $(p,q) = 1$ and $q > 0$, the inequality*

$$\left|\alpha - \frac{p}{q}\right| > \frac{c(\alpha)}{q^n}$$

holds.

Proof. Let $f(x) = a_n x^n + a_{n-1}x^{n-1} + \cdots + a_0$ be $\in \mathbb{Z}[x]$ whose degree equals that of α and for which α is a root. (So $\deg(f) \geq 2$). Notice that

$$\begin{aligned}
\left|f(\alpha) - f\left(\frac{p}{q}\right)\right| &= \left|f\left(\frac{p}{q}\right)\right| \\
&= \left|a_n\left(\frac{p}{q}\right)^n + a_{n-1}\left(\frac{p}{q}\right)^{n-1} + \cdots + a_0\right| \\
&= \left|\frac{a_n p^n + a_{n-1}p^{n-1}q + \cdots + a_0 q^n}{q^n}\right| \\
&\geq \frac{1}{q^n}.
\end{aligned}$$

If $\alpha = \alpha_1, \ldots, \alpha_n$ are the roots of f, let M be the maximum of the values $|\alpha_i|$, $1 \leq i \leq n$. If $|p/q|$ is greater than $2M$, then

$$\left|\alpha - \frac{p}{q}\right| \geq M \geq \frac{M}{q^n}.$$

If $|p/q| \leq M$, then

$$\left|\alpha_i - \frac{p}{q}\right| \leq 3M$$

so that

$$\left|\alpha - \frac{p}{q}\right| \geq \frac{1}{|a_n|q^n \prod_{j=2}^{n} |\alpha_j - p/q|} \geq \frac{1}{|a_n|(3M)^{n-1}q^n}.$$

Hence, the theorem holds with

$$c(\alpha) = \min\left(M, \frac{1}{|a_n|(3M)^{n-1}}\right).$$

□

Using this theorem, Liouville was able to give specific examples of transcendental numbers.

Example 3.2.2 Show that

$$\sum_{n=0}^{\infty} \frac{1}{10^{n!}}$$

is transcendental.

Solution. Suppose not, and call the sum α. Look at the partial sum

$$\sum_{n=0}^{k} \frac{1}{10^{n!}} = \frac{p_k}{q_k},$$

with $q_k = 10^{k!}$. Thus,

$$\begin{aligned}
\left|\alpha - \frac{p_k}{q_k}\right| &= \left|\sum_{n=k+1}^{\infty} \frac{1}{10^{n!}}\right| \\
&= \frac{1}{10^{(k+1)!}} + \left(\frac{1}{10^{(k+1)!}}\right)^{k+2} + \left(\frac{1}{10^{(k+1)!}}\right)^{(k+2)(k+3)} + \cdots \\
&\leq \frac{1}{10^{(k+1)!}}\left[1 + \frac{1}{10^2} + \frac{1}{10^3} + \cdots\right] \\
&= \left(\frac{1}{10^{(k+1)!}}\right) S,
\end{aligned}$$

where $S = 1 + 1/10^2 + 1/10^3 + \cdots$, an infinite geometric series which has a finite sum. So

$$\left|\sum_{n=k+1}^{\infty} \frac{1}{10^{n!}}\right| \leq \frac{S}{10^{(k+1)!}} = \frac{S}{q_k^{k+1}}.$$

If α were algebraic of degree of n then, by Liouville's theorem, there exists a constant $c(\alpha)$ such that

$$\left|\alpha - \frac{p_k}{q_k}\right| \geq \frac{c(\alpha)}{q_k^n},$$

so we have

$$\frac{S}{q_k^{k+1}} \geq \frac{c(\alpha)}{q_k^n}.$$

However, we can choose k to be as large as we want to obtain a contradiction. So α is transcendental.

It is easy to see that this argument can be generalized to show that $\sum_{n=0}^{\infty} a^{-n!}$ is transcendental for all positive integers a. We will prove this fact in the Supplementary Exercises for this chapter.

In 1873, Hermite showed the number e is transcendental and in 1882, Lindemann proved the transcendency of π. In fact, he showed more generally that for an algebraic number α, e^{α} is transcendental. This implies that π is transcendental since $e^{\pi i} = -1$.

In 1909, Thue was able to improve Liouville's inequality. He proved that if α is algebraic of degree n, then there exists a constant $c(\alpha)$ so that for all $p/q \in \mathbb{Q}$,

$$\left|\alpha - \frac{p}{q}\right| \geq \frac{c(\alpha)}{q^{n/2+1}}.$$

This theorem has immediate Diophantine applications.

Example 3.2.3 Let $f(x, y)$ be an irreducible polynomial of binary form of degree $n \geq 3$. Assuming Thue's theorem, show that $f(x, y) = m$ for any fixed $m \in \mathbb{Z}^*$ has only finitely many solutions.

Solution. Suppose $f(x, y) = m$ has infinitely many solutions, and write it in the form

$$f(x, y) = \prod_{i=1}^{n}(x - \alpha_i y) = m,$$

where α_i is an algebraic number of degree ≥ 3 $\forall i = 1, \ldots, n$.

Without loss of generality, we can suppose that for an infinite number of pairs (x, y), we have

$$\left|\frac{x}{y} - \alpha_1\right| \leq \left|\frac{x}{y} - \alpha_i\right| \qquad \text{for } i = 2, \ldots, n.$$

Further, by the triangle inequality,

$$\begin{aligned}\left|\frac{x}{y} - \alpha_i\right| &\geq \frac{1}{2}\left(\left|\frac{x}{y} - \alpha_i\right| + \left|\frac{x}{y} - \alpha_1\right|\right)\\ &\geq \frac{1}{2}|\alpha_i - \alpha_1| \qquad \text{for } i = 2, \ldots, n.\end{aligned}$$

Hence,

$$\begin{aligned} |f(x,y)| &= |y^n|\left|\frac{x}{y}-\alpha_1\right|\cdots\left|\frac{x}{y}-\alpha_n\right|, \\ |m| &\geq k|y|^n\left|\frac{x}{y}-\alpha_1\right|, \\ \frac{|m|}{k|y|^n} &\geq \left|\frac{x}{y}-\alpha_1\right|, \end{aligned}$$

where

$$k = \frac{1}{2^{n-1}}\prod_{i=2}^{n}|\alpha_i-\alpha_1|.$$

However, by Thue's theorem, this implies

$$\frac{c}{y^{n/2+1}} \leq \frac{m}{ky^n} \quad \Leftrightarrow \quad \frac{1}{y^{n/2+1}} \leq \frac{m(ck)^{-1}}{y^n}.$$

However, for $n \geq 3$, this holds for only finitely many (x, y), contradicting our assumption. Thus $f(x, y)$ has only finitely many solutions.

Over a long series of improvements upon Liouville's theorem, in 1955 Roth was able to show the inequality can be strengthened to

$$\left|\alpha - \frac{p}{q}\right| \geq \frac{c(\alpha,\varepsilon)}{q^{2+\varepsilon}},$$

for any $\varepsilon > 0$. This improved inequality gives us a new family of transcendental numbers.

Exercise 3.2.4 Show that $\sum_{n=1}^{\infty} 2^{-3^n}$ is transcendental.

Exercise 3.2.5 Show that, in fact, $\sum_{n=1}^{\infty} 2^{-f(n)}$ is transcendental whenever

$$\lim_{n\to\infty}\frac{f(n+1)}{f(n)} > 2.$$

3.3 Algebraic Number Fields

The theory of algebraic number fields is vast and rich. We will collect below the rudimentary facts of the theory. We begin with

Example 3.3.1 Let α be an algebraic number and define

$$\mathbb{Q}[\alpha] = \{f(\alpha) : f \in \mathbb{Q}[x]\},$$

a subring of $\mathbb{C}$. Show that $\mathbb{Q}[\alpha]$ is a field.

Solution. Let f be the minimal polynomial of α, and consider the map $\phi : \mathbb{Q}[x] \to \mathbb{Q}[\alpha]$ such that

$$\sum_{i=0}^{n} a_i x^i \to \sum_{i=0}^{n} a_i \alpha^i.$$

Notice that

$$\phi(g) + \phi(h) = \sum_{i=0}^{n} a_i \alpha^i + \sum_{i=0}^{m} b_i \alpha^i = \phi(g+h)$$

and

$$\phi(g)\phi(h) = \left(\sum_{i=0}^{n} a_i \alpha^i\right)\left(\sum_{j=0}^{m} b_j \alpha^j\right) = \sum_{0 \le i+j \le n+m} a_i b_j \alpha^{i+j} = \phi(gh).$$

So ϕ is a homomorphism. Furthermore, it is clear that $\ker \phi = (f)$, the ideal generated by f (see Theorem 3.1.4). Thus, by the ring homomorphism theorems,

$$\mathbb{Q}[x]/(f) \simeq \mathbb{Q}[\alpha].$$

Let g be a polynomial in $\mathbb{Q}[x]$ such that f does not divide g. From Chapter 2, we know that $\mathbb{Q}[x]$ is a Euclidean domain and is therefore also a PID. We also learned in Chapter 2 that the ideal generated by any irreducible element in a PID is a maximal ideal. Since f is irreducible, $\mathbb{Q}[x]/(f)$ is a field and so $\mathbb{Q}[\alpha]$ is a field, as desired.

From now on, we will denote $\mathbb{Q}[\alpha]$ by $\mathbb{Q}(\alpha)$.

A field $K \subseteq \mathbb{C}$ is called an *algebraic number field* if its dimension over $\mathbb{Q}$ is finite. The dimension of K over $\mathbb{Q}$ is called the *degree* of K and is denoted $[K : \mathbb{Q}]$. Notice that if α is an algebraic number of degree n, then $\mathbb{Q}(\alpha)$ is an algebraic number field of degree n over $\mathbb{Q}$.

Let α and β be algebraic numbers. $\mathbb{Q}(\alpha, \beta)$ is a field since it is the intersection of all the subfields of $\mathbb{C}$ containing $\mathbb{Q}$, α, and β. The intersection of a finite number of subfields in a fixed field is again a field.

Theorem 3.3.2 (Theorem of the Primitive Element) *If α and β are algebraic numbers, then $\exists\, \theta$, an algebraic number, such that $\mathbb{Q}(\alpha, \beta) = \mathbb{Q}(\theta)$.*

Proof. Let f be the minimal polynomial of α and let g be the minimal polynomial of β. We want to show that we can find $\lambda \in \mathbb{Q}$ such that $\theta = \alpha + \lambda\beta$ and $\mathbb{Q}(\alpha, \beta) = \mathbb{Q}(\theta)$. We will denote $\mathbb{Q}(\theta)$ by L. Clearly $L = \mathbb{Q}(\theta) \subseteq \mathbb{Q}(\alpha, \beta)$.

Define $\phi(x) = f(\theta - \lambda x) \in L[x]$. Notice that $\phi(\beta) = f(\theta - \lambda\beta) = f(\alpha) = 0$. So β is a root of ϕ. Choose $\lambda \in \mathbb{Q}$ in such a way that β is the only common root of ϕ and g. This can be done since only a finite number of choices of λ are thus ruled out. So $\gcd(\phi(x), g(x)) = c(x - \beta)$, $c \in \mathbb{C}^*$. Then $c(x - \beta) \in L[x]$ which implies that $c, c\beta \in L$, and so $\beta \in L$.

Now, $\theta = \alpha + \lambda\beta \in L$ which means that $\alpha \in L$. So $\mathbb{Q}(\alpha, \beta) \subseteq L = \mathbb{Q}(\theta)$. Thus, we have the desired equality: $\mathbb{Q}(\alpha, \beta) = \mathbb{Q}(\theta)$. □

This theorem can be generalized quite easily using induction: for a set $\alpha_1, \ldots, \alpha_n$ of algebraic numbers, there exists an algebraic number θ such that $\mathbb{Q}(\alpha_1, \ldots, \alpha_n) = \mathbb{Q}(\theta)$. Therefore, any algebraic number field K is $\mathbb{Q}(\theta)$ for some algebraic number θ.

Exercise 3.3.3 Let α be an algebraic number and let $p(x)$ be its minimal polynomial. Show that $p(x)$ has no repeated roots.

The roots of the minimal polynomial $p(x)$ of α are called the *conjugate roots* or *conjugates* of α. Thus, if n is the degree of $p(x)$, then α has n conjugates.

Exercise 3.3.4 Let α, β be algebraic numbers such that β is conjugate to α. Show that β and α have the same minimal polynomial.

If $\theta = \theta^{(1)}$ and $\theta^{(2)}, \ldots, \theta^{(n)}$ are the conjugates of θ, then $\mathbb{Q}(\theta^{(i)})$, for $i = 2, \ldots, n$, is called a conjugate field to $\mathbb{Q}(\theta)$. Further, the maps $\theta \to \theta^{(i)}$ are monomorphisms of $K = \mathbb{Q}(\theta) \to \mathbb{Q}(\theta^{(i)})$ (referred to as *embeddings* of K into $\mathbb{C}$).

We can partition the conjugates of θ into real roots and nonreal roots (called complex roots).

K is called a *normal* extension (or Galois extension) of $\mathbb{Q}$ if all the conjugate fields of K are identical to K. For example, any quadratic extension of $\mathbb{Q}$ is normal. However, $\mathbb{Q}(\sqrt[3]{2})$ is not since the two conjugate fields $\mathbb{Q}(\rho\sqrt[3]{2})$ and $\mathbb{Q}(\rho^2\sqrt[3]{2})$ are distinct from $\mathbb{Q}(\sqrt[3]{2})$. (Here ρ is a primitive cube root of unity.)

We also define the *normal closure* of any field K as the extension $\tilde{K}$ of smallest degree containing all the conjugate fields of K. Clearly this is well-defined for if there were two such fields, $\tilde{K_1}$ and $\tilde{K_2}$, then $\tilde{K_1} \cap \tilde{K_2}$ would have the same property and have smaller degree if $\tilde{K_1} \neq \tilde{K_2}$. In the above example, the normal closure of $\mathbb{Q}(\sqrt[3]{2})$ is clearly $\mathbb{Q}(\sqrt[3]{2}, \rho)$.

Example 3.3.5 Show that Liouville's theorem holds for α where α is a complex algebraic number of degree $n \geq 2$.

Solution. First we note that if α is algebraic, then so is $\overline{\alpha}$ (the complex conjugate of α), since they satisfy the same minimal polynomial. Also, every element in an algebraic number field is algebraic, since if the field $\mathbb{Q}(\gamma)$ has degree n over $\mathbb{Q}$, then for any $\beta \in \mathbb{Q}(\gamma)$ the elements $1, \beta, \ldots, \beta^n$ are surely linearly dependent. This implies that $\alpha + \overline{\alpha} = 2\,\mathrm{Re}(\alpha)$ and $\alpha - \overline{\alpha} = 2i\,\mathrm{Im}(\alpha)$ are algebraic, since they are both in the field $\mathbb{Q}(\alpha, \overline{\alpha})$.

We can apply Liouville's theorem to $\text{Re}(\alpha)$ to get a constant $c = c(\text{Re}(\alpha))$ such that

$$\left|\text{Re}(\alpha) - \frac{p}{q}\right| \geq \frac{c}{q^m},$$

where $\text{Re}(\alpha)$ has degree m. Now,

$$\begin{aligned}\left|\alpha - \frac{p}{q}\right| &= \sqrt{\left(\text{Re}(\alpha) - \frac{p}{q}\right)^2 + (\text{Im}(\alpha))^2} \\ &\geq \left|\text{Re}(\alpha) - \frac{p}{q}\right| \\ &\geq \frac{c}{q^m}.\end{aligned}$$

To prove the result, it remains only to show that if the degree of α is n, then the degree of $\text{Re}(\alpha) \leq n$. Consider the polynomial

$$g(x) = \prod_{i=1}^{n}[2x - (\alpha^{(i)} + \overline{\alpha}^{(i)})],$$

where $\alpha = \alpha^{(1)}, \alpha^{(2)}, \ldots, \alpha^{(n)}$ are the algebraic conjugates of α. Certainly $\text{Re}(\alpha)$ satisfies this equation, so we must verify that its coefficients are in $\mathbb{Q}$.

To prove this, we need some Galois Theory. Let f be the minimal polynomial of α over $\mathbb{Q}$, and let F be the splitting field of this polynomial (i.e., the normal closure of $\mathbb{Q}(\alpha)$). Recall that f is also the minimal polynomial of $\overline{\alpha}$, and so F contains $\alpha^{(i)}$ and $\overline{\alpha}^{(i)}$ for $i = 1, \ldots, n$. Consider the Galois group of F, that is, all automorphisms of F leaving $\mathbb{Q}$ fixed. These automorphisms permute the roots of f, which are simply the conjugates of α. It is easy to see that the coefficients of $g(x)$ will remain unchanged under a permutation of the $\alpha^{(i)}$'s, and so they must lie in the fixed field of the Galois group, which is $\mathbb{Q}$.

Since $\text{Re}(\alpha)$ satisfies a polynomial with coefficients in $\mathbb{Q}$ of degree n, it follows that the minimal polynomial of $\text{Re}(\alpha)$ must divide this polynomial, and so have degree less than or equal to n. This proves Liouville's theorem for complex algebraic numbers.

Exercise 3.3.6 Let $K = \mathbb{Q}(\theta)$ be of degree n over $\mathbb{Q}$. Let $\omega_1, \ldots, \omega_n$ be a basis of K as a vector space over $\mathbb{Q}$. Show that the matrix $\Omega = (\omega_i^{(j)})$ is invertible.

Exercise 3.3.7 Let α be an algebraic number. Show that there exists $m \in \mathbb{Z}$ such that $m\alpha$ is an algebraic integer.

Exercise 3.3.8 Show that $\mathbb{Z}[x]$ is not (a) Euclidean or (b) a PID.

The following theorem gives several characterizations of algebraic integers. Of these, (c) and (d) are the most useful for they supply us with an immediate tool to test whether a given number is an algebraic integer or not.

Theorem 3.3.9 *Prove that the following statements are equivalent:*

(a) *α is an algebraic integer.*

(b) *The minimal polynomial of α is monic $\in \mathbb{Z}[x]$.*

(c) *$\mathbb{Z}[\alpha]$ is a finitely generated $\mathbb{Z}$-module.*

(d) *$\exists$ a finitely generated $\mathbb{Z}$-submodule $M \neq \{0\}$ of $\mathbb{C}$ such that $\alpha M \subseteq M$.*

Proof. (a) $\Rightarrow$ (b) Let $f(x)$ be a monic polynomial in $\mathbb{Z}[x]$, $f(\alpha) = 0$. Let $\phi(x)$ be the minimal polynomial of α.

Recall the definition of primitive polynomials given in Chapter 2: a polynomial $f(x) = a_n x^n + \cdots + a_0 \in \mathbb{Z}[x]$ is said to be *primitive* if the gcd of the coefficients of f is 1. In particular, a monic polynomial is primitive. By Theorem 3.1.4, we know $f(x) = \phi(x)\psi(x)$, for some $\psi(x) \in \mathbb{Q}[x]$. By the proof of Theorem 2.1.10, we know we can write

$$\begin{aligned} \phi(x) &= \frac{a}{b}\phi_1(x), \quad \phi_1(x) \text{ primitive, } a, b \in \mathbb{Z}, \qquad \phi_1(x) \in \mathbb{Z}[x], \\ \psi(x) &= \frac{c}{d}\psi_1(x), \quad \psi_1(x) \text{ primitive, } c, d \in \mathbb{Z}, \qquad \psi_1(x) \in \mathbb{Z}[x]. \end{aligned}$$

So $b\,df(x) = ac\phi_1(x)\psi_1(x)$. But by Gauss' lemma (see Theorem 2.1.9), $\phi_1(x)\psi_1(x)$ is primitive, and $f(x)$ is primitive, so $bd = \pm ac$ and $f(x) = \pm\phi_1(x)\psi_1(x)$. Thus the leading term of both $\phi_1(x)$ and $\psi_1(x)$ is ± 1. Further, $\phi(\alpha) = 0 \Rightarrow \phi_1(\alpha) = 0$. So in fact $\phi(x) = \pm\phi_1(x)$ which is monic in $\mathbb{Z}[x]$.

(b) $\Rightarrow$ (c) Let $\phi(x) = x^n + a_{n-1}x^{n-1} + \cdots + a_0 \in \mathbb{Z}[x]$ be the minimal polynomial of α. Recall $\mathbb{Z}[\alpha] = \{f(\alpha) : f(x) \in \mathbb{Z}[x]\}$. In order to prove (c), it is enough to find a finite basis for $\mathbb{Z}[\alpha]$.

Claim: $\{1, \alpha, \ldots, \alpha^{n-1}\}$ *generates* $\mathbb{Z}[\alpha]$ *(as a $\mathbb{Z}$-module).*

Proof: It suffices to show that α^N, for any $N \in \mathbb{Z}^+$, is a linear combination of $\{1, \alpha, \ldots, \alpha^{n-1}\}$ with coefficients in $\mathbb{Z}$. We proceed inductively. Clearly this holds for $N \leq n-1$. For $N \geq n$, suppose this holds $\forall \alpha^j$, $j < N$.

$$\begin{aligned} \alpha^N &= \alpha^{N-n}\alpha^n \\ &= \alpha^{N-n}[-(a_0 + a_1\alpha + \cdots + a_{n-1}\alpha^{n-1})] \\ &= (-\alpha^{N-n}a_0)1 + (-\alpha^{N-n}a_1)\alpha + \cdots + (-\alpha^{N-n}a_{n-1})\alpha^{n-1}. \end{aligned}$$

By our inductive hypothesis, $-\alpha^{N-n}a_i \in \mathbb{Z}[\alpha]$ $\forall i = 0, 1, \ldots, n-1$. Then $\mathbb{Z}[\alpha]$ is a $\mathbb{Z}$-module generated by $\{1, \alpha, \ldots, \alpha^{n-1}\}$.

(c) $\Rightarrow$ (d) Let $M = \mathbb{Z}[\alpha]$. Clearly $\alpha\mathbb{Z}[\alpha] \subseteq \mathbb{Z}[\alpha]$.

(d) $\Rightarrow$ (a) Let $x_1, \ldots, x_r$ generate M over $\mathbb{Z}$. So $M \subseteq \mathbb{Z}x_1 + \cdots + \mathbb{Z}x_r$. By assumption $\alpha x_i \in M\ \forall i = 1, \ldots, r$. It follows that there exists a set of $c_{ij} \in \mathbb{Z}$ such that $\alpha x_i = \sum_{j=1}^{n} c_{ij}x_j \forall i = 1, \ldots, r$. Let $C = (c_{ij})$. Then

$$C\begin{pmatrix} x_1 \\ \vdots \\ x_r \end{pmatrix} = \alpha\begin{pmatrix} x_1 \\ \vdots \\ x_r \end{pmatrix},$$

$$\Leftrightarrow \qquad (C - \alpha I)\begin{pmatrix} x_1 \\ \vdots \\ x_r \end{pmatrix} = 0.$$

Since not all of $x_1, \ldots, x_r$ can vanish, $\det(C - \alpha I) = 0$. In other words,

$$\begin{vmatrix} c_{11} - x & c_{12} & \cdots & c_{1n} \\ c_{21} & c_{22} - x & \cdots & c_{2n} \\ \vdots & \vdots & & \vdots \\ c_{n1} & c_{n2} & \cdots & c_{nn} - x \end{vmatrix} = 0 \quad \text{when} \quad x = \alpha.$$

This is a polynomial equation in $\mathbb{Z}[x]$ of degree n whose leading coefficient is $(-1)^n$. Take

$$f(x) = \begin{cases} \det(C - xI) & \text{for } n \text{ even}, \\ -\det(C - xI) & \text{for } n \text{ odd}. \end{cases}$$

Then $f(x)$ is a monic polynomial in $\mathbb{Z}[x]$ such that $f(\alpha) = 0$. Thus α is an algebraic integer. □

Example 3.3.10 Let K be an algebraic number field. Let $\mathcal{O}_K$ be the set of all algebraic integers in K. Show that $\mathcal{O}_K$ is a ring.

Solution. From the above theorem, we know that for α, β, algebraic integers, $\mathbb{Z}[\alpha]$, $\mathbb{Z}[\beta]$ are finitely generated $\mathbb{Z}$-modules. Thus $M = \mathbb{Z}[\alpha, \beta]$ is also a finitely generated $\mathbb{Z}$-module. Moreover,

$$(\alpha \pm \beta)M \subseteq M,$$

and

$$(\alpha\beta)M \subseteq M.$$

So $\alpha \pm \beta$ and $\alpha\beta$ are algebraic integers; i.e., $\alpha \pm \beta$ and $\alpha\beta$ are in $\mathcal{O}_K$. So $\mathcal{O}_K$ is a ring.

Exercise 3.3.11 Let $f(x) = x^n + a_{n-1}x^{n-1} + \cdots + a_1x + a_0$, and assume that for p prime $p \mid a_i$ for $0 \leq i < k$ and $p^2 \nmid a_0$. Show that $f(x)$ has an irreducible factor of degree at least k. (The case $k = n$ is referred to as Eisenstein's criterion for irreducibility.)

Exercise 3.3.12 Show that $f(x) = x^5 + x^4 + 3x^3 + 9x^2 + 3$ is irreducible over $\mathbb{Q}$.

3.4 Supplementary Problems

Exercise 3.4.1 Show that

$$\sum_{n=0}^{\infty} \frac{1}{a^{n!}}$$

is transcendental for $a \in \mathbb{Z}, a \geq 2$.

Exercise 3.4.2 Show that

$$\sum_{n=1}^{\infty} \frac{1}{a^{3^n}}$$

is transcendental for $a \in \mathbb{Z}, a \geq 2$.

Exercise 3.4.3 Show that

$$\sum_{n=1}^{\infty} \frac{1}{a^{f(n)}}$$

is transcendental when

$$\lim_{n\to\infty} \frac{f(n+1)}{f(n)} > 2.$$

Exercise 3.4.4 Prove that $f(x) = x^6 + 7x^5 - 12x^3 + 6x + 2$ is irreducible over $\mathbb{Q}$.

Exercise 3.4.5 Using Thue's theorem, show that $f(x,y) = x^6 + 7x^5y - 12x^3y^3 + 6xy^5 + 8y^6 = m$ has only a finite number of solutions for $m \in \mathbb{Z}^*$.

Exercise 3.4.6 Let ζ_m be a primitive mth root of unity. Show that

$$\prod_{\substack{0 \leq i,j \leq m-1 \\ i \neq j}} (\zeta_m^i - \zeta_m^j) = (-1)^{m-1} m^m.$$

Exercise 3.4.7 Let

$$\phi_m(x) = \prod_{\substack{1 \leq i \leq m \\ (i,m)=1}} (x - \zeta_m^i)$$

denote the mth cyclotomic polynomial. Prove that

$$x^m - 1 = \prod_{d|m} \phi_d(x).$$

Exercise 3.4.8 Show that $\phi_m(x) \in \mathbb{Z}[x]$.

Exercise 3.4.9 Show that $\phi_m(x)$ is irreducible in $\mathbb{Q}[x]$ for every $m \geq 1$.

Exercise 3.4.10 Let I be a subset of the positive integers $\leq m$ which are coprime to m. Set

$$f(x) = \prod_{i \in I} (x - \zeta_m^i).$$

Suppose that $f(\zeta_m) = 0$ and $f(\zeta_m^p) \neq 0$ for some prime p. Show that $p \mid m$. (This observation gives an alternative proof for the irreducibility of $\phi_m(x)$.)

Exercise 3.4.11 Consider the equation $x^3 + 3x^2y + xy^2 + y^3 = m$. Using Thue's theorem, deduce that there are only finitely many integral solutions to this equation.

Exercise 3.4.12 Assume that n is an odd integer, $n \geq 3$. Show that $x^n + y^n = m$ has only finitely many integral solutions.

Exercise 3.4.13 Let ζ_m denote a primitive mth root of unity. Show that $\mathbb{Q}(\zeta_m)$ is normal over $\mathbb{Q}$.

Exercise 3.4.14 Let a be squarefree and greater than 1, and let p be prime. Show that the normal closure of $\mathbb{Q}(a^{1/p})$ is $\mathbb{Q}(a^{1/p}, \zeta_p)$.

Chapter 4

Integral Bases

In this chapter, we look more closely at the algebraic structure of $\mathcal{O}_K$, the ring of integers of an algebraic number field K. In particular, we show that $\mathcal{O}_K$ is always a finitely generated $\mathbb{Z}$-module admitting a $\mathbb{Q}$-basis for K as a generating set (where K is viewed as a $\mathbb{Q}$-vector space). We will define the trace and norm of an element of any number field. We will also define an important invariant of a number field called the discriminant which arises in many calculations within the number field. Finally, ideals in the ring of integers of a number field will be briefly discussed at the end of the chapter.

4.1 The Norm and the Trace

We begin by defining two important rational numbers associated with an element of an algebraic number field K. Recall that if K is an algebraic number field, then K can be viewed as a finite-dimensional vector space over $\mathbb{Q}$. Then if $\alpha \in K$, the map from K to K defined by $\Phi_\alpha : v \to \alpha v$ defines a linear operator on K. We define the *trace* of α by $\mathrm{Tr}_K(\alpha) := \mathrm{Tr}(\Phi_\alpha)$ and the *norm* of α by $\mathrm{N}_K(\alpha) := \det(\Phi_\alpha)$ (where Tr and det are the usual trace and determinant of a linear map). We sometimes also use the notation $\mathrm{Tr}_{K/\mathbb{Q}}$ for Tr_K and $\mathrm{N}_{K/\mathbb{Q}}$ for N_K.

Thus, to find $\mathrm{Tr}_K(\alpha)$, we choose any $\mathbb{Q}$-basis $\omega_1, \omega_2, \ldots, \omega_n$ of K and write

$$\alpha\omega_i = \sum a_{ij}\omega_j \quad \forall\, i,$$

so $\mathrm{Tr}_K(\alpha) = \mathrm{Tr} A$ and $\mathrm{N}_K(\alpha) = \det A$ where A is the matrix (a_{ij}).

Lemma 4.1.1 *If K is an algebraic number field of degree n over $\mathbb{Q}$, and $\alpha \in \mathcal{O}_K$ its ring of integers, then $\mathrm{Tr}_K(\alpha)$ and $\mathrm{N}_K(\alpha)$ are in $\mathbb{Z}$.*

Proof. We begin by writing $\alpha\omega_i = \sum_{j=1}^n a_{ij}\omega_j \ \forall i$. Then we have

$$\alpha^{(k)}\omega_i^{(k)} = \sum_{j=1}^n a_{ij}\omega_j^{(k)} \qquad \forall i, k,$$

where $\alpha^{(k)}$ is the kth conjugate of α. We rewrite the above by introducing the Kronecker delta function to get

$$\sum_{j=1}^n \delta_{jk}\alpha^{(j)}\omega_i^{(j)} = \sum_{j=1}^n a_{ij}\omega_j^{(k)},$$

where $\delta_{ij} = \begin{cases} 0 & \text{if } i \neq j, \\ 1 & \text{if } i = j. \end{cases}$ Now, if we define the matrices

$$A_0 = (\alpha^{(i)}\delta_{ij}), \qquad \Omega = (\omega_i^{(j)}), \qquad A = (a_{ij}),$$

the preceding statement tells us that $\Omega A_0 = A\Omega$ or $A_0 = \Omega^{-1}A\Omega$, so we conclude that $\mathrm{Tr}A = \mathrm{Tr}A_0$ and $\det A = \det A_0$. But $\mathrm{Tr}A_0$ is just the sum of the conjugates of α and is thus (up to sign) the coefficient of the x^{n-1} term in the minimal polynomial for α; similarly, $\det A_0$ is just the product of the conjugates of α and is thus equal (up to sign) to the constant term in the minimal polynomial for α. Thus $\mathrm{Tr}_K(\alpha)$ and $\mathrm{N}_K(\alpha)$ are in $\mathbb{Z}$. □

Exercise 4.1.2 Let $K = \mathbb{Q}(i)$. Show that $i \in \mathcal{O}_K$ and verify that $\mathrm{Tr}_K(i)$ and $\mathrm{N}_K(i)$ are integers.

Exercise 4.1.3 Determine the algebraic integers of $K = \mathbb{Q}(\sqrt{-5})$.

Given an algebraic number field K and $\omega_1, \omega_2, \dots, \omega_n$ a $\mathbb{Q}$-basis for K, consider the correspondence from K to $M_n(\mathbb{Q})$ given by $\alpha \to (a_{ij})$ where $\alpha\omega_i = \sum a_{ij}\omega_j$. This is readily seen to give a homomorphism from K to $M_n(\mathbb{Q})$. From this we can deduce that $\mathrm{Tr}_K(\cdot)$ is in fact a $\mathbb{Q}$-linear map from K to $\mathbb{Q}$.

Lemma 4.1.4 *The bilinear pairing given by $B(x, y) : K \times K \to \mathbb{Q}$ such that $(x, y) \to \mathrm{Tr}_K(xy)$ is nondegenerate.*

Proof. We recall that if V is a finite-dimensional vector space over a field F with basis $e_1, e_2, \dots, e_n$ and $B : V \times V \to F$ is a bilinear map, we can associate a matrix to B as follows. Write

$$\begin{aligned} v &= \sum a_i e_i \quad \text{with } a_i \in F, \\ u &= \sum b_i e_i \quad \text{with } b_i \in F. \end{aligned}$$

Then

$$\begin{aligned} B(v,u) &= \sum_i B(a_i e_i, u) \\ &= \sum_i a_i B(e_i, u) \\ &= \sum_{i,j} a_i b_j B(e_i, e_j) \end{aligned}$$

and we associate to B the matrix $(B(e_i, e_j))$. B is said to be nondegenerate if the matrix associated to it is nonsingular. This definition is independent of the choice of basis (see Exercise 4.1.5 below).

Now, if $\omega_1, \omega_2, \ldots, \omega_n$ is a $\mathbb{Q}$-basis for K, then the matrix associated to $B(x,y)$ with respect to this basis is just

$$(B(\omega_i, \omega_j)) = (\mathrm{Tr}_K(\omega_i \omega_j)),$$

but $\mathrm{Tr}_K(\omega_i \omega_j) = \sum \omega_i^{(k)} \omega_j^{(k)}$ and thus we see that

$$(B(\omega_i, \omega_j)) = \Omega \Omega^T,$$

where Ω is nonsingular because $\omega_1, \omega_2, \ldots, \omega_n$ form a basis for K. Thus $B(x,y)$ is indeed nondegenerate. □

Exercise 4.1.5 Show that the definition of nondegeneracy above is independent of the choice of basis.

4.2 Existence of an Integral Basis

Let K be an algebraic number field of degree n over $\mathbb{Q}$, and $\mathcal{O}_K$ its ring of integers. We say that $\omega_1, \omega_2, \ldots, \omega_n$ is an *integral basis* for K if $\omega_i \in \mathcal{O}_K$ for all i, and $\mathcal{O}_K = \mathbb{Z}\omega_1 + \mathbb{Z}\omega_2 + \cdots + \mathbb{Z}\omega_n$.

Exercise 4.2.1 Show that $\exists \omega_1^*, \omega_2^*, \ldots, \omega_n^* \in K$ such that

$$\mathcal{O}_K \subseteq \mathbb{Z}\omega_1^* + \mathbb{Z}\omega_2^* + \cdots + \mathbb{Z}\omega_n^*.$$

Theorem 4.2.2 *Let $\alpha_1, \alpha_2, \ldots, \alpha_n$ be a set of generators for a finitely generated $\mathbb{Z}$-module M, and let N be a submodule.*

(a) *$\exists \beta_1, \beta_2, \ldots, \beta_m$ in N with $m \leq n$ such that*

$$N = \mathbb{Z}\beta_1 + \mathbb{Z}\beta_2 + \cdots + \mathbb{Z}\beta_m$$

and $\beta_i = \sum_{j \geq i} p_{ij} \alpha_j$ with $1 \leq i \leq m$ and $p_{ij} \in \mathbb{Z}$.

(b) *If $m = n$, then $[M : N] = p_{11} p_{22} \cdots p_{nn}$.*

Proof. (a) We will proceed by induction on the number of generators of a $\mathbb{Z}$-module. This is trivial when $n = 0$. We can assume that we have proved the above statement to be true for all $\mathbb{Z}$-modules with $n-1$ or fewer generators, and proceed to prove it for n. We define M' to be the submodule generated by $\alpha_2, \alpha_3, \ldots, \alpha_n$ over $\mathbb{Z}$, and define N' to be $N \cap M'$. Now, if $n = 1$, then $M' = 0$ and there is nothing to prove. If $N = N'$, then the statement is true by our induction hypothesis.

So we assume that $N \neq N'$ and consider A, the set of all integers k such that $\exists k_2, k_3, \ldots, k_n$ with $k\alpha_1 + k_2\alpha_2 + \cdots + k_n\alpha_n \in N$. Since N is a submodule, we deduce that A is a subgroup of $\mathbb{Z}$. All additive subgroups of $\mathbb{Z}$ are of the form $m\mathbb{Z}$ for some integer m, and so $A = k_{11}\mathbb{Z}$ for some k_{11}. Then let $\beta_1 = k_{11}\alpha_1 + k_{12}\alpha_2 + \cdots + k_{1n}\alpha_n \in N$. If we have some $\alpha \in N$, then

$$\alpha = \sum_{i=1}^{n} h_i\alpha_i,$$

with $h_i \in \mathbb{Z}$ and $h_1 \in A$ so $h_1 = ak_{11}$. Therefore, $\alpha - a\beta_1 \in N'$. By the induction hypothesis, there exist

$$\beta_i = \sum_{j \geq i} k_{ij}\alpha_j,$$

$i = 2, 3 \ldots, m$, which generate N' over $\mathbb{Z}$ and which satisfy all the conditions above. It is clear that adding β_1 to this list gives us a set of generators of N.

(b) Consider α, an arbitrary element of M. Then $\alpha = \sum c_i\alpha_i$. Recalling that

$$\beta_i = \sum_{j \geq i} p_{ij}\alpha_j,$$

we write $c_1 = p_{11}q_1 + r_1$, with $0 \leq r_1 < p_{11}$. Then $\alpha - q_1\beta_1 = \sum c_i'\alpha_i$ where $0 \leq c_1' < p_{11}$. Note that $\alpha \equiv \alpha - q_1\beta_1 \pmod{N}$. Next we write $c_2' = p_{22}q_2 + r_2$, where $0 \leq r_2 < p_{22}$, and note that

$$\alpha \equiv \alpha - q_1\beta_1 - q_2\beta_2 \pmod{N}.$$

It is clear by induction that we can continue this process to arrive at an expression $\alpha' = \sum k_i\alpha_i$ with $0 \leq k_i < p_{ii}$ and $\alpha \equiv \alpha' \pmod{N}$.

It remains only to show that if we have $\alpha = \sum c_i\alpha_i$ and $\beta = \sum d_i\alpha_i$ where $c_i \neq d_i$ for at least one i and $0 \leq c_i, d_i < p_{ii}$, then α and β are distinct mod N. Suppose that this is not true, and that

$$\sum c_i\alpha_i \equiv \sum d_i\alpha_i \pmod{N},$$

where $c_i \neq d_i$ for at least one i. Suppose $c_i = d_i$ for $i < r$ and $c_r \neq d_r$. Then $\sum(c_i - d_i)\alpha_i \in N$, so

$$\sum_{i \geq r}(c_i - d_i)\alpha_i = \sum_{i \geq r} k_i\beta_i = \sum_{i \geq r} k_i \left(\sum_{j \geq i} p_{ij}\alpha_j\right).$$

Since c_r, d_r are both less than p_{rr}, we have $c_r = d_r$, a contradiction. Thus, each coset in M/N has a unique representative

$$\alpha = \sum c_i \alpha_i,$$

with $0 \leq c_i < p_{ii}$, and there are $p_{11}p_{22}\cdots p_{nn}$ of them. So $[M : N] = p_{11}p_{22}\cdots p_{nn}$. □

Exercise 4.2.3 Show that $\mathcal{O}_K$ has an integral basis.

Exercise 4.2.4 Show that $\det(\mathrm{Tr}(\omega_i\omega_j))$ is independent of the choice of integral basis.

We are justified now in making the following definition. If K is an algebraic number field of degree n over $\mathbb{Q}$, define the *discriminant* of K as

$$d_K := \det(\omega_i^{(j)})^2,$$

where $\omega_1, \omega_2, \ldots, \omega_n$ is an integral basis for K.

Exercise 4.2.5 Show that the discriminant is well-defined. In other words, show that given $\omega_1, \omega_2, \ldots, \omega_n$ and $\theta_1, \theta_2, \ldots, \theta_n$, two integral bases for K, we get the same discriminant for K.

We can generalize the notion of a discriminant for arbitrary elements of K. Let $K/\mathbb{Q}$ be an algebraic number field, a finite extension of $\mathbb{Q}$ of degree n. Let $\sigma_1, \sigma_2, \ldots, \sigma_n$ be the embeddings of K. For $a_1, a_2, \ldots, a_n \in K$ we can define $d_{K/\mathbb{Q}}(a_1, \ldots, a_n) = \left[\det(\sigma_i(a_j))\right]^2$.

Exercise 4.2.6 Show that

$$d_{K/\mathbb{Q}}(1, a, \ldots, a^{n-1}) = \prod_{i>j} \left(\sigma_i(a) - \sigma_j(a)\right)^2.$$

We denote $d_{K/\mathbb{Q}}(1, a, \ldots, a^{n-1})$ by $d_{K/\mathbb{Q}}(a)$.

Exercise 4.2.7 Suppose that $u_i = \sum_{j=1}^n a_{ij}v_j$ with $a_{ij} \in \mathbb{Q}, v_j \in K$. Show that $d_{K/\mathbb{Q}}(u_1, u_2, \ldots, u_n) = \left(\det(a_{ij})\right)^2 d_{K/\mathbb{Q}}(v_1, v_2, \ldots, v_n)$.

For a module M with submodule N, we can define the *index* of N in M to be the number of elements in M/N, and denote this by $[M : N]$. Suppose α is an algebraic integer of degree n, generating a field K. We define the *index* of α to be the index of $\mathbb{Z} + \mathbb{Z}\alpha + \cdots + \mathbb{Z}\alpha^{n-1}$ in $\mathcal{O}_K$.

Exercise 4.2.8 Let $a_1, a_2, \ldots, a_n \in \mathcal{O}_K$ be linearly independent over $\mathbb{Q}$. Let $N = \mathbb{Z}a_1 + \mathbb{Z}a_2 + \cdots + \mathbb{Z}a_n$ and $m = [\mathcal{O}_K : N]$. Prove that

$$d_{K/\mathbb{Q}}(a_1, a_2, \ldots, a_n) = m^2 d_K.$$

4.3 Examples

Example 4.3.1 Suppose that the minimal polynomial of α is Eisensteinian with respect to a prime p, i.e., α is a root of the polynomial

$$x^n + a_{n-1}x^{n-1} + \cdots + a_1 x + a_0,$$

where $p \mid a_i$, $0 \le i \le n-1$ and $p^2 \nmid a_0$. Show that the index of α is not divisible by p.

Solution. Let $M = \mathbb{Z} + \mathbb{Z}\alpha + \cdots + \mathbb{Z}\alpha^{n-1}$. First observe that since

$$\alpha^n + a_{n-1}\alpha^{n-1} + \cdots + a_1\alpha + a_0 = 0,$$

then $\alpha^n/p \in M \subseteq \mathcal{O}_K$. Also, $|\mathrm{N}_K(\alpha)| = a_0 \not\equiv 0 \pmod{p^2}$.

We will proceed by contradiction. Suppose $p \mid [\mathcal{O}_K : M]$. Then there is an element of order p in the group $\mathcal{O}_K/M$, meaning $\exists \xi \in \mathcal{O}_K$ such that $\xi \notin M$ but $p\xi \in M$. Then

$$p\xi = b_0 + b_1\alpha + \cdots + b_{n-1}\alpha^{n-1},$$

where not all the b_i are divisible by p, for otherwise $\xi \in M$. Let j be the least index such that $p \nmid b_j$. Then

$$\begin{aligned}\eta &= \xi - \left(\frac{b_0}{p} + \frac{b_1}{p}\alpha + \cdots + \frac{b_{j-1}}{p}\alpha^{j-1}\right)\\ &= \frac{b_j}{p}\alpha^j + \frac{b_{j+1}}{p}\alpha^{j+1} + \cdots + \frac{b_n}{p}\alpha^n\end{aligned}$$

is in $\mathcal{O}_K$, since both ξ and

$$\frac{b_0}{p} + \frac{b_1}{p}\alpha + \cdots + \frac{b_n}{p}\alpha^{j-1}$$

are in $\mathcal{O}_K$.

If $\eta \in \mathcal{O}_K$, then of course $\eta\alpha^{n-j-1}$ is also in $\mathcal{O}_K$, and

$$\eta\alpha^{n-j-1} = \frac{b_j}{p}\alpha^{n-1} + \frac{\alpha^n}{p}(b_{j+1} + b_{j+2}\alpha + \cdots + b_n\alpha^{n-j-2}).$$

Since both α^n/p and $(b_{j+1} + b_{j+2}\alpha + \cdots + b_n\alpha^{n-j-2})$ are in $\mathcal{O}_K$, we conclude that $(b_j\alpha^{n-1})/p \in \mathcal{O}_K$.

We know from Lemma 4.1.1 that the norm of an algebraic integer is always a rational integer, so

$$\begin{aligned}\mathrm{N}_K\left(\frac{b_j}{p}\alpha^{n-1}\right) &= \frac{b_j^n \mathrm{N}_K(\alpha)^{n-1}}{p^n}\\ &= \frac{b_j^n a_0^{n-1}}{p^n}\end{aligned}$$

must be an integer. But p does not divide b_j, and p^2 does not divide a_0, so this is impossible. This proves that we do not have an element of order p, and thus $p \nmid [\mathcal{O}_K : M]$.

Exercise 4.3.2 Let $m \in \mathbb{Z}, \alpha \in \mathcal{O}_K$. Prove that $d_{K/\mathbb{Q}}(\alpha + m) = d_{K/\mathbb{Q}}(\alpha)$.

Exercise 4.3.3 Let α be an algebraic integer, and let $f(x)$ be the minimal polynomial of α. If f has degree n, show that $d_{K/\mathbb{Q}}(\alpha) = (-1)^{\binom{n}{2}} \prod_{i=1}^{n} f'(\alpha^{(i)})$.

Example 4.3.4 Let $K = \mathbb{Q}(\sqrt{D})$ with D a squarefree integer. Find an integral basis for $\mathcal{O}_K$.

Solution. An arbitrary element α of K is of the form $\alpha = r_1 + r_2\sqrt{D}$ with $r_1, r_2 \in \mathbb{Q}$. Since $[K : \mathbb{Q}] = 2$, α has only one conjugate: $r_1 - r_2\sqrt{D}$. From Lemma 4.1.1 we know that if α is an algebraic integer, then $\mathrm{Tr}_K(\alpha) = 2r_1$ and

$$\begin{aligned} \mathrm{N}_K(\alpha) &= (r_1 + r_2\sqrt{D})(r_1 - r_2\sqrt{D}) \\ &= r_1^2 - Dr_2^2 \end{aligned}$$

are both integers. We note also that since α satisfies the monic polynomial $x^2 - 2r_1x + r_1^2 - Dr_2^2$, if $\mathrm{Tr}_K(\alpha)$ and $\mathrm{N}_K(\alpha)$ are integers, then α is an algebraic integer. If $2r_1 \in \mathbb{Z}$ where $r_1 \in \mathbb{Q}$, then the denominator of r_1 can be at most 2. We also need $r_1^2 - Dr_2^2$ to be an integer, so the denominator of r_2 can be no more than 2. Then let $r_1 = g_1/2, r_2 = g_2/2$, where $g_1, g_2 \in \mathbb{Z}$. The second condition amounts to

$$\frac{g_1^2 - Dg_2^2}{4} \in \mathbb{Z},$$

which means that $g_1^2 - Dg_2^2 \equiv 0 \pmod 4$, or $g_1^2 \equiv Dg_2^2 \pmod 4$.

We will discuss two cases:

Case 1. $D \equiv 1 \pmod 4$.

If $D \equiv 1 \pmod 4$, and $g_1^2 \equiv Dg_2^2 \pmod 4$, then g_1 and g_2 are either both even or both odd. So if $\alpha = r_1 + r_2\sqrt{D}$ is an algebraic integer of $\mathbb{Q}(\sqrt{D})$, then either r_1 and r_2 are both integers, or they are both fractions with denominator 2.

We recall from Chapter 3 that if $4 \mid (-D + 1)$, then $(1 + \sqrt{D})/2$ is an algebraic integer. This suggests that we use $1, (1 + \sqrt{D})/2$ as a basis; it is clear from the discussion above that this is in fact an integral basis.

Case 2. $D \equiv 2, 3 \pmod 4$.

If $g_1^2 \equiv Dg_2^2 \pmod 4$, then both g_1 and g_2 must be even. Then a basis for $\mathcal{O}_K$ is $1, \sqrt{D}$; again it is clear that this is an integral basis.

Exercise 4.3.5 If $D \equiv 1 \pmod 4$, show that every integer of $\mathbb{Q}(\sqrt{D})$ can be written as $(a + b\sqrt{D})/2$ where $a \equiv b \pmod 2$.

Example 4.3.6 Let $K = \mathbb{Q}(\alpha)$ where $\alpha = r^{1/3}$, $r = ab^2 \in \mathbb{Z}$ where ab is squarefree. If $3 \mid r$, assume that $3 \mid a, 3 \nmid b$. Find an integral basis for K.

Solution. The minimal polynomial of α is $f(x) = x^3 - r$, and α's conjugates are $\alpha, \omega\alpha$, and $\omega^2\alpha$ where ω is a primitive cube root of unity. By Exercise 4.3.3,

$$d_{K/\mathbb{Q}}(\alpha) = -\prod_{i=1}^{3} f'(\alpha^{(i)}) = -3^3 r^2.$$

So $-3^3 r^2 = m^2 d_K$ where $m = [\mathcal{O}_K : \mathbb{Z} + \mathbb{Z}\alpha + \mathbb{Z}\alpha^2]$. We note that $f(x)$ is Eisensteinian for every prime divisor of a so by Example 4.3.1 if $p \mid a$, $p \nmid m$. Thus if $3 \mid a$, $27a^2 \mid d_K$, and if $3 \nmid a$, then $3a^2 \mid d_K$.

We now consider $\beta = \alpha^2/b$, which is a root of the polynomial $x^3 - a^2 b$. This polynomial is Eisensteinian for any prime which divides b. Therefore $b^2 \mid d_K$. We conclude that $d_K = -3^n(ab)^2$ where $n = 3$ if $3 \mid r$ and $n = 1$ or 3 otherwise. We will consider three cases: $r \not\equiv 1, 8 \pmod 9$, $r \equiv 1 \pmod 9$ and $r \equiv 8 \pmod 9$.

Case 1. If $r \not\equiv 1, 8 \pmod 9$, then $r^3 \not\equiv r \pmod 9$.

Then the polynomial $(x + r)^3 - r$ is Eisensteinian with respect to the prime 3. A root of this polynomial is $\alpha - r$ and $d_{K/\mathbb{Q}}(\alpha - r) = d_{K/\mathbb{Q}}(\alpha) = -27r^2$. This implies that $3 \nmid m$ and so $m = b$.

We can choose as our integral basis $1, \alpha, \alpha^2/b$, all of which are algebraic integers. We verify that this is an integral basis by checking the index of $\mathbb{Z} + \mathbb{Z}\alpha + \mathbb{Z}\alpha^2$ in $\mathbb{Z} + \mathbb{Z}\alpha + \mathbb{Z}\alpha^2/b$, which is clearly b. Thus

$$\mathcal{O}_K = \mathbb{Z} + \mathbb{Z}\alpha + \mathbb{Z}\frac{\alpha^2}{b}.$$

Case 2. If $r \equiv 1 \pmod 9$, then $c = (1 + \alpha + \alpha^2)/3$ is an algebraic integer.

In fact, since $\mathrm{Tr}_K(\alpha) = \mathrm{Tr}_K(\alpha^2) = 0$, then $\mathrm{Tr}_K(c) = 1 \in \mathbb{Z}$ and

$$\mathrm{N}_K(c) = \frac{\mathrm{N}_K(1 + \alpha + \alpha^2)}{27} = \frac{\mathrm{N}_K(\alpha^3 - 1)}{27\mathrm{N}_K(\alpha - 1)} = \frac{(r-1)^2}{27}$$

because the minimal polynomial for $1 - \alpha$ is $x^3 + 3x^2 + 3x + 1 - r$. The other coefficient for the minimal polynomial of c is $(1 - r)/3$ which is an integer. If c is in $\mathcal{O}_K$, then $\mathcal{O}_K/(\mathbb{Z} + \mathbb{Z}\alpha + \mathbb{Z}\alpha^2)$ has an element of order 3, and so $3 \mid m$. Then $d_K = -3(ab)^2$, so $m = 3b$. We will choose as our integral basis $\alpha, \alpha^2/b, c$, noting that since

$$\begin{aligned} 1 &= 3c - \alpha - b\frac{\alpha^2}{b}, \\ \alpha &= 0 + \alpha + 0, \\ \alpha^2 &= 0 + 0 + b\frac{\alpha^2}{b}, \end{aligned}$$

then Theorem 4.2.2 tells us that the index of $\mathbb{Z}+\mathbb{Z}\alpha+\mathbb{Z}\alpha^2$ in $\mathbb{Z}\alpha+\mathbb{Z}\alpha^2/b+\mathbb{Z}c$ is $3b$. Therefore

$$\mathcal{O}_K = \mathbb{Z}\alpha + \mathbb{Z}\frac{\alpha^2}{b} + \mathbb{Z}\frac{1+\alpha+\alpha^2}{3}.$$

Case 3. If $r \equiv 8 \pmod 9$, consider $d = (1-\alpha+\alpha^2)/3$.

This is an algebraic integer. $\mathrm{Tr}_K(d) = 1, \mathrm{N}_K(d) = (1+r)^2/27 \in \mathbb{Z}$, and the remaining coefficient for the minimal polynomial of d is $(1+r)/3 \in \mathbb{Z}$. By the same reasoning as above, we conclude that $3 \mid m$ and so $m = 3b$. We choose $\alpha, \alpha^2/b, d$ as an integral basis, noting that

$$\begin{aligned} 1 &= 3d + \alpha - b\frac{\alpha^2}{b}, \\ \alpha &= 0 + \alpha + 0, \\ \alpha^2 &= 0 + 0 + b\frac{\alpha^2}{b}, \end{aligned}$$

so that the index of $\mathbb{Z}+\mathbb{Z}\alpha+\mathbb{Z}\alpha^2$ in $\mathbb{Z}\alpha+\mathbb{Z}\alpha^2/b+\mathbb{Z}d$ is $3b$. We conclude that

$$\mathcal{O}_K = \mathbb{Z}\alpha + \mathbb{Z}\frac{\alpha^2}{b} + \mathbb{Z}\frac{1-\alpha+\alpha^2}{b}.$$

Exercise 4.3.7 Let ζ be any primitive pth root of unity, and $K = \mathbb{Q}(\zeta)$. Show that $1, \zeta, \ldots, \zeta^{p-2}$ form an integral basis of K.

4.4 Ideals in $\mathcal{O}_K$

At this point, we have shown that $\mathcal{O}_K$ is indeed much like $\mathbb{Z}$ in its algebraic structure. It turns out that we are only halfway to the final step in our generalization of an integer in a number field. We may think of the ideals in $\mathcal{O}_K$ as the most general integers in K, and we remark that when this set of ideals is endowed with the usual operations of ideal addition and multiplication, we recover an arithmetic most like that of $\mathbb{Z}$. We prove now several properties of the ideals in $\mathcal{O}_K$.

Exercise 4.4.1 Let $\mathfrak{a}$ be a nonzero ideal of $\mathcal{O}_K$. Show that $\mathfrak{a} \cap \mathbb{Z} \neq \{0\}$.

Exercise 4.4.2 Show that $\mathfrak{a}$ has an integral basis.

Exercise 4.4.3 Show that if $\mathfrak{a}$ is a nonzero ideal in $\mathcal{O}_K$, then $\mathfrak{a}$ has finite index in $\mathcal{O}_K$.

Exercise 4.4.4 Show that every nonzero prime ideal in $\mathcal{O}_K$ contains exactly one integer prime.

We define the *norm* of a nonzero ideal in $\mathcal{O}_K$ to be its index in $\mathcal{O}_K$. We will denote the norm of an ideal by $N(\mathfrak{a})$.

Exercise 4.4.5 Let $\mathfrak{a}$ be an integral ideal with basis $\alpha_1, \ldots, \alpha_n$. Show that

$$[\det(\alpha_i^{(j)})]^2 = (N\mathfrak{a})^2 d_K.$$

4.5 Supplementary Problems

Exercise 4.5.1 Let K be an algebraic number field. Show that $d_K \in \mathbb{Z}$.

Exercise 4.5.2 Let $K/\mathbb{Q}$ be an algebraic number field of degree n. Show that $d_K \equiv 0$ or $1 \pmod 4$. This is known as Stickelberger's criterion.

Exercise 4.5.3 Let $f(x) = x^n + a_{n-1}x^{n-1} + \cdots + a_1 x + a_0$ with $(a_i \in \mathbb{Z})$ be the minimal polynomial of θ. Let $K = \mathbb{Q}(\theta)$. If for each prime p such that $p^2 \mid d_{K/\mathbb{Q}}(\theta)$ we have $f(x)$ Eisensteinian with respect to p, show that $\mathcal{O}_K = \mathbb{Z}[\theta]$.

Exercise 4.5.4 If the minimal polynomial of α is $f(x) = x^n + ax + b$, show that for $K = \mathbb{Q}(\alpha)$,

$$d_{K/\mathbb{Q}}(\alpha) = (-1)^{\binom{n}{2}} \left(n^n b^{n-1} + a^n (1-n)^{n-1}\right).$$

Exercise 4.5.5 Determine an integral basis for $K = \mathbb{Q}(\theta)$ where $\theta^3 + 2\theta + 1 = 0$.

Exercise 4.5.6 (Dedekind) Let $K = \mathbb{Q}(\theta)$ where $\theta^3 - \theta^2 - 2\theta - 8 = 0$.

(a) Show that $f(x) = x^3 - x^2 - 2x - 8$ is irreducible over $\mathbb{Q}$.

(b) Consider $\beta = (\theta^2 + \theta)/2$. Show that $\beta^3 - 3\beta^2 - 10\beta - 8 = 0$. Hence β is integral.

(c) Show that $d_{K/\mathbb{Q}}(\theta) = -4(503)$, and $d_{K/\mathbb{Q}}(1, \theta, \beta) = -503$. Deduce that $1, \theta, \beta$ is a $\mathbb{Z}$-basis of $\mathcal{O}_K$.

(d) Show that every integer x of K has an even discriminant.

(e) Deduce that $\mathcal{O}_K$ has no integral basis of the form $\mathbb{Z}[\alpha]$.

Exercise 4.5.7 Let $m = p^a$, with p prime and $K = \mathbb{Q}(\zeta_m)$. Show that

$$(1 - \zeta_m)^{\varphi(m)} = p\mathcal{O}_K.$$

Exercise 4.5.8 Let $m = p^a$, with p prime, and $K = \mathbb{Q}(\zeta_m)$. Show that

$$d_{K/\mathbb{Q}}(\zeta_m) = \frac{(-1)^{\varphi(m)/2} m^{\varphi(m)}}{p^{m/p}}.$$

Exercise 4.5.9 Let $m = p^a$, with p prime. Show that $\{1, \zeta_m, \ldots, \zeta_m^{\varphi(m)-1}\}$ is an integral basis for the ring of integers of $K = \mathbb{Q}(\zeta_m)$.

Exercise 4.5.10 Let $K = \mathbb{Q}(\zeta_m)$ where $m = p^a$. Show that

$$d_K = \frac{(-1)^{\varphi(m)/2} m^{\varphi(m)}}{p^{m/p}}.$$

Exercise 4.5.11 Show that $\mathbb{Z}[\zeta_n + \zeta_n^{-1}]$ is the ring of integers of $\mathbb{Q}(\zeta_n + \zeta_n^{-1})$, where ζ_n denotes a primitive nth root of unity, and $n = p^\alpha$.

Exercise 4.5.12 Let K and L be algebraic number fields of degree m and n, respectively, over $\mathbb{Q}$. Let $d = \gcd(d_K, d_L)$. Show that if $[KL : \mathbb{Q}] = mn$, then $\mathcal{O}_{KL} \subseteq 1/d\mathcal{O}_K\mathcal{O}_L$.

Exercise 4.5.13 Let K and L be algebraic number fields of degree m and n, respectively, with $\gcd(d_K, d_L) = 1$. If $\{\alpha_1, \ldots, \alpha_m\}$ is an integral basis of $\mathcal{O}_K$ and $\{\beta_1, \ldots, \beta_n\}$ is an integral basis of $\mathcal{O}_L$, show that $\mathcal{O}_{KL}$ has an integral basis $\{\alpha_i\beta_j\}$ given that $[KL : \mathbb{Q}] = mn$. (In a later chapter, we will see that $\gcd(d_K, d_L) = 1$ implies that $[KL : \mathbb{Q}] = mn$.)

Exercise 4.5.14 Find an integral basis for $\mathbb{Q}(\sqrt{2}, \sqrt{-3})$.

Exercise 4.5.15 Let p and q be distinct primes $\equiv 1 \pmod 4$. Let $K = \mathbb{Q}(\sqrt{p})$, $L = \mathbb{Q}(\sqrt{q})$. Find a $\mathbb{Z}$-basis for $\mathbb{Q}(\sqrt{p}, \sqrt{q})$.

Exercise 4.5.16 Let K be an algebraic number field of degree n over $\mathbb{Q}$. Let $a_1, \ldots, a_n \in \mathcal{O}_K$ be linearly independent over $\mathbb{Q}$. Set

$$\Delta = d_{K/\mathbb{Q}}(a_1, \ldots, a_n).$$

Show that if $\alpha \in \mathcal{O}_K$, then $\Delta\alpha \in \mathbb{Z}[a_1, \ldots, a_n]$.

Exercise 4.5.17 (Explicit Construction of Integral Bases) Suppose K is an algebraic number field of degree n over $\mathbb{Q}$. Let $a_1, \ldots, a_n \in \mathcal{O}_K$ be linearly independent over $\mathbb{Q}$ and set

$$\Delta = d_{K/\mathbb{Q}}(a_1, \ldots, a_n).$$

For each i, choose the least natural number d_{ii} so that for some $d_{ij} \in \mathbb{Z}$, the number

$$w_i = \Delta^{-1} \sum_{j=1}^{i} d_{ij} a_j \in \mathcal{O}_K.$$

Show that $w_1, \ldots, w_n$ is an integral basis of $\mathcal{O}_K$.

Exercise 4.5.18 If K is an algebraic number field of degree n over $\mathbb{Q}$ and $a_1, \ldots, a_n \in \mathcal{O}_K$ are linearly independent over $\mathbb{Q}$, then there is an integral basis $w_1, \ldots, w_n$ of $\mathcal{O}_K$ such that

$$a_j = c_{j1}w_1 + \cdots + c_{jj}w_j,$$

$c_{ij} \in \mathbb{Z}, j = 1, \ldots, n$.

Exercise 4.5.19 If $\mathbb{Q} \subseteq K \subseteq L$ and K, L are algebraic number fields, show that $d_K \mid d_L$.

Exercise 4.5.20 (The Sign of the Discriminant) Suppose K is a number field with r_1 real embeddings and $2r_2$ complex embeddings so that

$$r_1 + 2r_2 = [K : \mathbb{Q}] = n$$

(say). Show that d_K has sign $(-1)^{r_2}$.

Exercise 4.5.21 Show that only finitely many imaginary quadratic fields K are Euclidean.

Exercise 4.5.22 Show that $\mathbb{Z}[(1+\sqrt{-19})/2]$ is not Euclidean. (Recall that in Exercise 2.5.6 we showed this ring is not Euclidean for the norm map.)

Exercise 4.5.23 (a) Let $A = (a_{ij})$ be an $m \times m$ matrix, $B = (b_{ij})$ an $n \times n$ matrix. We define the (Kronecker) tensor product $A \otimes B$ to be the $mn \times mn$ matrix obtained as

$$\begin{pmatrix} Ab_{11} & Ab_{12} & \cdots & Ab_{1n} \\ Ab_{21} & Ab_{22} & \cdots & Ab_{2n} \\ \vdots & \vdots & & \vdots \\ Ab_{n1} & Ab_{n2} & \cdots & Ab_{nn} \end{pmatrix},$$

where each block Ab_{ij} has the form

$$\begin{pmatrix} a_{11}b_{ij} & a_{12}b_{ij} & \cdots & a_{1m}b_{ij} \\ a_{21}b_{ij} & a_{22}b_{ij} & \cdots & a_{2m}b_{ij} \\ \vdots & \vdots & & \vdots \\ a_{m1}b_{ij} & a_{m2}b_{ij} & \cdots & a_{mm}b_{ij} \end{pmatrix}.$$

If C and D are $m \times m$ and $n \times n$ matrices, respectively, show that

$$(A \otimes B)(C \otimes D) = (AC) \otimes (BD).$$

(b) Prove that $\det(A \otimes B) = (\det A)^n (\det B)^m$.

Exercise 4.5.24 Let K and L be algebraic number fields of degree m and n, respectively, with $\gcd(d_K, d_L) = 1$. Show that

$$d_{KL} = d_K^n \cdot d_L^m.$$

If we set

$$\delta(M) = \frac{\log |d_M|}{[M : \mathbb{Q}]},$$

deduce that $\delta(KL) = \delta(K) + \delta(L)$ whenever $\gcd(d_K, d_L) = 1$.

Exercise 4.5.25 Let ζ_m denote a primitive mth root of unity and let $K = \mathbb{Q}(\zeta_m)$. Show that $\mathcal{O}_K = \mathbb{Z}[\zeta_m]$ and

$$d_K = \frac{(-1)^{\phi(m)/2} m^{\varphi(m)}}{\prod_{p|m} p^{\phi(m)/(p-1)}}.$$

Exercise 4.5.26 Let K be an algebraic number field. Suppose that $\theta \in \mathcal{O}_K$ is such that $d_{K/\mathbb{Q}}(\theta)$ is squarefree. Show that $\mathcal{O}_K = \mathbb{Z}[\theta]$.

Chapter 5

Dedekind Domains

5.1 Integral Closure

The notion of a Dedekind domain is the concept we need in order to establish the unique factorization of ideals as a product of prime ideals. En route to this goal, we will also meet the fundamental idea of a Noetherian ring. It turns out that Dedekind domains can be studied in the wider context of Noetherian rings. Even though a theory of factorization of ideals can also be established for Noetherian rings, we do not pursue it here.

Exercise 5.1.1 Show that a nonzero commutative ring R with identity is a field if and only if it has no nontrivial ideals.

Theorem 5.1.2 *Let R be a commutative ring with identity. Then:*

(a) *$\mathfrak{m}$ is a maximal ideal if and only if $R/\mathfrak{m}$ is a field.*

(b) *$\wp$ is a prime ideal if and only if $R/\wp$ is an integral domain.*

(c) *Let $\mathfrak{a}$ and $\mathfrak{b}$ be ideals of R. If $\wp$ is a prime ideal containing $\mathfrak{ab}$, then $\wp \supseteq \mathfrak{a}$ or $\wp \supseteq \mathfrak{b}$.*

(d) *If $\wp$ is a prime ideal containing the product $\mathfrak{a}_1\mathfrak{a}_2 \cdots \mathfrak{a}_r$ of r ideals of R, then $\wp \supseteq \mathfrak{a}_i$, for some i.*

Proof. (a) By the correspondence between ideals of R containing $\mathfrak{m}$ and ideals of $R/\mathfrak{m}$, $R/\mathfrak{m}$ has a nontrivial ideal if and only if there is an ideal $\mathfrak{a}$ of R strictly between $\mathfrak{m}$ and R. Thus,

$$\begin{aligned} & \mathfrak{m} \text{ is maximal,} \\ \Leftrightarrow\ & R/\mathfrak{m} \text{ has no nontrivial ideals,} \\ \Leftrightarrow\ & R/\mathfrak{m} \text{ is a field.} \end{aligned}$$

(b) $\wp$ is a prime ideal

$\Leftrightarrow$ $ab \in \wp \Rightarrow a \in \wp$ or $b \in \wp$,

$\Leftrightarrow$ $ab + \wp = 0 + \wp$ in $R/\wp \Rightarrow a + \wp = 0 + \wp$ or $b + \wp = 0 + \wp$ in $R/\wp$,

$\Leftrightarrow$ $R/\wp$ has no zero-divisors,

$\Leftrightarrow$ $R/\wp$ is an integral domain.

(c) Suppose that $\wp \supseteq \mathfrak{ab}$, $\wp \not\supseteq \mathfrak{a}$. Let $a \in \mathfrak{a}$, $a \notin \wp$. We know that $ab \in \wp$ for all $b \in \mathfrak{b}$ since $\mathfrak{ab} \subseteq \wp$. But, $a \notin \wp$. Thus, $b \in \wp$ for all $b \in \mathfrak{b}$, since $\wp$ is prime. Thus, $\mathfrak{b} \subseteq \wp$.

(d) (By induction on r). The base case $r = 1$ is trivial. Suppose $r > 1$ and $\wp \supseteq \mathfrak{a}_1\mathfrak{a}_2 \cdots \mathfrak{a}_r$. Then from part (c), $\wp \supseteq \mathfrak{a}_1\mathfrak{a}_2 \cdots \mathfrak{a}_{r-1}$ or $\wp \supseteq \mathfrak{a}_r$. If $\wp \supseteq \mathfrak{a}_1\mathfrak{a}_2 \cdots \mathfrak{a}_{r-1}$, then the induction hypothesis implies that $\wp \supseteq \mathfrak{a}_i$ for some $i \in \{1, \ldots, r-1\}$. In either case, $\wp \supseteq \mathfrak{a}_i$ for some $i \in \{1, \ldots, r\}$. □

Exercise 5.1.3 Show that a finite integral domain is a field.

Exercise 5.1.4 Show that every nonzero prime ideal $\wp$ of $\mathcal{O}_K$ is maximal.

Let R be an integral domain. We can always find a field containing R. As an example of such a field, take $Q(R) := \{[a, b] : a, b \in R, b \neq 0\}$ such that we identify elements $[a, b]$ and $[c, d]$ if $ad - bc = 0$. We define addition and multiplication on $Q(R)$ by the following rules: $[a, b] \cdot [c, d] = [ac, bd]$ and $[a, b] + [c, d] = [ad + bc, bd]$.

We can show that this makes $Q(R)$ into a commutative ring with $[a, b] \cdot [b, a] = 1$, for $a, b \neq 0$, so that any nonzero element is invertible (i.e., $Q(R)$ is a field). It contains R in the sense that the map taking a to $[a, 1]$ is a one-to-one homomorphism from R into $Q(R)$. The field $Q(R)$ is called the *quotient field of* R. We will usually write a/b rather than $[a, b]$.

For any field L containing R, we say that $\alpha \in L$ is *integral over* R if α satisfies a monic polynomial equation $f(\alpha) = 0$ with $f(x) \in R[x]$.

R is said to be *integrally closed* if every element in the quotient field of R which is integral over R, already lies in R.

Exercise 5.1.5 Show that every unique factorization domain is integrally closed.

Theorem 5.1.6 *For $\alpha \in \mathbb{C}$, the following are equivalent:*

(1) *α is integral over $\mathcal{O}_K$;*

(2) *$\mathcal{O}_K[\alpha]$ is a finitely generated $\mathcal{O}_K$-module;*

(3) *There is a finitely generated $\mathcal{O}_K$-module $M \subseteq \mathbb{C}$ such that $\alpha M \subseteq M$.*

Proof. (1) $\Rightarrow$ (2) Let $\alpha \in \mathbb{C}$ be integral over $\mathcal{O}_K$. Say α satisfies a monic polynomial of degree n over $\mathcal{O}_K$. Then

$$\mathcal{O}_K[\alpha] = \mathcal{O}_K + \mathcal{O}_K\alpha + \mathcal{O}_K\alpha^2 + \cdots + \mathcal{O}_K\alpha^{n-1}$$

and so is a finitely generated $\mathcal{O}_K$-module.

(2) $\Rightarrow$ (3) Certainly, $\alpha\mathcal{O}_K[\alpha] \subseteq \mathcal{O}_K[\alpha]$, so if $\mathcal{O}_K[\alpha]$ is a finitely generated $\mathcal{O}_K$-module, then (3) is satisfied with $M = \mathcal{O}_K[\alpha]$.

(3) $\Rightarrow$ (1) Let $u_1, u_2, \dots, u_n$ generate M as an $\mathcal{O}_K$-module. Then $\alpha u_i \in M$ for all $i = 1, 2, \dots, n$ since $\alpha M \subseteq M$. Let

$$\alpha u_i = \sum_{j=1}^{n} a_{ij} u_j$$

for $i \in \{1, 2, \dots, n\}$, $a_{ij} \in \mathcal{O}_K$. Let $A = (a_{ij})$, $B = \alpha I_n - A = (b_{ij})$. Then

$$\begin{aligned} \sum_{j=1}^{n} b_{ij} u_j &= \sum_{j=1}^{n} (\alpha\delta_{ij} - a_{ij}) u_{ij} \\ &= \alpha u_i - \sum_{j=1}^{n} a_{ij} u_j \\ &= 0 \quad \text{for all } i. \end{aligned}$$

Thus, $B(u_1, u_2, \dots, u_n)^T = (0, 0, \dots, 0)^T$. But

$$(0, 0, \dots, 0)^T \neq (u_1, u_2, \dots, u_n)^T \in \mathbb{C}^n.$$

Thus, $\det(B) = 0$. But, the determinant of B is a monic polynomial in $\mathcal{O}_K[\alpha]$, so α is integral over $\mathcal{O}_K$. □

Note that this theorem, and its proof, were exactly the same as Theorem 3.3.9, with $\mathcal{O}_K$ replacing $\mathbb{Z}$.

Theorem 5.1.7 *$\mathcal{O}_K$ is integrally closed.*

Proof. If $\alpha \in K$ is integral over $\mathcal{O}_K$, then let

$$M = \mathcal{O}_K u_1 + \cdots + \mathcal{O}_K u_n, \qquad \alpha M \subseteq M.$$

Let $\mathcal{O}_K = \mathbb{Z}v_1 + \cdots + \mathbb{Z}v_m$, where $\{v_1, \dots, v_m\}$ is a basis for K over $\mathbb{Q}$. Then $M = \sum_{i=1}^{m} \sum_{j=1}^{n} \mathbb{Z}v_i u_j$ is a finitely generated $\mathbb{Z}$-module with $\alpha M \subseteq M$, so α is integral over $\mathbb{Z}$. By definition, $\alpha \in \mathcal{O}_K$. □

5.2 Characterizing Dedekind Domains

A ring is called *Noetherian* if every ascending chain $\mathfrak{a}_1 \subseteq \mathfrak{a}_2 \subseteq \mathfrak{a}_3 \subseteq \cdots$ of ideals terminates, i.e., if there exists n such that $\mathfrak{a}_n = \mathfrak{a}_{n+k}$ for all $k \geq 0$.

Exercise 5.2.1 If $\mathfrak{a} \subsetneq \mathfrak{b}$ are ideals of $\mathcal{O}_K$, show that $N(\mathfrak{a}) > N(\mathfrak{b})$.

Exercise 5.2.2 Show that $\mathcal{O}_K$ is Noetherian.

Theorem 5.2.3 *For any commutative ring R with identity, the following are equivalent:*

(1) *R is Noetherian;*

(2) *every nonempty set of ideals contains a maximal element; and*

(3) *every ideal of R is finitely generated.*

Proof. (1) $\Rightarrow$ (2) Suppose that S is a nonempty set of ideals of R that does not contain a maximal element. Let $\mathfrak{a}_1 \in S$. $\mathfrak{a}_1$ is not maximal in S, so there is an $\mathfrak{a}_2 \in S$ with $\mathfrak{a}_1 \subsetneq \mathfrak{a}_2$. $\mathfrak{a}_2$ is not a maximal element of S, so there exists an $\mathfrak{a}_3 \in S$ with $\mathfrak{a}_1 \subsetneq \mathfrak{a}_2 \subsetneq \mathfrak{a}_3$. Continuing in this way, we find an infinite ascending chain of ideals of R. This contradicts R being Noetherian, so every nonempty set of ideals contains a maximal element.

(2) $\Rightarrow$ (3) Let $\mathfrak{b}$ be an ideal of R. Let A be the set of ideals contained in $\mathfrak{b}$ which are finitely generated. A is nonempty, since $(0) \in A$. Thus, A has a maximal element, say $\mathfrak{a} = (x_1, \ldots, x_n)$. If $\mathfrak{a} \neq \mathfrak{b}$, then $\exists x \in \mathfrak{b} \setminus \mathfrak{a}$. But then $\mathfrak{a} + (x) = (x_1, x_2, \ldots, x_n, x)$ is a larger finitely generated ideal contained in $\mathfrak{b}$, contradicting the maximality of $\mathfrak{b}$. Thus, $\mathfrak{b} = \mathfrak{a}$, so $\mathfrak{b}$ is finitely generated. Thus, every ideal of R is finitely generated.

(3) $\Rightarrow$ (1) Let $\mathfrak{a}_1 \subseteq \mathfrak{a}_2 \subseteq \mathfrak{a}_3 \subseteq \cdots$ be an ascending chain of ideals of R. Then $\mathfrak{a} = \bigcup_{i=1}^{\infty} \mathfrak{a}_i$ is also an ideal of R, and so is finitely generated, say $\mathfrak{a} = (x_1, \ldots, x_n)$. Then $x_1 \in \mathfrak{a}_{i_1}, \ldots, x_n \in \mathfrak{a}_{i_n}$. Let $m = \max(i_1, \ldots, i_n)$. Then $\mathfrak{a} \subseteq \mathfrak{a}_m$, so $\mathfrak{a} = \mathfrak{a}_m$. Thus, $\mathfrak{a}_m = \mathfrak{a}_{m+1} = \cdots$, and the chain does terminate. Thus, R is Noetherian. $\square$

Thus, we have proved that:

(1) $\mathcal{O}_K$ is integrally closed;

(2) every nonzero prime ideal of $\mathcal{O}_K$ is maximal; and

(3) $\mathcal{O}_K$ is Noetherian.

A commutative integral domain which satisfies these three conditions is called a *Dedekind domain.* We have thus seen that $\mathcal{O}_K$ is a Dedekind domain.

Exercise 5.2.4 Show that any principal ideal domain is a Dedekind domain.

Exercise 5.2.5 Show that $\mathbb{Z}[\sqrt{-5}]$ is a Dedekind domain, but not a principal ideal domain.

5.3 Fractional Ideals and Unique Factorization

Our next goal is to show that, in $\mathcal{O}_K$, every ideal can be written as a product of prime ideals uniquely. In fact, this is true in any Dedekind domain.

A *fractional ideal* $\mathcal{A}$ of $\mathcal{O}_K$ is an $\mathcal{O}_K$-module contained in K such that there exists $m \in \mathbb{Z}$ with $m\mathcal{A} \subseteq \mathcal{O}_K$. Of course, any ideal of $\mathcal{O}_K$ is a fractional ideal by taking $m = 1$.

Exercise 5.3.1 Show that any fractional ideal is finitely generated as an $\mathcal{O}_K$-module.

Exercise 5.3.2 Show that the sum and product of two fractional ideals are again fractional ideals.

Lemma 5.3.3 *Any proper ideal of* $\mathcal{O}_K$ *contains a product of nonzero prime ideals.*

Proof. Let S be the set of all proper ideals of $\mathcal{O}_K$ that do not contain a product of prime ideals. We need to show that S is empty. If not, then since $\mathcal{O}_K$ is Noetherian, S has a maximal element, say $\mathfrak{a}$. Then, $\mathfrak{a}$ is not prime since $\mathfrak{a} \in S$, so there exist $a, b \in \mathcal{O}_K$, with $ab \in \mathfrak{a}$, $a \notin \mathfrak{a}$, $b \notin \mathfrak{a}$. Then, $(\mathfrak{a}, a) \supsetneq \mathfrak{a}$, $(\mathfrak{a}, b) \supsetneq \mathfrak{a}$. Thus, $(\mathfrak{a}, a) \notin S$, $(\mathfrak{a}, b) \notin S$, by the maximality of $\mathfrak{a}$.

Thus, $(\mathfrak{a}, a) \supseteq \wp_1 \cdots \wp_r$ and $(\mathfrak{a}, b) \supseteq \wp_1' \cdots \wp_s'$, with the $\wp_i$ and $\wp_j'$ nonzero prime ideals. But $ab \in \mathfrak{a}$, so $(\mathfrak{a}, ab) = \mathfrak{a}$.

Thus, $\mathfrak{a} = (\mathfrak{a}, ab) \supseteq (\mathfrak{a}, a)(\mathfrak{a}, b) \supseteq \wp_1 \cdots \wp_r \wp_1' \cdots \wp_s'$. Therefore $\mathfrak{a}$ contains a product of prime ideals. This contradicts $\mathfrak{a}$ being in S, so S must actually be empty.

Thus, any proper ideal of $\mathcal{O}_K$ contains a product of nonzero prime ideals.

□

Lemma 5.3.4 *Let* $\wp$ *be a prime ideal of* $\mathcal{O}_K$. *There exists* $z \in K$, $z \notin \mathcal{O}_K$, *such that* $z\wp \subseteq \mathcal{O}_K$.

Proof. Take $x \in \wp$. From the previous lemma, $(x) = x\mathcal{O}_K$ contains a product of prime ideals. Let r be the least integer such that (x) contains a product of r prime ideals, and say $(x) \supseteq \wp_1 \cdots \wp_r$, with the $\wp_i$ nonzero prime ideals.

Since $\wp \supseteq \wp_1 \cdots \wp_r$, $\wp \supseteq \wp_i$, for some i, from Theorem 5.1.2 (d). Without loss of generality, we can assume that $i = 1$, so $\wp \supseteq \wp_1$. But $\wp_1$ is a nonzero prime ideal of $\mathcal{O}_K$, and so is maximal. Thus, $\wp = \wp_1$.

Now, $\wp_2 \cdots \wp_r \not\subseteq (x)$, since r was chosen to be minimal. Choose an element $b \in \wp_2 \cdots \wp_r$, $b \notin x\mathcal{O}_K$. Then

$$\begin{aligned} bx^{-1}\wp = bx^{-1}\wp_1 &\subseteq (\wp_2 \cdots \wp_r)(x^{-1}\wp_1) \\ &= x^{-1}(\wp_1 \cdots \wp_r) \\ &\subseteq x^{-1}x\mathcal{O}_K \\ &= \mathcal{O}_K. \end{aligned}$$

Put $z = bx^{-1}$. Then $z\wp \subseteq \mathcal{O}_K$. Now, if z were in $\mathcal{O}_K$, we would have $bx^{-1} \in \mathcal{O}_K$, and so $b \in x\mathcal{O}_K$. But this is not the case, so $z \notin \mathcal{O}_K$. Thus, we have found $z \in K, z \notin \mathcal{O}_K$ with $z\wp \subseteq \mathcal{O}_K$. □

Let $\wp$ be a prime ideal. Define

$$\wp^{-1} = \{x \in K : x\wp \subseteq \mathcal{O}_K\}.$$

Lemma 5.3.4 implies, in particular, that $\wp^{-1} \neq \mathcal{O}_K$.

Theorem 5.3.5 *Let $\wp$ be a prime ideal of $\mathcal{O}_K$. Then $\wp^{-1}$ is a fractional ideal and $\wp\wp^{-1} = \mathcal{O}_K$.*

Proof. It is easily seen that $\wp^{-1}$ is an $\mathcal{O}_K$-module. Now, $\wp \cap \mathbb{Z} \neq (0)$, from Exercise 4.4.1, so let $n \in \wp \cap \mathbb{Z}$, $n \neq 0$. Then, $n\wp^{-1} \subseteq \wp\wp^{-1} \subseteq \mathcal{O}_K$, by definition. Thus, $\wp^{-1}$ is a fractional ideal.

It remains to show that $\wp\wp^{-1} = \mathcal{O}_K$. Since $1 \in \wp^{-1}$, $\wp \subseteq \wp\wp^{-1} \subseteq \mathcal{O}_K$. $\wp\wp^{-1}$ is an ideal of $\mathcal{O}_K$, since it is an $\mathcal{O}_K$-module contained in $\mathcal{O}_K$. But $\wp$ is maximal, so either $\wp\wp^{-1} = \mathcal{O}_K$, in which case we are done, or $\wp\wp^{-1} = \wp$.

Suppose that $\wp\wp^{-1} = \wp$. Then $x\wp \subseteq \wp$ $\forall x \in \wp^{-1}$. Since $\wp$ is a finitely generated $\mathbb{Z}$-module (from Exercise 4.4.2), $x \in \mathcal{O}_K$ for all $x \in \wp^{-1}$, by Theorem 5.1.6. Thus, $\wp^{-1} \subseteq \mathcal{O}_K$. But $1 \in \wp^{-1}$, so $\wp^{-1} = \mathcal{O}_K$. From the comments above, and by the previous lemma, we know this is not true, so $\wp\wp^{-1} \neq \wp$. Thus, $\wp\wp^{-1} = \mathcal{O}_K$. □

Theorem 5.3.6 *Any ideal of $\mathcal{O}_K$ can be written as a product of prime ideals uniquely.*

Proof.

Existence. Let S be the set of ideals of $\mathcal{O}_K$ that cannot be written as a product of prime ideals. If S is nonempty, then S has a maximal element, since $\mathcal{O}_K$ is Noetherian. Let $\mathfrak{a}$ be a maximal element of S. Then $\mathfrak{a} \subseteq \wp$ for some maximal ideal $\wp$, since $\mathcal{O}_K$ is Noetherian. Recall that every maximal ideal of $\mathcal{O}_K$ is prime. Since $\mathfrak{a} \in S$, $\mathfrak{a} \neq \wp$ and therefore $\mathfrak{a}$ is not prime.

Consider $\wp^{-1}\mathfrak{a}$. $\wp^{-1}\mathfrak{a} \subset \wp^{-1}\wp = \mathcal{O}_K$. Since $\mathfrak{a} \subsetneq \wp$,

$$\wp^{-1}\mathfrak{a} \subsetneq \wp^{-1}\wp = \mathcal{O}_K,$$

since for any $x \in \wp \backslash \mathfrak{a}$,

$$\wp^{-1} x \subseteq \wp^{-1} \mathfrak{a} \quad \Rightarrow \quad x \in \wp\wp^{-1}\mathfrak{a} = \mathcal{O}_K \mathfrak{a} = \mathfrak{a},$$

which is not true. Thus, $\wp^{-1}\mathfrak{a}$ is a proper ideal of $\mathcal{O}_K$, and contains $\mathfrak{a}$ properly since $\wp^{-1}$ contains $\mathcal{O}_K$ properly. Thus, $\wp^{-1}\mathfrak{a} \notin S$, since $\mathfrak{a}$ is a maximal element of S. Thus, $\wp^{-1}\mathfrak{a} = \wp_1 \cdots \wp_r$, for some prime ideals $\wp_i$. Then, $\wp\wp^{-1}\mathfrak{a} = \wp\wp_1 \cdots \wp_r$, so $\mathfrak{a} = \wp\wp^{-1}\mathfrak{a} = \wp\wp_1 \cdots \wp_r$. But then $\mathfrak{a} \notin S$, a contradiction.

Thus, S is empty, so every ideal of $\mathcal{O}_K$ can be written as a product of prime ideals.

Uniqueness. Suppose that $\mathfrak{a} = \wp_1 \cdots \wp_r = \wp_1' \cdots \wp_s'$ are two factorizations of $\mathfrak{a}$ as a product of prime ideals.

Then, $\wp_1' \supseteq \wp_1' \cdots \wp_s' = \wp_1 \cdots \wp_r$, so $\wp_1' \supseteq \wp_i$, for some i, say $\wp_1' \supseteq \wp_1$. But $\wp_1$ is maximal, so $\wp_1' = \wp_1$. Thus, multiplying both sides by $(\wp_1')^{-1}$ and cancelling using $(\wp_1')^{-1}\wp_1' = \mathcal{O}_K$, we obtain

$$\wp_2' \cdots \wp_s' = \wp_2 \cdots \wp_r.$$

Thus, continuing in this way, we see that $r = s$ and the primes are unique up to reordering. □

It is possible to show that any integral domain in which every non-zero ideal can be factored as a product of prime ideals is necessarily a Dedekind domain. We refer the reader to p. 82 of [Mat] for the details of the proof. This fact gives us an interesting characterization of Dedekind domains.

When $\wp$ and $\wp'$ are prime ideals, we will write $\wp/\wp'$ for $(\wp')^{-1}\wp$. We will also write

$$\frac{\wp_1\wp_2 \cdots \wp_r}{\wp_1'\wp_2' \cdots \wp_s'}$$

to mean $(\wp_1')^{-1}(\wp_2')^{-1} \cdots (\wp_s')^{-1}\wp_1\wp_2 \cdots \wp_r$.

Exercise 5.3.7 Show that any fractional ideal $\mathcal{A}$ can be written uniquely in the form

$$\frac{\wp_1 \cdots \wp_r}{\wp_1' \cdots \wp_s'},$$

where the $\wp_i$ and $\wp_j'$ may be repeated, but no $\wp_i = \wp_j'$.

Exercise 5.3.8 Show that, given any fractional ideal $\mathcal{A} \neq 0$ in K, there exists a fractional ideal $\mathcal{A}^{-1}$ such that $\mathcal{A}\mathcal{A}^{-1} = \mathcal{O}_K$.

For $\mathfrak{a}$ and $\mathfrak{b}$ ideals of $\mathcal{O}_K$, we say $\mathfrak{a}$ divides $\mathfrak{b}$ (denoted $\mathfrak{a} \mid \mathfrak{b}$), if $\mathfrak{a} \supseteq \mathfrak{b}$.

Exercise 5.3.9 Show that if $\mathfrak{a}$ and $\mathfrak{b}$ are ideals of $\mathcal{O}_K$, then $\mathfrak{b} \mid \mathfrak{a}$ if and only if there is an ideal $\mathfrak{c}$ of $\mathcal{O}_K$ with $\mathfrak{a} = \mathfrak{b}\mathfrak{c}$.

Define $\mathfrak{d}$ to be the *greatest common divisor* of $\mathfrak{a}, \mathfrak{b}$ if:

(i) $\mathfrak{d} \mid \mathfrak{a}$ and $\mathfrak{d} \mid \mathfrak{b}$; and

(ii) $\mathfrak{e} \mid \mathfrak{a}$ and $\mathfrak{e} \mid \mathfrak{b} \Rightarrow \mathfrak{e} \mid \mathfrak{d}$.

Denote $\mathfrak{d}$ by $\gcd(\mathfrak{a}, \mathfrak{b})$.

Similarly, define $\mathfrak{m}$ to be the *least common multiple* of $\mathfrak{a}$, $\mathfrak{b}$ if:

(i) $\mathfrak{a} \mid \mathfrak{m}$ and $\mathfrak{b} \mid \mathfrak{m}$; and

(ii) $\mathfrak{a} \mid \mathfrak{n}$ and $\mathfrak{b} \mid \mathfrak{n} \Rightarrow \mathfrak{m} \mid \mathfrak{n}$.

Denote $\mathfrak{m}$ by $\operatorname{lcm}(\mathfrak{a}, \mathfrak{b})$.

In the next two exercises, we establish the existence (and uniqueness) of the gcd and lcm of two ideals of $\mathcal{O}_K$.

Let $\mathfrak{a} = \prod_{i=1}^r \wp_i^{e_i}$, $\mathfrak{b} = \prod_{i=1}^r \wp_i^{f_i}$, with $e_i, f_i \in \mathbb{Z}_{\geq 0}$.

Exercise 5.3.10 Show that $\gcd(\mathfrak{a}, \mathfrak{b}) = \mathfrak{a} + \mathfrak{b} = \prod_{i=1}^r \wp_i^{\min(e_i, f_i)}$.

Exercise 5.3.11 Show that $\operatorname{lcm}(\mathfrak{a}, \mathfrak{b}) = \mathfrak{a} \cap \mathfrak{b} = \prod_{i=1}^r \wp_i^{\max(e_i, f_i)}$.

Exercise 5.3.12 Suppose $\mathfrak{a}, \mathfrak{b}, \mathfrak{c}$ are ideals of $\mathcal{O}_K$. Show that if $\mathfrak{a}\mathfrak{b} = \mathfrak{c}^g$ and $\gcd(\mathfrak{a}, \mathfrak{b}) = 1$, then $\mathfrak{a} = \mathfrak{d}^g$ and $\mathfrak{b} = \mathfrak{e}^g$ for some ideals $\mathfrak{d}$ and $\mathfrak{e}$ of $\mathcal{O}_K$. (This generalizes Exercise 1.2.1.)

Theorem 5.3.13 (Chinese Remainder Theorem) (a) *Let $\mathfrak{a}$, $\mathfrak{b}$ be ideals so that* $\gcd(\mathfrak{a}, \mathfrak{b}) = 1$, *i.e.,* $\mathfrak{a} + \mathfrak{b} = \mathcal{O}_K$. *Given $a, b \in \mathcal{O}_K$, we can solve*

$$\begin{aligned} x &\equiv a \pmod{\mathfrak{a}}, \\ x &\equiv b \pmod{\mathfrak{b}}. \end{aligned}$$

(b) *Let $\wp_1, \ldots, \wp_r$ be r distinct prime ideals in $\mathcal{O}_K$. Given $a_i \in \mathcal{O}_K$, $e_i \in \mathbb{Z}_{>0}$, $\exists x$ such that $x \equiv a_i \pmod{\wp_i^{e_i}}$ for all $i \in \{1, \ldots, r\}$.*

Proof. (a) Since $\mathfrak{a} + \mathfrak{b} = \mathcal{O}_K$, $\exists x_1 \in \mathfrak{a}$, $x_2 \in \mathfrak{b}$ with $x_1 + x_2 = 1$. Let $x = bx_1 + ax_2 \equiv ax_2 \pmod{\mathfrak{a}}$. But, $x_2 = 1 - x_1 \equiv 1 \pmod{\mathfrak{a}}$. Thus, we have found an x such that $x \equiv a \pmod{\mathfrak{a}}$. Similarly, $x \equiv b \pmod{\mathfrak{b}}$.

(b) We proceed by induction on r. If $r = 1$, there is nothing to show. Suppose $r > 1$, and that we can solve $x \equiv a_i \pmod{\wp_i^{e_i}}$ for $i = 1, \ldots, r-1$, say $a \equiv a_i \pmod{\wp_i^{e_i}}$ for $i = 1, \ldots, r-1$. From part (a), we can solve

$$\begin{aligned} x &\equiv a \pmod{\wp_1^{e_1} \cdots \wp_{r-1}^{e_{r-1}}}, \\ x &\equiv a_r \pmod{\wp_r^{e_r}}. \end{aligned}$$

Then $x - a_i \in \wp_1^{e_1} \cdots \wp_{r-1}^{e_{r-1}}$, $x \equiv a_r \pmod{\wp_r^{e_r}}$. Thus, $x - a_i \in \wp_i^{e_i}$ $\forall i$, i.e., $x \equiv a_i \pmod{\wp_i^{e_i}}$ $\forall i$. $\square$

We define the *order* of $\mathfrak{a}$ in $\wp$ by $\operatorname{ord}_\wp(\mathfrak{a}) = t$ if $\wp^t \mid \mathfrak{a}$ and $\wp^{t+1} \nmid \mathfrak{a}$.

Exercise 5.3.14 Show that $\text{ord}_\wp(\mathfrak{a}\mathfrak{b}) = \text{ord}_\wp(\mathfrak{a}) + \text{ord}_\wp(\mathfrak{b})$, where $\wp$ is a prime ideal.

Exercise 5.3.15 Show that, for $\alpha \neq 0$ in $\mathcal{O}_K$, $N((\alpha)) = |\text{N}_K(\alpha)|$.

Theorem 5.3.16 (a) *If* $\mathfrak{a} = \prod_{i=1}^r \wp_i^{e_i}$, *then*

$$N(\mathfrak{a}) = \prod_{i=1}^r N(\wp_i^{e_i}).$$

(b) $\mathcal{O}_K/\wp \simeq \wp^{e-1}/\wp^e$, *and*

$$N(\wp^e) = (N(\wp))^e$$

for any integer $e \geq 0$.

Proof. (a) Consider the map

$$\begin{aligned} \phi : \mathcal{O}_K &\longrightarrow \bigoplus_{i=1}^r (\mathcal{O}_K/\wp_i^{e_i}), \\ x &\longrightarrow (x_1, \ldots, x_r), \end{aligned}$$

where $x_i \equiv x \pmod{\wp_i^{e_i}}$.

The function ϕ is surjective by the Chinese Remainder Theorem, and ϕ is a homomorphism since each of the r components $x \longrightarrow x_i \pmod{\wp_i^{e_i}}$ is a homomorphism.

Next, we show by induction that $\bigcap_{i=1}^r \wp_i^{e_i} = \prod_{i=1}^r \wp_i^{e_i}$. The base case $r = 1$ is trivial. Suppose $r > 1$, and that the result is true for numbers smaller than r.

$$\begin{aligned} \bigcap_{i=1}^r \wp_i^{e_i} &= \text{lcm}\left(\bigcap_{i=1}^{r-1} \wp_i^{e_i}, \wp_r^{e_r}\right) \\ &= \text{lcm}\left(\prod_{i=1}^{r-1} \wp_i^{e_i}, \wp_r^{e_r}\right) \\ &= \prod_{i=1}^r \wp_i^{e_i}. \end{aligned}$$

Thus, $\ker(\phi) = \bigcap_{i=1}^r \wp_i^{e_i} = \prod_{i=1}^r \wp_i^{e_i}$, which implies that

$$\mathcal{O}_K/\mathfrak{a} \simeq \oplus(\mathcal{O}_K/\wp_i^{e_i}).$$

Hence, $N(\mathfrak{a}) = \prod_{i=1}^r N(\wp_i^{e_i})$.

(b) Since $\wp^e \subsetneq \wp^{e-1}$, we can find an element $\alpha \in \wp^{e-1}/\wp^e$, so that $\text{ord}_\wp(\alpha) = e - 1$. Then $\wp^e \subseteq (\alpha) + \wp^e \subseteq \wp^{e-1}$. So $\wp^{e-1} \mid (\alpha) + \wp^e$. But

$(\alpha)+\wp^e \neq \wp^e$, so $\wp^{e-1} = (\alpha)+\wp^e$, by unique factorization. Define the map $\phi : \mathcal{O}_K \longrightarrow \wp^{e-1}/\wp^e$ by $\phi(\gamma) = \gamma\alpha + \wp^e$. This is clearly a homomorphism and is surjective since $\wp^{e-1} = (\alpha) + \wp^e$.

Now,

$$\begin{aligned} \gamma \in \ker(\phi) &\Leftrightarrow \gamma\alpha \in \wp^e \\ &\Leftrightarrow \mathrm{ord}_\wp(\gamma\alpha) \geq e \\ &\Leftrightarrow \mathrm{ord}_\wp(\gamma) + \mathrm{ord}_\wp(\alpha) \geq e \\ &\Leftrightarrow \mathrm{ord}_\wp(\gamma) + e - 1 \geq e \\ &\Leftrightarrow \mathrm{ord}_\wp(\gamma) \geq 1 \\ &\Leftrightarrow \gamma \in \wp. \end{aligned}$$

Thus, $\mathcal{O}_K/\wp \simeq \wp^{e-1}/\wp^e$. Also,

$$(\mathcal{O}_K/\wp^e)/(\wp^{e-1}/\wp^e) \simeq \mathcal{O}_K/\wp^{e-1}$$

since the map from $\mathcal{O}_K/\wp^e$ to $\mathcal{O}_K/\wp^{e-1}$ taking $x + \wp^e$ to $x + \wp^{e-1}$ is a surjective homomorphism with kernel $\wp^{e-1}/\wp^e$. Thus,

$$\begin{aligned} N(\wp^e) = |\mathcal{O}_K/\wp^e| &= |\mathcal{O}_K/\wp^{e-1}|\,|\wp^{e-1}/\wp^e| \\ &= N(\wp^{e-1})N(\wp) \\ &= N(\wp)^{e-1}N(\wp) \quad \text{by the induction hypothesis} \\ &= N(\wp)^e. \end{aligned}$$

□

Thus, the norm function is multiplicative. Also, we can extend the definition of norm to fractional ideals, in the following way. Since any fractional ideal can be written uniquely in the form $\mathfrak{a}\mathfrak{b}^{-1}$ where $\mathfrak{a}, \mathfrak{b}$ are ideals of $\mathcal{O}_K$, we can put

$$N(\mathfrak{a}\mathfrak{b}^{-1}) = \frac{N(\mathfrak{a})}{N(\mathfrak{b})}.$$

Let $\mathcal{O}_K = \mathbb{Z}\omega_1 + \cdots + \mathbb{Z}\omega_n$. Then, if p is a prime number, we have $p\mathcal{O}_K = \mathbb{Z}p\omega_1 + \cdots + \mathbb{Z}p\omega_n$, and so $N((p)) = p^n$, where $n = [K : \mathbb{Q}]$.

Exercise 5.3.17 If we write $p\mathcal{O}_K$ as its prime factorization,

$$p\mathcal{O}_K = \wp_1^{e_1} \cdots \wp_g^{e_g},$$

show that $N(\wp_i)$ is a power of p and that if $N(\wp_i) = p^{f_i}$, $\sum_{i=1}^g e_i f_i = n$.

5.4 Dedekind's Theorem

The number e_i found in the previous exercise is called the *ramification degree* of $\wp_i$. (We sometimes write $e_{\wp_i}$ for e_i.) We say the prime number p *ramifies in* K if some $e_i \geq 2$. If all the f_i's are 1, we say p *splits completely.*

Our next goal is to show *Dedekind's Theorem*: If p is a prime number that ramifies in K, then $p \mid d_K$. Recall that $d_K = \det(\mathrm{Tr}(\omega_i\omega_j))$, where $\omega_1, \ldots, \omega_n$ is any integral basis for $\mathcal{O}_K$.

Let $\mathcal{D}^{-1} = \{x \in K : \mathrm{Tr}(x\mathcal{O}_K) \subseteq \mathbb{Z}\}$.

Exercise 5.4.1 Show that $\mathcal{D}^{-1}$ is a fractional ideal of K and find an integral basis.

Exercise 5.4.2 Let $\mathcal{D}$ be the fractional ideal inverse of $\mathcal{D}^{-1}$. We call $\mathcal{D}$ the *different* of K. Show that $\mathcal{D}$ is an ideal of $\mathcal{O}_K$.

Theorem 5.4.3 *Let $\mathcal{D}$ be the different of an algebraic number field K. Then $N(\mathcal{D}) = |d_K|$.*

Proof. For some $m > 0$, $m\mathcal{D}^{-1}$ is an ideal of $\mathcal{O}_K$. Now,

$$m\mathcal{D}^{-1} = \mathbb{Z}m\omega_1^* + \cdots + \mathbb{Z}m\omega_n^*.$$

Let

$$m\omega_i^* = \sum_{j=1}^{n} a_{ij}\omega_j,$$

so

$$\omega_i^* = \sum_{j=1}^{n} \frac{a_{ij}}{m}\omega_j,$$

and

$$\omega_i = \sum_{j=1}^{n} b_{ij}\omega_j^*.$$

Thus, (b_{ij}) is the matrix inverse of (a_{ij}/m). But

$$\begin{aligned} \mathrm{Tr}(\omega_i\omega_j) &= \mathrm{Tr}\left(\sum_{k=1}^{n} b_{ik}\omega_k^*\omega_j\right) \\ &= \sum_{k=1}^{n} b_{ik}\ \mathrm{Tr}(\omega_k^*\omega_j) \\ &= b_{ij}. \end{aligned}$$

Thus, $\det(b_{ij}) = d_K$.

However, by Exercise 4.2.8, since $m\mathfrak{D}^{-1}$ is an ideal of $\mathcal{O}_K$ with integral basis $m\omega_1^*, \ldots, m\omega_n^*$, we know that

$$d_{K/\mathbb{Q}}(m\omega_1^*, \ldots, m\omega_n^*) = N(m\mathfrak{D}^{-1})^2 d_K$$

and from Exercise 4.2.7 we have

$$d_{K/\mathbb{Q}}(m\omega_1^*, \ldots, m\omega_n^*) = (\det(a_{ij}))^2 d_K,$$

which shows that

$$|\det(a_{ij})| = N(m\mathfrak{D}^{-1}) = m^n N(\mathfrak{D}^{-1}),$$

and thus

$$|\det(\frac{a_{ij}}{m})| = N(\mathfrak{D}^{-1}) = N(\mathfrak{D})^{-1}.$$

Hence, $|d_K| = |\det(b_{ij})| = |\det(a_{ij}/m)|^{-1} = N(\mathfrak{D})$. □

Theorem 5.4.4 *Let $p \in \mathbb{Z}$ be prime, $\wp \subseteq \mathcal{O}_K$, a prime ideal and $\mathfrak{D}$ the different of K. If $\wp^e \mid (p)$, then $\wp^{e-1} \mid \mathfrak{D}$.*

Proof. We may assume that e is the highest power of $\wp$ dividing (p). So let $(p) = \wp^e \mathfrak{a}$, $\gcd(\mathfrak{a}, \wp) = 1$. Let $x \in \wp\mathfrak{a}$. Then $x = \sum_{i=1}^n p_i a_i$, $p_i \in \wp$, $a_i \in \mathfrak{a}$. Hence,

$$x^p \equiv \sum_{i=1}^n p_i^p a_i^p \pmod p,$$

and

$$x^{p^m} \equiv \sum_{i=1}^n p_i^{p^m} a_i^{p^m} \pmod p.$$

For sufficiently large m, $p_i^{p^m} \in \wp^e$, so $x^{p^m} \in \wp^e$ and thus, $x^{p^m} \in \wp^e \mathfrak{a} = (p)$.
Therefore, $\mathrm{Tr}(x^{p^m}) \in p\mathbb{Z}$, which implies that

$$\begin{aligned}
& (\mathrm{Tr}(x))^{p^m} \in p\mathbb{Z} \\
\Rightarrow\ & \mathrm{Tr}(x) \in p\mathbb{Z} \\
\Rightarrow\ & \mathrm{Tr}(p^{-1}\wp\mathfrak{a}) \subseteq \mathbb{Z} \\
\Rightarrow\ & p^{-1}\wp\mathfrak{a} \subseteq \mathfrak{D}^{-1} \\
\Rightarrow\ & \mathfrak{D}p^{-1}\wp\mathfrak{a} \subseteq \mathfrak{D}\mathfrak{D}^{-1} = \mathcal{O}_K \\
\Rightarrow\ & \mathfrak{D} \subseteq p\wp^{-1}\mathfrak{a}^{-1} = \wp^e\mathfrak{a}\wp^{-1}\mathfrak{a}^{-1} = \wp^{e-1} \\
\Rightarrow\ & \wp^{e-1} \mid \mathfrak{D}.
\end{aligned}$$

□

Exercise 5.4.5 Show that if p is ramified, $p \mid d_K$.

Dedekind also proved that if $p \mid d_K$, then p ramifies. We do not prove this here. In the Supplementary Problems, some special cases are derived.

5.5 Factorization in $\mathcal{O}_K$

The following theorem gives an important connection between factoring polynomials mod p and factoring ideals in number fields:

Theorem 5.5.1 *Suppose that there is a $\theta \in K$ such that $\mathcal{O}_K = \mathbb{Z}[\theta]$. Let $f(x)$ be the minimal polynomial of θ over $\mathbb{Z}[x]$. Let p be a rational prime, and suppose*

$$f(x) \equiv f_1(x)^{e_1} \cdots f_g(x)^{e_g} \pmod{p},$$

where each $f_i(x)$ is irreducible in $\mathbb{F}_p[x]$. Then $p\mathcal{O}_K = \wp_1^{e_1} \cdots \wp_g^{e_g}$ where $\wp_i = (p, f_i(\theta))$ are prime ideals, with $N(\wp_i) = p^{\deg f_i}$.

Proof. We first note that $(p, f_1(\theta))^{e_1} \cdots (p, f_g(\theta))^{e_g} \subseteq p\mathcal{O}_K$. Thus it suffices to show that $(p, f_i(\theta))$ is a prime ideal of norm p^{d_i} where d_i is the degree of f_i.

Now, since $f_i(x)$ is irreducible over $\mathbb{F}_p$, then $\mathbb{F}_p[x]/(f_i(x))$ is a field. Also,

$$\mathbb{Z}[x]/(p) \simeq \mathbb{F}_p[x], \quad \Rightarrow \quad \mathbb{Z}[x]/(p, f_i(x)) \simeq \mathbb{F}_p[x]/(f_i(x)),$$

and so $\mathbb{Z}[x]/(p, f_i(x))$ is a field.

Consider the map $\varphi : \mathbb{Z}[x] \to \mathbb{Z}[\theta]/(p, f_i(\theta))$. Clearly

$$(p, f_i(x)) \subseteq \ker(\varphi) = \{n(x) : n(\theta) \in (p, f_i(\theta))\}.$$

If $n(x) \in \ker(\varphi)$, we can divide by $f_i(x)$ to get

$$n(x) = q(x) f_i(x) + r_i(x), \qquad \deg(r_i) < \deg(f_i).$$

We assume that r_i is nonzero, for otherwise the result is trivial. Since $n(\theta) \in (p, f_i(\theta))$, then $r_i(\theta) \in (p, f_i(\theta))$, so $r_i(\theta) = pa(\theta) + f_i(\theta)b(\theta)$. Here we have used the fact that $\mathcal{O}_K = \mathbb{Z}[\theta]$.

Now define the polynomial $h(x) = r_i(x) - pa(x) - f_i(x)b(x)$. Since $h(\theta) = 0$ and f is the minimal polynomial of θ, then $h(x) = g(x)f(x)$ for some polynomial $g(x) \in \mathbb{Z}[x]$. We conclude that $r_i(x) = p\tilde{a}(x) + f_i(x)\tilde{b}(x)$ for some $\tilde{a}(x), \tilde{b}(x) \in \mathbb{Z}[x]$. Therefore $r_i(x) \in (p, f_i(x))$.

Thus,

$$\mathbb{Z}[\theta]/(p, f_i(\theta)) \simeq \mathbb{Z}[x]/(p, f_i(x)) \simeq \mathbb{F}_p[x]/(f_i(x))$$

and is therefore a field. Hence, $(p, f_i(\theta))$ is a maximal ideal and is therefore prime.

Now, let e_i' be the ramification index of $\wp_i$, so that $p\mathcal{O}_K = \wp_1^{e_1'} \cdots \wp_g^{e_g'}$, and let $d_i = [\mathcal{O}_K/\wp_i : \mathbb{Z}/p]$. Clearly d_i is the degree of the polynomial $f_i(x)$. Since $f(\theta) = 0$, and since $f(x) - f_1(x)^{e_1} \cdots f_g(x)^{e_g} \in p\mathbb{Z}[x]$, it follows that $f_1(\theta)^{e_1} \cdots f_g(\theta)^{e_g} \in p\mathcal{O}_K = p\mathbb{Z}[\theta]$. Also, $\wp_i^{e_i} \subset p\mathcal{O}_K + (f_i(\theta)^{e_i})$ and so

$$\wp_1^{e_1} \cdots \wp_g^{e_g} \subseteq p\mathcal{O}_K = \wp_1^{e_1'} \cdots \wp_g^{e_g'}.$$

Therefore, $e_i \geq e_i'$ for all i. But

$$\sum e_i d_i = \deg f = [K : \mathbb{Q}] = \sum e_i' d_i.$$

Thus, $e_i = e_i'$ for all i. $\square$

Exercise 5.5.2 If in the previous theorem we do not assume that $\mathcal{O}_K = \mathbb{Z}[\theta]$ but instead that $p \nmid [\mathcal{O}_K : \mathbb{Z}[\theta]]$, show that the same result holds.

Exercise 5.5.3 Suppose that $f(x)$ in the previous exercise is Eisensteinian with respect to the prime p. Show that p ramifies totally in K. That is, $p\mathcal{O}_K = (\theta)^n$ where $n = [K : \mathbb{Q}]$.

Exercise 5.5.4 Show that $(p) = (1 - \zeta_p)^{p-1}$ when $K = \mathbb{Q}(\zeta_p)$.

5.6 Supplementary Problems

Exercise 5.6.1 Show that if a ring R is a Dedekind domain and a unique factorization domain, then it is a principal ideal domain.

Exercise 5.6.2 Using Theorem 5.5.1, find a prime ideal factorization of $5\mathcal{O}_K$ and $7\mathcal{O}_K$ in $\mathbb{Z}[(1 + \sqrt{-3})/2]$.

Exercise 5.6.3 Find a prime ideal factorization of $(2), (5), (11)$ in $\mathbb{Z}[i]$.

Exercise 5.6.4 Compute the different $\mathcal{D}$ of $K = \mathbb{Q}(\sqrt{-2})$.

Exercise 5.6.5 Compute the different $\mathcal{D}$ of $K = \mathbb{Q}(\sqrt{-3})$.

Exercise 5.6.6 Let $K = \mathbb{Q}(\alpha)$ be an algebraic number field of degree n over $\mathbb{Q}$. Suppose $\mathcal{O}_K = \mathbb{Z}[\alpha]$ and that $f(x)$ is the minimal polynomial of α. Write

$$f(x) = (x - \alpha)(b_0 + b_1 x + \cdots + b_{n-1} x^{n-1}), \qquad b_i \in \mathcal{O}_K.$$

Prove that the dual basis to $1, \alpha, \ldots, \alpha^{n-1}$ is

$$\frac{b_0}{f'(\alpha)}, \ldots, \frac{b_{n-1}}{f'(\alpha)}.$$

Deduce that

$$\mathcal{D}^{-1} = \frac{1}{f'(\alpha)}(\mathbb{Z}b_0 + \cdots + \mathbb{Z}b_{n-1}).$$

Exercise 5.6.7 Let $K = \mathbb{Q}(\alpha)$ be of degree n over $\mathbb{Q}$. Suppose that $\mathcal{O}_K = \mathbb{Z}[\alpha]$. Prove that $\mathcal{D} = (f'(\alpha))$.

Exercise 5.6.8 Compute the different $\mathcal{D}$ of $\mathbb{Q}[\zeta_p]$ where ζ_p is a primitive pth root of unity.

Exercise 5.6.9 Let p be a prime, $p \nmid m$, and $a \in \mathbb{Z}$. Show that $p \mid \phi_m(a)$ if and only if the order of $a \pmod p$ is n. (Here $\phi_m(x)$ is the mth cyclotomic polynomial.)

Exercise 5.6.10 Suppose $p \nmid m$ is prime. Show that $p \mid \phi_m(a)$ for some $a \in \mathbb{Z}$ if and only if $p \equiv 1 \pmod m$. Deduce from Exercise 1.2.5 that there are infinitely many primes congruent to $1 \pmod m$.

Exercise 5.6.11 Show that $p \nmid m$ splits completely in $\mathbb{Q}(\zeta_m)$ if and only if $p \equiv 1 \pmod m$.

Exercise 5.6.12 Let p be prime and let a be squarefree and coprime to p. Set $\theta = a^{1/p}$ and consider $K = \mathbb{Q}(\theta)$. Show that $\mathcal{O}_K = \mathbb{Z}[\theta]$ if and only if $a^{p-1} \not\equiv 1 \pmod{p^2}$.

Exercise 5.6.13 Suppose that $K = \mathbb{Q}(\theta)$ and $\mathcal{O}_K = \mathbb{Z}[\theta]$. Show that if $p \mid d_K$, p ramifies.

Exercise 5.6.14 Let $K = \mathbb{Q}(\theta)$ and suppose that $p \mid d_{K/\mathbb{Q}}(\theta)$, $p^2 \nmid d_{K/\mathbb{Q}}(\theta)$. Show that $p \mid d_K$ and p ramifies in K.

Exercise 5.6.15 Let K be an algebraic number field of discriminant d_K. Show that the normal closure of K contains a quadratic field of the form $\mathbb{Q}(\sqrt{d_K})$.

Exercise 5.6.16 Show that if p ramifies in K, then it ramifies in each of the conjugate fields of K. Deduce that if p ramifies in the normal closure of K, then it ramifies in K.

Exercise 5.6.17 Deduce the following special case of Dedekind's theorem: if $p^{2m+1} \| d_K$ show that p ramifies in K.

Exercise 5.6.18 Determine the prime ideal factorization of (7), (29), and (31) in $K = \mathbb{Q}(\sqrt[3]{2})$.

Exercise 5.6.19 If L/K is a finite extension of algebraic number field, we can view L as a finite dimensional vector space over K. If $\alpha \in L$, the map $v \mapsto \alpha v$ is a linear mapping and one can define, as before, the *relative norm* $N_{L/K}(\alpha)$ and *relative trace* $Tr_{L/K}(\alpha)$ as the determinant and trace, respectively, of this linear map. If $\alpha \in \mathcal{O}_L$, show that $Tr_{L/K}(\alpha)$ and $N_{L/K}(\alpha)$ lie in $\mathcal{O}_K$.

Exercise 5.6.20 If $K \subseteq L \subseteq M$ are finite extensions of algebraic number fields, show that $N_{M/K}(\alpha) = N_{L/K}(N_{M/L}(\alpha))$ and $Tr_{M/K}(\alpha) = Tr_{L/K}(Tr_{M/L}(\alpha))$ for any $\alpha \in M$. (We refer to this as the *transitivity property* of the norm and trace map, respectively.)

Exercise 5.6.21 Let L/K be a finite extension of algebraic number fields. Show that the map

$$Tr_{L/K} : L \times L \to K$$

is non-degenerate.

Exercise 5.6.22 Let L/K be a finite extension of algebraic number fields. Let $\mathfrak{a}$ be a finitely generated $\mathcal{O}_K$-module contained in L. The set

$$\mathcal{D}_{L/K}^{-1}(\mathfrak{a}) = \{x \in L : \quad Tr_{L/K}(x\mathfrak{a}) \subseteq \mathcal{O}_K\}$$

is called *codifferent* of $\mathfrak{a}$ over K. If $\mathfrak{a} \neq 0$, show that $\mathcal{D}_{L/K}^{-1}(\mathfrak{a})$ is a finitely generated $\mathcal{O}_K$-module. Thus, it is a fractional ideal of L.

Exercise 5.6.23 If in the previous exercise $\mathfrak{a}$ is an ideal of $\mathcal{O}_L$, show that the fractional ideal inverse, denoted $\mathcal{D}_{L/K}(\mathfrak{a})$ of $\mathcal{D}_{L/K}^{-1}(\mathfrak{a})$ is an integral ideal of $\mathcal{O}_L$. (We call $\mathcal{D}_{L/K}(\mathfrak{a})$ the *different* of $\mathfrak{a}$ over K. In the case $\mathfrak{a}$ is $\mathcal{O}_L$, we call it the *relative different* of L/K and denote it by $\mathcal{D}_{L/K}$.)

Exercise 5.6.24 Let $K \subseteq L \subseteq M$ be algebraic number fields of finite degree over the rationals. Show that

$$\mathcal{D}_{M/K} = \mathcal{D}_{M/L}(\mathcal{D}_{L/K}\mathcal{O}_M).$$

Exercise 5.6.25 Let L/K be a finite extension of algebraic number fields. We define the *relative discriminant* of L/K, denoted $d_{L/K}$ as $N_{L/K}(\mathcal{D}_{L/K})$. This is an integral ideal of $\mathcal{O}_K$. If $K \subseteq L \subseteq M$ are as in Exercise 5.6.24, show that

$$d_{M/K} = d_{L/K}^{[M:L]} N_{L/K}(d_{M/L}).$$

Exercise 5.6.26 Let L/K be a finite extension of algebraic number fields. Suppose that $\mathcal{O}_L = \mathcal{O}_K[\alpha]$ for some $\alpha \in L$. If $f(x)$ is the minimal polynomial of α over $\mathcal{O}_K$, show that $\mathcal{D}_{L/K} = (f'(\alpha))$.

Exercise 5.6.27 Let K_1, K_2 be algebraic number fields of finite degree over K. If L/K is the compositum of K_1/K and K_2/K, show that the set of prime ideals dividing $d_{L/K}$ and $d_{K_1/K}d_{K_2/K}$ are the same.

Exercise 5.6.28 Let L/K be a finite extension of algebraic number fields. If $\tilde{L}$ denotes the normal closure, show that a prime $\mathfrak{p}$ of $\mathcal{O}_K$ is unramified in L if and only if it is unramified in $\tilde{L}$.

Chapter 6

The Ideal Class Group

This chapter mainly discusses the concept of the *ideal class group*, and some of its applications to Diophantine equations. We will prove that the ideal class group of an algebraic number field is finite, and establish some related results.

As in all other chapters, we shall let K be an algebraic number field with degree n over $\mathbb{Q}$, and let $\mathcal{O}_K$ be the ring of algebraic integers in K.

6.1 Elementary Results

This section serves as preparation and introduction to the remainder of the chapter. We start by a number of standard results.

Recall that the ring $\mathcal{O}_K$ is Euclidean if given $\alpha \in K$, $\exists \beta \in \mathcal{O}_K$ such that $|N(\alpha - \beta)| < 1$. Indeed, given $\theta, \gamma \in \mathcal{O}_K$, the fact that there exist $q, r \in \mathcal{O}_K$ with $r = \theta - q\gamma$ and

$$|N(r)| < |N(\gamma)|$$

is equivalent to the fact that there exists $q \in \mathcal{O}_K$ such that

$$|N(\theta/\gamma - q)| < 1.$$

Let $\alpha = \theta/\gamma$, let $\beta = q$, and we have

$$|N(\alpha - \beta)| < 1.$$

In general, $\mathcal{O}_K$ is not Euclidean, but the following results always hold:

Lemma 6.1.1 *There is a constant H_K such that given $\alpha \in K$, $\exists \beta \in \mathcal{O}_K$, and a non-zero integer t, with $|t| \leq H_K$, such that*

$$|N(t\alpha - \beta)| < 1.$$

Proof. Let $\{\omega_1, \omega_2, \ldots, \omega_n\}$ be an integral basis of $\mathcal{O}_K$. Given any $\alpha \in K$, there exists $m \in \mathbb{Z}$ such that $m\alpha \in \mathcal{O}_K$, so α can be written as

$$\alpha = \sum_{i=1}^{n} c_i \omega_i,$$

with $c_i \in \mathbb{Q}$ for all $i = 1, 2, \ldots, n$.

Let L be a natural number. Partition the interval $[0, 1]$ into L parts, each of length $1/L$. This induces a subdivision of $[0, 1]^n$ into L^n subcubes. Consider the map $\varphi : \alpha\mathbb{Z} \longrightarrow [0, 1]^n$ defined by

$$t\alpha \xrightarrow{\varphi} (\{tc_1\}, \{tc_2\}, \ldots, \{tc_n\}),$$

where $t \in \mathbb{Z}$, and $\{a\}$ denotes the fractional part of $a \in \mathbb{R}$. Let t run from 0 to L^n (the number of subcubes in $[0, 1]^n$). The number of choices for t is then $L^n + 1$, which is one more than the number of subcubes. There must be two distinct values of t, say t_1 and t_2, so that $t_1\alpha$ and $t_2\alpha$ get mapped to the same subcube of $[0, 1]^n$. Let

$$\beta = \sum_{i=1}^{n} ([t_1 c_i] - [t_2 c_i])\omega_i,$$

where $[a]$ denotes the integer part of $a \in \mathbb{R}$. Then,

$$(t_1 - t_2)\alpha - \beta = \sum_{i=1}^{n} (\{t_1 c_i\} - \{t_2 c_i\})\omega_i.$$

Let $t = t_1 - t_2$, then

$$|N(t\alpha - \beta)| = \left| N\left(\sum_{i=1}^{n} (\{t_1 c_i\} - \{t_2 c_i\})\omega_i \right) \right|.$$

Since $|(\{t_1 c_i\} - \{t_2 c_i\})| \leq 1/L$, we then have

$$|N(t\alpha - \beta)| \leq \frac{1}{L^n} \prod_{j=1}^{n} \left(\sum_{i=1}^{n} |\omega_i^{(j)}| \right),$$

where $\omega_i^{(j)}$ is the jth conjugate of ω_i. If we take $L^n > \prod_{j=1}^{n} (\sum_{i=1}^{n} |\omega_i^{(j)}|) = H_K$ (say), then

$$|N(t\alpha - \beta)| < 1.$$

Furthermore, since $0 \leq t_1, t_2 \leq L^n$, we have $|t| \leq L^n$. Thus, if we choose $L = H_K^{1/n}$, we are done. □

Let us call H_K as defined above the Hurwitz constant, since the lemma is due to A. Hurwitz.

Exercise 6.1.2 Show that given $\alpha, \beta \in \mathcal{O}_K$, there exist $t \in \mathbb{Z}, |t| \leq H_K$, and $w \in \mathcal{O}_K$ so that $|N(\alpha t - \beta w)| < |N(\beta)|$.

6.2 Finiteness of the Ideal Class Group

The concept of the ideal class group arose from Dedekind's work in establishing the unique factorization theory for ideals in the ring of algebraic integers of a number field. Our main aim of this section is to prove that the ideal class group is finite. We start by introducing an equivalence relation on ideals.

We proved in Exercise 5.3.7 that any fractional ideal $\mathcal{A}$ can be written uniquely in the form

$$\mathcal{A} = \frac{\wp_1 \cdots \wp_s}{\wp_1' \cdots \wp_r'},$$

where the $\wp_i, \wp_i'$ are primes in $\mathcal{O}_K$, and no $\wp_i$ is a $\wp_j'$ (recall that we write $1/\wp = \wp^{-1}$). In particular, we can always write any fractional ideal $\mathcal{A}$ in the form

$$\mathcal{A} = \frac{\mathfrak{b}}{\mathfrak{c}},$$

where $\mathfrak{b}, \mathfrak{c}$ are two integral ideals.

Two fractional ideals $\mathcal{A}$ and $\mathcal{B}$ in K are said to be equivalent if there exist $\alpha, \beta \in \mathcal{O}_K$ such that $(\alpha)\mathcal{A} = (\beta)\mathcal{B}$. In this case, we write $\mathcal{A} \sim \mathcal{B}$.

Notice that if $\mathcal{O}_K$ is a principal ideal domain then any two ideals are equivalent.

Exercise 6.2.1 Show that the relation $\sim$ defined above is an equivalence relation.

Theorem 6.2.2 *There exists a constant C_K such that every ideal $\mathfrak{a} \subseteq \mathcal{O}_K$ is equivalent to an ideal $\mathfrak{b} \subseteq \mathcal{O}_K$ with $N(\mathfrak{b}) \leq C_K$.*

Proof. Suppose $\mathfrak{a}$ is an ideal of $\mathcal{O}_K$. Let $\beta \in \mathfrak{a}$ be a non-zero element such that $|N(\beta)|$ is minimal.

For each $\alpha \in \mathfrak{a}$, by Exercise 6.1.2, we can find $t \in \mathbb{Z}, |t| \leq H_K$, and $w \in \mathcal{O}_K$ such that

$$|N(t\alpha - w\beta)| < |N(\beta)|.$$

Moreover, since $\alpha, \beta \in \mathfrak{a}$, so $t\alpha - w\beta \in \mathfrak{a}$; and therefore, by the minimality of $|N(\beta)|$, we must have $t\alpha = w\beta$. Thus, we have shown that for any $\alpha \in \mathfrak{a}$, there exist $t \in \mathbb{Z}, |t| \leq H_K$, and $w \in \mathcal{O}_K$ such that $t\alpha = w\beta$.

Let

$$M = \prod_{|t| \leq H_K} t,$$

and we have $M\mathfrak{a} \subseteq (\beta)$. This means that (β) divides $(M)\mathfrak{a}$, and so

$$(M)\mathfrak{a} = (\beta)\mathfrak{b},$$

for some ideal $\mathfrak{b} \subseteq \mathcal{O}_K$.

Observe that $\beta \in \mathfrak{a}$, so $M\beta \in (\beta)\mathfrak{b}$, and hence $(M) \subseteq \mathfrak{b}$. This implies that $|N(\mathfrak{b})| \leq N((M)) = C_K$. Hence, $\mathfrak{a} \sim \mathfrak{b}$, and $C_K = N((M))$ satisfies the requirements. □

Exercise 6.2.3 Show that each equivalence class of ideals has an integral ideal representative.

Exercise 6.2.4 Prove that for any integer $x > 0$, the number of integral ideals $\mathfrak{a} \subseteq \mathcal{O}_K$ for which $N(\mathfrak{a}) \leq x$ is finite.

Theorem 6.2.5 *The number of equivalence classes of ideals is finite.*

Proof. By Exercise 6.2.3, each equivalence class of ideals can be represented by an integral ideal. This integral ideal, by Theorem 6.2.2, is equivalent to another integral ideal with norm less than or equal to a given constant C_K. Apply Exercise 6.2.4, and we are done. □

As we did in the proof of Exercise 6.2.3, it is sufficient to consider only integral representatives when dealing with equivalence classes of ideals.

Let $\mathcal{H}$ be the set of all the equivalence classes of ideals of K. Given $\mathcal{C}_1$ and $\mathcal{C}_2$ in $\mathcal{H}$, we define the *product* of $\mathcal{C}_1$ and $\mathcal{C}_2$ to be the equivalence class of $\mathcal{AB}$, where $\mathcal{A}$ and $\mathcal{B}$ are two representatives of $\mathcal{C}_1$ and $\mathcal{C}_2$, respectively.

Exercise 6.2.6 Show that the product defined above is well defined, and that $\mathcal{H}$ together with this product form a group, of which the equivalence class containing the principal ideals is the identity element.

Theorem 6.2.5 and Exercise 6.2.6 give rise to the notion of class number. Given an algebraic number field K, we denote by $h(K)$ the cardinality of the group of equivalence classes of ideals ($h(K) = |\mathcal{H}|$), and call it the class number of the field K. The group of equivalence classes of ideals is called the *ideal class group*.

With the establishment of the ideal class group, the result in Theorem 6.2.2 can be improved as follows:

Exercise 6.2.7 Show that the constant C_K in Theorem 6.2.2 could be taken to be the greatest integer less than or equal to H_K, the Hurwitz constant.

The improvement on the bound enables us to determine the class number of many algebraic number fields. We demonstrate this by looking at the following example:

Example 6.2.8 Show that the class number of $K = \mathbb{Q}(\sqrt{-5})$ is 2.

Solution. We proved in Exercise 4.1.3 that the integers in K are $\mathbb{Z}[\sqrt{-5}]$, so that

$$\omega_1^{(1)} = 1, \quad \omega_2^{(1)} = \sqrt{-5},$$
$$\omega_1^{(2)} = 1, \quad \omega_2^{(2)} = -\sqrt{-5},$$

and the Hurwitz constant is $(1+\sqrt{5})^2 = 10.45\cdots$. Thus, $C_K = 10$. This implies that every equivalence class of ideals $\mathcal{C} \in \mathcal{H}$ has an integral representative $\mathfrak{a}$ such that $N(\mathfrak{a}) \leq 10$. $\mathfrak{a}$ has a factorization into a product of primes, say,

$$\mathfrak{a} = \wp_1\wp_2\cdots\wp_m,$$

where $\wp_i$ is prime in $\mathcal{O}_K$ for all $i = 1, \ldots, m$.

Consider $\wp_1$. There exists, by Exercise 4.4.4, a unique prime number $p \in \mathbb{Z}$ such that $p \in \wp_1$. This implies that $\wp_1$ is in the factorization of (p) into product of primes in $\mathcal{O}_K$. Thus, $N(\wp_1)$ is a power of p. Since $N(\mathfrak{a}) = \prod_{i=1}^m N(\wp_i)$, and $N(\mathfrak{a}) \leq 10$, we deduce that $N(\wp_i) \leq 10$ for all i. And so, in particular, $N(\wp_1) \leq 10$. Therefore, $p \leq 10$. Thus, p could be 2, 3, 5, or 7.

For $p = 2$, 3, 5, and 7, (p) factors in $\mathbb{Z}[\sqrt{-5}]$ as follows:

$$\begin{aligned}(2) &= (2, 1+\sqrt{-5})(2, 1-\sqrt{-5}),\\ (3) &= (3, 1+\sqrt{-5})(3, 1-\sqrt{-5}),\\ (7) &= (7, 3+\sqrt{-5})(7, 3-\sqrt{-5}),\end{aligned}$$

and

$$(5) = (\sqrt{-5})^2.$$

Thus, $\wp_1$ can only be $(2, 1+\sqrt{-5}), (2, 1-\sqrt{-5}), (3, 1+\sqrt{-5}), (3, 1-\sqrt{-5}), (7, 3+\sqrt{-5}), (7, 3-\sqrt{-5})$, or $(\sqrt{-5})$. The same conclusion holds for any $\wp_i$ for $i = 2, \ldots, m$. Moreover, it can be seen that $(\sqrt{-5})$ is principal, and all the others are not principal (by taking the norms), but are pairwise equivalent by the following relations:

$$\begin{aligned}(2, 1+\sqrt{-5}) &= (2, 1-\sqrt{-5}),\\ (3, 1+\sqrt{-5})(1-\sqrt{-5}) &= (3)(2, 1-\sqrt{-5}),\\ (3, 1-\sqrt{-5})(1+\sqrt{-5}) &= (3)(2, 1+\sqrt{-5}),\\ (7, 3+\sqrt{-5})(3-\sqrt{-5}) &= (7)(2, 1-\sqrt{-5}),\\ (7, 3-\sqrt{-5})(3+\sqrt{-5}) &= (7)(2, 1+\sqrt{-5}).\end{aligned}$$

Therefore, $\mathfrak{a}$ is equivalent to either the class of principal ideals or the class of those primes listed above.

Hence, the class number of $K = \mathbb{Q}(\sqrt{-5})$ is 2, and the problem is solved. In the supplementary problems we will derive a sharper constant than C_K.

6.3 Diophantine Equations

In this section, we look at the equation

$$x^2 + k = y^3, \tag{6.2}$$

which was first introduced by Bachet in 1621, and has played a fundamental role in the development of number theory. When $k = 2$, the only integral solutions to this equation are given by $y = 3$ (see Exercise 2.4.3); and this result is due to Fermat. It is known that the equation has no integral solution for many different values of k. There are various methods for discussing integral solutions of equation (6.2). We shall present, here, the one that uses applications of the quadratic field $\mathbb{Q}(\sqrt{-k})$, and the concept of ideal class group. This method is usually referred to as Minkowski's method. We start with a simple case, when $k = 5$.

Example 6.3.1 Show that the equation $x^2 + 5 = y^3$ has no integral solution.

Solution. Observe that if y is even, then x is odd, and $x^2+5 \equiv 0 \pmod 4$, and hence $x^2 \equiv 3 \pmod 4$, which is a contradiction. Therefore, y is odd. Also, if a prime $p \mid (x, y)$, then $p \mid 5$, so $p = 5$; and hence, by dividing both sides of the equation by 5, we end up with $1 \equiv 0 \pmod 5$, which is absurd. Thus, x and y are coprime.

Suppose now that (x, y) is an integral solution to the given equation. We consider the factorization

$$(x + \sqrt{-5})(x - \sqrt{-5}) = y^3, \tag{6.3}$$

in the ring of integers $\mathbb{Z}[\sqrt{-5}]$.

Suppose a prime $\wp$ divides the gcd of $(x + \sqrt{-5})$ and $(x - \sqrt{-5})$ (which implies $\wp$ divides (y)). Then $\wp$ divides $(2x)$. Also, since y is odd, $\wp$ does not divide (2). Thus, $\wp$ divides (x). This is a contradiction to the fact that x and y are coprime. Hence, $(x + \sqrt{-5})$ and $(x - \sqrt{-5})$ are coprime ideals. This and equation (6.3) ensure (by Exercise 5.3.12) that

$$(x + \sqrt{-5}) = \mathfrak{a}^3 \qquad \text{and} \qquad (x - \sqrt{-5}) = \mathfrak{b}^3,$$

for some ideals $\mathfrak{a}$ and $\mathfrak{b}$.

Since the class number of $\mathbb{Q}(\sqrt{-5})$ was found in Example 6.2.8 to be 2, $\mathfrak{c}^2$ is principal for any ideal $\mathfrak{c}$. Thus, since $\mathfrak{a}^3$ and $\mathfrak{b}^3$ are principal, we deduce that $\mathfrak{a}$ and $\mathfrak{b}$ are also principal. Moreover, since the units of $\mathbb{Q}(\sqrt{-5})$ are 1 and -1, which are both cubes, we conclude that

$$x + \sqrt{-5} = (a + b\sqrt{-5})^3,$$

for some integers a and b. This implies that

$$1 = b(3a^2 - 5b^2).$$

It is easy to see that $b \mid 1$, so $b = \pm 1$. Therefore, $3a^2 - 5 = \pm 1$. Both cases lead to contradiction with the fact that $a \in \mathbb{Z}$.

Hence, the given equation does not have an integral solution.

The discussion for many, but by no means all, values of k goes through without any great change. For instance, one can show that when $k = 13$ and $k = 17$, the only integral solutions to equation (6.2) are given by $y = 17$ and $y = 5234$, respectively.

We now turn to a more general result.

Exercise 6.3.2 Let $k > 0$ be a squarefree positive integer. Suppose that $k \equiv 1, 2 \pmod 4$, and k does not have the form $k = 3a^2 \pm 1$ for an integer a. Consider the equation

$$x^2 + k = y^3. \tag{6.4}$$

Show that if 3 does not divide the class number of $\mathbb{Q}(\sqrt{-k})$, then this equation has no integral solution.

6.4 Exponents of Ideal Class Groups

The study of class groups of quadratic fields is a fascinating one with many conjectures and few results. For instance, it was proved in 1966 by H. Stark and A. Baker (independently) that there are exactly nine imaginary quadratic fields of class number one. They are $\mathbb{Q}(\sqrt{-d})$ with $d = 1, 2, 3, 7, 11, 19, 43, 67, 163$.

By combining Dirichlet's class number formula (see Chapter 10, Exercise 10.5.12) with analytic results due to Siegel, one can show that the class number of $\mathbb{Q}(\sqrt{-d})$ grows like $\sqrt{d}$. More precisely, if $h(-d)$ denotes the class number,

$$\log h(-d) \sim \tfrac{1}{2} \log d$$

as $d \to \infty$.

The study of the growth of class numbers of real quadratic fields is more complicated. For example, it is a classical conjecture of Gauss that there are infinitely many real quadratic fields of class number 1. Related to the average behaviour of class numbers of real quadratic fields, C. Hooley formulated some interesting conjectures in 1984.

Around the same time, Cohen and Lenstra formulated general conjectures about the distribution of class groups of quadratic fields. A particular case of these conjectures is illustrated by the following. Let p be prime $\neq 2$. They predict that the probability that p divides the order of the class group of an imaginary quadratic field is

$$1 - \prod_{i=1}^{\infty}\left(1 - \frac{1}{p^i}\right).$$

A similar conjecture is made in the real quadratic case.

These conjectures suggest that as a first step, it might be worthwhile to investigate the exponents of class groups of imaginary quadratic fields.

A similar analysis for the real quadratic fields is more difficult and is postponed to Exercise 8.3.17 in Chapter 8.

Exercise 6.4.1 Fix a positive integer $g > 1$. Suppose that n is odd, greater than 1, and $n^g - 1 = d$ is squarefree. Show that the ideal class group of $\mathbb{Q}(\sqrt{-d})$ has an element of order g.

Exercise 6.4.2 Let g be odd and greater than 1. If $d = 3^g - x^2$ is squarefree with x odd and satisfying $x^2 < 3^g/2$, show that $\mathbb{Q}(\sqrt{-d})$ has an element of order g in the class group.

Exercise 6.4.3 Let g be odd. Let N be the number of squarefree integers of the form $3^g - x^2$, x odd, $0 < x^2 < 3^g/2$. For g sufficiently large, show that $N \gg 3^{g/2}$. Deduce that there are infinitely many imaginary quadratic fields whose class number is divisible by g.

6.5 Supplementary Problems

Exercise 6.5.1 Show that the class number of $K = \mathbb{Q}(\sqrt{-19})$ is 1.

We define the *volume* of a domain $C \subseteq \mathbb{R}^n$ to be

$$\operatorname{vol}(C) = \int_C \chi(\underline{x})d\underline{x}$$

where $\chi(\underline{x})$ is the characteristic function of C:

$$\chi(\underline{x}) = \begin{cases} 1, & \underline{x} \in C, \\ 0, & \underline{x} \notin C. \end{cases}$$

Exercise 6.5.2 (Siegel) Let C be a symmetric, bounded domain in $\mathbb{R}^n$. (That is, C is bounded and if $x \in C$ so is $-x$.) If $\operatorname{vol}(C) > 1$, then there are two distinct points $P, Q \in C$ such that $P - Q$ is a lattice point.

Exercise 6.5.3 If C is any convex, bounded, symmetric domain of volume $> 2^n$, show that C contains a non-zero lattice point. (C is said to be *convex* if $x, y \in C$ implies $\lambda x + (1 - \lambda)y \in C$ for $0 \leq \lambda \leq 1$.)

Exercise 6.5.4 Show in the previous question if the volume $\geq 2^n$, the result is still valid, if C is closed.

Exercise 6.5.5 Show that there exist bounded, symmetric convex domains with volume $< 2^n$ that do not contain a lattice point.

Exercise 6.5.6 (Minkowski) For $x = (x_1, \ldots, x_n)$, let

$$L_i(x) = \sum_{j=1}^{n} a_{ij} x_j, \qquad 1 \le i \le n,$$

be n linear forms with real coefficients. Let C be the domain defined by

$$|L_i(x)| \le \lambda_i, \qquad 1 \le i \le n.$$

Show that if $\lambda_1 \cdots \lambda_n \ge |\det A|$ where $A = (a_{ij})$, then C contains a nonzero lattice point.

Exercise 6.5.7 Suppose that among the n linear forms above, $L_i(x)$, $1 \le i \le r_1$ are real (i.e., $a_{ij} \in \mathbb{R}$), and $2r_2$ are not real (i.e., some a_{ij} may be nonreal). Further assume that

$$L_{r_1+r_2+j} = \overline{L_{r_1+j}}, \qquad 1 \le j \le r_2.$$

That is,

$$L_{r_1+r_2+j}(x) = \sum_{k=1}^{n} \overline{a}_{r_1+j,k} x_k, \qquad 1 \le j \le r_2.$$

Now let C be the convex, bounded symmetric domain defined by

$$|L_i(x)| \le \lambda_i, \qquad 1 \le i \le n,$$

with $\lambda_{r_1+j} = \lambda_{r_1+r_2+j}$, $1 \le j \le r_2$. Show that if $\lambda_1 \cdots \lambda_n \ge |\det A|$, then C contains a nonzero lattice point.

Exercise 6.5.8 Using the previous result, deduce that if K is an algebraic number field with discriminant d_K, then every ideal class contains an ideal $\mathfrak{b}$ satisfying $N\mathfrak{b} \le \sqrt{|d_K|}$.

Exercise 6.5.9 Let X_t consist of points

$$(x_1, \ldots, x_r, y_1, z_1, \ldots, y_s, z_s)$$

in $\mathbb{R}^{r+2s}$ where the coordinates satisfy

$$|x_1| + \cdots + |x_r| + 2\sqrt{y_1^2 + z_1^2} + \cdots + 2\sqrt{y_s^2 + z_s^2} < t.$$

Show that X_t is a bounded, convex, symmetric domain.

Exercise 6.5.10 In the previous question, show that the volume of X_t is

$$\frac{2^{r-s} \pi^s t^n}{n!},$$

where $n = r + 2s$.

Exercise 6.5.11 Let C be a bounded, symmetric, convex domain in $\mathbb{R}^n$. Let $a_1, \ldots, a_n$ be linearly independent vectors in $\mathbb{R}^n$. Let A be the $n \times n$ matrix whose rows are the a_i's. If

$$\text{vol}(C) > 2^n |\det A|,$$

show that there exist rational integers $x_1, \ldots, x_n$ (not all zero) such that

$$x_1 a_1 + \cdots + x_n a_n \in C.$$

Exercise 6.5.12 (Minkowski's Bound) Let K be an algebraic number field of degree n over $\mathbb{Q}$. Show that each ideal class contains an ideal $\mathfrak{a}$ satisfying

$$N\mathfrak{a} \leq \frac{n!}{n^n} \left(\frac{4}{\pi}\right)^{r_2} |d_K|^{1/2},$$

where r_2 is the number of pairs of complex embeddings of K, and d_K is the discriminant.

Exercise 6.5.13 Show that if $K \neq \mathbb{Q}$, then $|d_K| > 1$. Thus, by Dedekind's theorem, in any nontrivial extension of K, some prime ramifies.

Exercise 6.5.14 If K and L are algebraic number fields such that d_K and d_L are coprime, show that $K \cap L = \mathbb{Q}$. Deduce that

$$[KL : \mathbb{Q}] = [K : \mathbb{Q}][L : \mathbb{Q}].$$

Exercise 6.5.15 Use Minkowski's bound to show that $\mathbb{Q}(\sqrt{5})$ has class number 1.

Exercise 6.5.16 Using Minkowski's bound, show that $\mathbb{Q}(\sqrt{-5})$ has class number 2.

Exercise 6.5.17 Compute the class numbers of the fields $\mathbb{Q}(\sqrt{2})$, $\mathbb{Q}(\sqrt{3})$, and $\mathbb{Q}(\sqrt{13})$.

Exercise 6.5.18 Compute the class number of $\mathbb{Q}(\sqrt{17})$.

Exercise 6.5.19 Compute the class number of $\mathbb{Q}(\sqrt{6})$.

Exercise 6.5.20 Show that the fields $\mathbb{Q}(\sqrt{-1})$, $\mathbb{Q}(\sqrt{-2})$, $\mathbb{Q}(\sqrt{-3})$, and $\mathbb{Q}(\sqrt{-7})$ each have class number 1.

Exercise 6.5.21 Let K be an algebraic number field of degree n over $\mathbb{Q}$. Prove that

$$|d_K| \geq \left(\frac{\pi}{4}\right)^n \left(\frac{n^n}{n!}\right)^2.$$

Exercise 6.5.22 Show that $|d_K| \to \infty$ as $n \to \infty$ in the preceding question.

Exercise 6.5.23 (Hermite) Show that there are only finitely many algebraic number fields with a given discriminant.

Exercise 6.5.24 Let p be a prime $\equiv 11 \pmod{12}$. If $p > 3^n$, show that the ideal class group of $\mathbb{Q}(\sqrt{-p})$ has an element of order greater than n.

Exercise 6.5.25 Let $K = \mathbb{Q}(\alpha)$ where α is a root of the polynomial $f(x) = x^5 - x + 1$. Prove that $\mathbb{Q}(\alpha)$ has class number 1.

Exercise 6.5.26 Determine the class number of $\mathbb{Q}(\sqrt{14})$.

Exercise 6.5.27 If K is an algebraic number field of finite degree over $\mathbb{Q}$ with d_K squarefree, show that K has no non-trivial subfields.

Chapter 7

Quadratic Reciprocity

The equation $x^2 \equiv a \pmod{p}$, where p is some prime, provides the starting point for our discussion on quadratic reciprocity. We can ask whether there exist solutions to the above equation. If yes, how do these solutions depend upon a? upon p? Gauss developed the theory of quadratic reciprocity to answer these questions. His solution is today called the Law of Quadratic Reciprocity. Gauss, however, christened his result *Theorema Auruem*, the Golden Theorem.

In this chapter, we will be examining this interesting facet of number theory. We will begin with some of the basic properties of reciprocity. We will then take a brief trip into the realm of Gauss sums, which will provide us with the necessary tools to prove the Law of Quadratic Reciprocity. Finally, once we have developed this Golden Theorem, we will show its usefulness in the study of quadratic fields, as well as primes in certain arithmetic progressions.

7.1 Preliminaries

In this section, we would like to search for solutions to equations of the form $x^2 \equiv a \pmod{p}$, where p is prime. We will discover that quadratic reciprocity gives us a means to determine if any solution exists.

In order to appreciate the usefulness of quadratic reciprocity, let us consider how we would tackle the congruence

$$x^2 \equiv -1 \pmod{5}.$$

The naive method would be to take all the residue classes in $(\mathbb{Z}/5\mathbb{Z})$ and square them. We would get $0^2 \equiv 0$, $1^2 \equiv 1$, $2^2 \equiv 4$, $3^2 \equiv 4$, and $4^2 \equiv 1$. Since $4 \equiv -1 \pmod{5}$, we have found two solutions to the above equation, namely 2 and 3. This *brute force* method works well for small primes but becomes impractical once the size of the numbers gets too large. Thus it would be nice to have a more accessible method to determine solutions. The following exercise shows us a way to determine if there exists a solution when p is fixed. However, determining solutions to the congruence is still a difficult problem.

Exercise 7.1.1 Let p be a prime and $a \neq 0$. Show that $x^2 \equiv a \pmod{p}$ has a solution if and only if $a^{(p-1)/2} \equiv 1 \pmod{p}$.

Notice that Exercise 7.1.1 merely provides us with a means of determining whether a solution exists and gives us no information on how to actually find a square root of $a \pmod{p}$.

Exercise 7.1.1 works very well for a fixed p. Suppose, however, we wish to fix a and vary p. What happens in this case? This question motivates the remainder of our discussion on quadratic reciprocity.

Definition. The *Legendre symbol* (a/p), with p prime, is defined as follows:

$$\left(\frac{a}{p}\right) = \begin{cases} 1 & \text{if } x^2 \equiv a \pmod{p} \text{ has a solution,} \\ -1 & \text{if } x^2 \equiv a \pmod{p} \text{ has no solution,} \\ 0 & \text{if } p \mid a. \end{cases}$$

If $(a/p) = 1$, we say that a is a quadratic residue mod p. If $(a/p) = -1$, a is said to be a quadratic nonresidue mod p.

Exercise 7.1.2 Using Wilson's theorem and the congruence

$$k(p-k) \equiv -k^2 \pmod{p},$$

compute $(-1/p)$ for all primes p.

Remark. One of the interesting results of this exercise is that we can now determine which finite fields $\mathbb{F}_p$, for p prime, have an element that acts like $\sqrt{-1}$. For example, if $p = 5$, then $p \equiv 1 \pmod{4}$, and so, $(-1/p) = 1$. So there exists an element $a \in \mathbb{F}_5$ such that $a^2 = -1$. However, $7 \equiv 3 \pmod{4}$, so $\mathbb{F}_7$ can have no element that is the square root of -1.

Before going any further, we will determine some properties of the Legendre symbol.

Exercise 7.1.3 Show that

$$a^{(p-1)/2} \equiv \left(\frac{a}{p}\right) \pmod{p}.$$

Exercise 7.1.4 Show that

$$\left(\frac{ab}{p}\right) = \left(\frac{a}{p}\right)\left(\frac{b}{p}\right).$$

Exercise 7.1.5 If $a \equiv b \pmod{p}$, then $(a/p) = (b/p)$.

Exercises 7.1.3 to 7.1.5 give some of the basic properties of the Legendre symbol that we will exploit throughout the remainder of this chapter. Notice that Exercise 7.1.4 shows us that the product of two residues mod p is again a residue mod p. As well, the product of two quadratic nonresidues mod p is a quadratic residue mod p. However, a residue mod p multiplied by a nonresidue mod p is a nonresidue mod p.

Theorem 7.1.6 *For all odd primes p,*

$$\left(\frac{2}{p}\right) = \begin{cases} 1 & \text{if } p \equiv \pm 1 \pmod 8, \\ -1 & \text{if } p \equiv 3, 5 \pmod 8. \end{cases}$$

Proof. To exhibit this result, we will work in the field $\mathbb{Q}(i)$, where $i = \sqrt{-1}$. Notice that the ring of integers of this field is $\mathbb{Z}[i]$. We wish to find when there exist solutions to $x^2 \equiv 2 \pmod p$. We will make use of Exercise 7.1.1 which tells us there exists a solution if and only if $2^{(p-1)/2} \equiv 1 \pmod p$.

Working in $\mathbb{Z}[i]$, we observe that

$$(1+i)^2 = 1 + 2i + i^2 = 2i.$$

Also, for p prime,

$$(1+i)^p = 1 + \binom{p}{1} i + \binom{p}{2} i^2 + \cdots + i^p.$$

Considering the above equation mod $p\mathbb{Z}[i]$, we get

$$(1+i)^p \equiv 1 + i^p \pmod{p\mathbb{Z}[i]}.$$

But we also observe that

$$\begin{aligned} (1+i)^p &= (1+i)(1+i)^{p-1} \\ &= (1+i)((1+i)^2)^{(p-1)/2} \\ &= (1+i)(2i)^{(p-1)/2} \\ &= i^{(p-1)/2}(1+i)2^{(p-1)/2}. \end{aligned}$$

So,

$$i^{(p-1)/2}(1+i)2^{(p-1)/2} \equiv 1 + i^p \pmod{p\mathbb{Z}[i]}. \tag{7.1}$$

We now consider the various possiblities for $p \pmod 8$.

If $p \equiv 1 \pmod 8$, then $i^p = i$. As well, $i^{(p-1)/2} = 1$. So, equation (7.1) becomes

$$(1+i) \equiv (1+i)2^{(p-1)/2} \pmod{p\mathbb{Z}[i]},$$

which implies

$$1 \equiv 2^{(p-1)/2} \pmod{p\mathbb{Z}[i]}.$$

So, $1 \equiv 2^{(p-1)/2} \pmod{p}$, and thus $(2/p) = 1$ by Exercise 7.1.1.

If $p \equiv -1 \pmod{8}$, then $i^p = -i$. As well, $i^{(p-1)/2} = -i$. So, (7.1) becomes

$$(1-i) \equiv -i(1+i)2^{(p-1)/2} \pmod{p\mathbb{Z}[i]},$$

which implies

$$1 \equiv 2^{(p-1)/2} \pmod{p\mathbb{Z}[i]}.$$

Again, we have $1 \equiv 2^{(p-1)/2} \pmod{p}$, and thus $(2/p)=1$.

If $p \equiv 3 \pmod{8}$, then $i^p = -i$. Also, $i^{(p-1)/2} = i$. So, the above equation becomes

$$\begin{aligned} (1-i) &\equiv i(1+i)2^{(p-1)/2} \pmod{p\mathbb{Z}[i]}, \\ -i(1+i) &\equiv i(1+i)2^{(p-1)/2} \pmod{p\mathbb{Z}[i]}, \\ -1 &\equiv 2^{(p-1)/2} \pmod{p\mathbb{Z}[i]}. \end{aligned}$$

Since $1 \not\equiv 2^{(p-1)/2} \pmod{p}$, $(2/p) = -1$.

Finally, if $p \equiv 5 \pmod{8}$, then $i^p = i$. As well, $i^{(p-1)/2} = -1$. From this, it follows that

$$\begin{aligned} (1+i) &\equiv -1(1+i)2^{(p-1)/2} \pmod{p\mathbb{Z}[i]}, \\ -1 &\equiv 2^{(p-1)/2} \pmod{p\mathbb{Z}[i]}. \end{aligned}$$

Hence, $(2/p) = -1$, thus completing the proof. □

The above result can be restated as

$$\left(\frac{2}{p}\right) = (-1)^{(p^2-1)/8}$$

for odd primes p.

Exercise 7.1.7 Show that the number of quadratic residues mod p is equal to the number of quadratic nonresidues mod p.

Exercise 7.1.8 Show that $\sum_{a=1}^{p-1}(a/p) = 0$ for any fixed prime p.

The proof of the Law of Quadratic Reciprocity to be given does not originate with Gauss, but is of a later date. The proof, however, makes use of Gauss sums, and as a result, we will make a brief detour to describe these functions.

7.2 Gauss Sums

Definition. Let p be a prime and let ζ_p be a primitive pth root of unity. We define the *Gauss Sum* as follows:

$$S = \sum_{a \bmod p} \left(\frac{a}{p}\right) \zeta_p^a,$$

where (a/p) is the Legendre symbol.

This sum has some interesting properties. We explore some of them below.

Theorem 7.2.1 *For S as defined above,*

$$S^2 = \left(\frac{-1}{p}\right)p.$$

Proof. From the definition of a Gauss sum, we have

$$S^2 = \left(\sum_{a \bmod p} \left(\frac{a}{p}\right)\zeta_p^a\right)\left(\sum_{b \bmod p} \left(\frac{b}{p}\right)\zeta_p^b\right).$$

By applying Exercise 7.1.4, we can simplify the above to get

$$S^2 = \sum_{a,b} \left(\frac{ab}{p}\right)\zeta_p^{a+b}.$$

We now make a substitution by letting $b = ca$, where $(c,p) = 1$. Thus,

$$S^2 = \sum_{(a,p)=1}\sum_{(c,p)=1} \left(\frac{a^2c}{p}\right)\zeta_p^{a(1+c)}.$$

Again, using Exercise 7.1.4, we get

$$\begin{aligned} S^2 &= \sum_{(a,p)=1}\sum_{(c,p)=1} \left(\frac{c}{p}\right)\zeta_p^{a(1+c)} \\ &= \sum_{(c,p)=1} \left(\frac{c}{p}\right)\left(\sum_{(a,p)=1} \zeta_p^{a(1+c)}\right). \end{aligned}$$

Observe that $(1+c,p) = 1$ or $(1+c,p) = p$. Since $(c,p) = 1$, the second case will only happen if $c = p-1$. But then, if $(1+c,p) = 1$, we will have

$$\sum_{(a,p)=1} \zeta_p^{a(1+c)} = \zeta_p^{(1+c)} + \zeta_p^{2(1+c)} + \cdots + \zeta_p^{(p-1)(1+c)} = -1.$$

But $(1+c,p) = p$ implies that

$$\sum_{(a,p)=1} \zeta_p^{a(1+c)} = 1 + 1^2 + \cdots + 1^{p-1} = p - 1.$$

Thus

$$\begin{aligned} S^2 &= \sum_{(c,p)=1} \left(\frac{c}{p}\right) \left(\sum_{(a,p)=1} \zeta_p^{a(1+c)} \right) \\ &= \sum_{1 \le c \le p-2} \left(\frac{c}{p}\right)(-1) + \left(\frac{p-1}{p}\right)(p-1) \\ &= (-1) \sum_{1 \le c \le p-2} \left(\frac{c}{p}\right) + \left(\frac{-1}{p}\right)(p-1). \end{aligned}$$

But

$$\sum_{1 \le c \le p-2} \left(\frac{c}{p}\right) = \sum_{1 \le c \le p-1} \left(\frac{c}{p}\right) - \left(\frac{-1}{p}\right).$$

From Exercise 7.1.8, we know that the first term on the right-hand side must be equal to 0. So,

$$\begin{aligned} S^2 &= (-1) \sum_{1 \le c \le p-2} \left(\frac{c}{p}\right) + \left(\frac{-1}{p}\right)(p-1) \\ &= (-1)\left[-\left(\frac{-1}{p}\right)\right] + \left(\frac{-1}{p}\right)(p-1) \\ &= \left(\frac{-1}{p}\right) + \left(\frac{-1}{p}\right)(p-1) \\ &= \left(\frac{-1}{p}\right) p. \end{aligned}$$

But now we have shown the desired result, namely, $S^2 = \left(\frac{-1}{p}\right) p$. □

In the next exercise, we are going to prove an important identity that we will utilize in proving the law of quadratic reciprocity.

Exercise 7.2.2 Show that

$$S^q \equiv \left(\frac{q}{p}\right) S \pmod q,$$

where q and p are odd primes.

7.3 The Law of Quadratic Reciprocity

We are now in a position to prove the *Theorema Auruem*, which we do in this section. We also demonstrate how to use this beautiful result.

Theorem 7.3.1 (Law of Quadratic Reciprocity) *Let p and q be odd primes. Then*

$$\left(\frac{p}{q}\right) = \left(\frac{q}{p}\right)(-1)^{\frac{p-1}{2}\cdot\frac{q-1}{2}}.$$

Proof. From Exercise 7.2.2, we have

$$S^q \equiv \left(\frac{q}{p}\right) S \pmod{q}.$$

Thus, cancelling out an S from both sides will give us

$$S^{q-1} \equiv \left(\frac{q}{p}\right) \pmod{q}.$$

Since q is odd, $q-1$ must be divisible by 2. So

$$S^{q-1} = (S^2)^{(q-1)/2} = \left[p\left(\frac{-1}{p}\right)\right]^{(q-1)/2}.$$

The last equality follows from Theorem 7.2.1. Thus,

$$\left(\frac{q}{p}\right) \equiv \left[p\left(\frac{-1}{p}\right)\right]^{(q-1)/2} \pmod{q}.$$

From Exercise 7.1.2, $(-1/p) = (-1)^{(p-1)/2}$. We substitute this into the above equation to get

$$\left(\frac{q}{p}\right) \equiv p^{\frac{q-1}{2}}(-1)^{\frac{p-1}{2}\cdot\frac{q-1}{2}} \pmod{q}.$$

Exercise 7.1.3 tells us that $p^{(q-1)/2} \equiv (p/q) \pmod{q}$. So,

$$\left(\frac{q}{p}\right) \equiv \left(\frac{p}{q}\right)(-1)^{\frac{p-1}{2}\cdot\frac{q-1}{2}} \pmod{q}.$$

But both sides only take on the value ± 1, and since $q \geq 3$, the congruence can be replaced by an equals sign. This gives us

$$\left(\frac{p}{q}\right) = \left(\frac{q}{p}\right)(-1)^{\frac{p-1}{2}\cdot\frac{q-1}{2}}.$$

□

With this result, we can answer the question we asked at the beginning of this chapter. That is, if we fix some a, for what primes p will $x^2 \equiv a \pmod{p}$ have a solution? Expressed in terms of the Legendre symbol, we want to know for which p will $(a/p) = 1$. We know from Exercise 7.1.4 that the Legendre symbol is multiplicative. So, we can factor a as $a = q_1^{e_1} \cdots q_n^{e_n}$. Thus

$$\left(\frac{a}{p}\right) = \left(\frac{q_1}{p}\right)^{e_1} \cdots \left(\frac{q_n}{p}\right)^{e_n}.$$

So the question is reduced to evaluating (q/p) for each prime q. Note that we already know how to solve $(-1/p)$ and $(2/p)$. Thus, all that needs to be done is to evaluate (q/p) where q is an odd prime. The next exercise helps us to determine this.

Exercise 7.3.2 Let q be an odd prime. Prove:

(a) If $q \equiv 1 \pmod 4$, then q is a quadratic residue mod p if and only if $p \equiv r \pmod q$, where r is a quadratic residue mod q.

(b) If $q \equiv 3 \pmod 4$, then q is a quadratic residue mod p if and only if $p \equiv \pm b^2 \pmod{4q}$, where b is an odd integer prime to q.

The next exercise will demonstrate how to use Exercise 7.3.2 to compute (q/p) in the special cases $q = 5, 7$.

Exercise 7.3.3 Compute $\left(\frac{5}{p}\right)$ and $\left(\frac{7}{p}\right)$.

7.4 Quadratic Fields

In this section, we will focus on quadratic fields, that is, all algebraic number fields K such that $[K : \mathbb{Q}] = 2$. It can be shown that all quadratic extensions can be written as $K = \mathbb{Q}(\sqrt{d})$, where d is some squarefree integer.

With every algebraic number field comes an associated ring of integers, $\mathcal{O}_K$. Suppose that $p \in \mathbb{Z}$, and p is prime. We can let $p\mathcal{O}_K$ be the ideal of $\mathcal{O}_K$ generated by p. Since $\mathcal{O}_K$ is a Dedekind domain (see Chapter 5), every ideal can be written as a product of prime ideals, i.e., $p\mathcal{O}_K = \wp_1^{e_1} \cdots \wp_r^{e_r}$. However, because K is a quadratic extension,

$$p^2 = N(p\mathcal{O}_K) = N(\wp_1)^{e_1} \cdots N(\wp_r)^{e_r}.$$

So $N(\wp) = p$, or $N(\wp) = p^2$. But then, we have three possibilities:

(1) $p\mathcal{O}_K = \wp\wp'$, where $\wp \neq \wp'$;

(2) $p\mathcal{O}_K = \wp^2$; and

(3) $p\mathcal{O}_K = \wp$.

If (1) is true, we say that p splits. When case (2) occurs, we say that p ramifies. Finally, if (3) occurs, we say that p is inert, i.e., it stays prime. In the next exercises, we will see that we can determine which case occurs by using quadratic reciprocity.

Exercise 7.4.1 Find the discriminant of $K = \mathbb{Q}(\sqrt{d})$ when:

(a) $d \equiv 2, 3 \pmod 4$; and

(b) $d \equiv 1 \pmod 4$.

Remark. From the above exercise it follows that if $m = d_K$ is the discriminant of a quadratic field K, then $1, (m + \sqrt{m})/2$ will always form an integral basis.

Theorem 7.4.2 *Assume p is an odd prime. Then $(d/p) = 1$ if and only if $p\mathcal{O}_K = \wp\wp'$, where $\wp \neq \wp'$, and $\wp$ prime.*

Proof. $\Rightarrow$ From our assumption, we have $a^2 \equiv d \pmod p$ for some a. Let $\wp = (p, a + \sqrt{d})$ and $\wp' = (p, a - \sqrt{d})$.

We claim that $p\mathcal{O}_K = \wp\wp'$.

$$\begin{aligned} \wp\wp' &= (p, a + \sqrt{d})(p, a - \sqrt{d}) \\ &= (p^2, p(a + \sqrt{d}), p(a - \sqrt{d}), a^2 - d) \\ &= (p)(p, a + \sqrt{d}, a - \sqrt{d}, (a^2 - d)/p) \\ &= (p). \end{aligned}$$

The last equality holds because $2a$ and p are both elements of the second ideal. But $(2a, p) = 1$. From this, it follows that $1 \in (p, a+\sqrt{d}, a-\sqrt{d}, (a^2 - d)/p)$. It is clear that $\wp \neq \wp'$ because if they were equal, then $2a$ and p would be in $\wp$, from which it follows that $\wp = \mathcal{O}_K$, which is false. Since the norm of $p\mathcal{O}_K$ is p^2, $N(\wp)$ must divide p^2. Since $\wp \neq (1)$, $N(\wp) \neq 1$. Also, it cannot be p^2 because then $N(\wp') = 1$, which is false. So, both $\wp$ and $\wp'$ have norm p, and thus, they must be prime.

$\Leftarrow$ In the comments after Exercise 7.4.1 we noted that $\{1, (m+\sqrt{m})/2\}$ always forms an integral basis of $\mathcal{O}_K$ where $m = d$ if $d \equiv 1 \pmod 4$ and $m = 4d$ if $d \equiv 2, 3 \pmod 4$. Since $\wp\wp' = p\mathcal{O}_K$, there must exist $a \in \wp$, but $a \notin p\mathcal{O}_K$. So, $a = x + y(m + \sqrt{m})/2$, where $x, y \in \mathbb{Z}$, but p does not divide both x and y. Now, consider $a\mathcal{O}_K$, the ideal generated by a. We can write $a\mathcal{O}_K = \wp\mathfrak{q}$, $\mathfrak{q} \subseteq \mathcal{O}_K$. Now, taking the norms of both sides, we discover that $N(\wp) = p$ must divide

$$N(a\mathcal{O}_K) = \left| \left(x + \frac{ym}{2}\right)^2 - \frac{y^2 m}{4} \right|.$$

So, $(2x + ym)^2 \equiv y^2 m \pmod p$. If $p \mid y$, then $p \mid (2x + ym)^2$. But then $p \mid 2x$, and since p is odd, $p \mid x$. This contradicts the fact that p did not divide both x and y. So p does not divide y, and since $\mathbb{Z}/p\mathbb{Z}$ is a field,

$$\frac{(2x + ym)^2}{y^2} \equiv m \pmod p.$$

But then we have found some z such that $z^2 \equiv m \pmod p$. Since $m = d$ or $m = 4d$, then $(d/p) = 1$. $\square$

Exercise 7.4.3 Assume that p is an odd prime. Show that $(d/p) = 0$ if and only if $p\mathcal{O}_K = \wp^2$, where $\wp$ is prime.

Exercise 7.4.4 Assume p is an odd prime. Then $(d/p) = -1$ if and only if $p\mathcal{O}_K = \wp$, where $\wp$ is prime.

What we have shown is a method for determining what happens to an odd prime in a quadratic field that utilizes the Legendre symbol. We have yet to answer what happens to p if $p = 2$. An analogous result holds for this case.

Theorem 7.4.5 *Suppose* $p = 2$. *Then:*

(a) $2\mathcal{O}_K = \wp^2, \wp$ *prime if and only if* $2 \mid d_K$*;*

(b) $2\mathcal{O}_K = \wp\wp', \wp$ *prime if and only if* $d \equiv 1 \pmod 8$ *and* $2 \nmid d_K$*; and*

(c) $2\mathcal{O}_K = \wp, \wp$ *prime if and only if* $d \equiv 5 \pmod 8$ *and* $2 \nmid d_K$.

Proof. (a) $\Leftarrow$ If $2 \mid d_K$, then $d \equiv 2, 3 \pmod 4$. If $d \equiv 2 \pmod 4$, then we claim that $(2) = (2, \sqrt{d})^2$. Note that

$$(2, \sqrt{d})^2 = (4, 2\sqrt{d}, d) = (2)(2, \sqrt{d}, d/2).$$

Since d is squarefree, then 2 and $d/2$ are relatively prime and thus the second ideal above is actually $\mathcal{O}_K$. So $(2) = (2, \sqrt{d})^2$.

If $d \equiv 3 \pmod 4$ we claim that $(2) = (2, 1 + \sqrt{d})^2$, since

$$(2, 1 + \sqrt{d})^2 = (4, 2 + 2\sqrt{d}, 1 + d + 2\sqrt{d}) = (2)\left(2, 1 + \sqrt{d}, \frac{1+d}{2} + \sqrt{d}\right).$$

Now we note that $1 + \sqrt{d}$ and $(1 + d)/2 + \sqrt{d}$ are relatively prime, and so the second ideal is $\mathcal{O}_K$.

$\Rightarrow$ We consider d_K, which we know is congruent to either 0 or 1 mod 4. Suppose that $d_K \equiv 1 \pmod 4$. Then $\mathcal{O}_K$ is generated as a $\mathbb{Z}$-module by $1, (1 + \sqrt{d})/2$. There must exist some element a in $\wp$ which is not in $\wp^2$. So $a = m + n(1 + \sqrt{d})/2$ where we can assume that m and n are either 0 or 1, since for any $\alpha \in \mathcal{O}_K$, $a + 2\alpha$ is in $\wp$ but not in $\wp^2$.

Now, if $n = 0$, then $m \neq 0$ because otherwise $a = 0$ and is obviously in (2). But if $m = 1$, then $a = 1$ and $a \notin \wp$. So, $n = 1$ and $m = 0$ or 1. We know that $a^2 \in (2)$, and

$$\begin{aligned} a^2 &= \left(m + \frac{1 + \sqrt{d}}{2}\right)^2 \\ &= m^2 + \frac{d-1}{4} + (2m+1)\frac{1+\sqrt{d}}{2} \in (2). \end{aligned}$$

But $2m + 1$ is odd and so $a^2 \notin (2)$, and we have arrived at a contradiction.

We conclude that $d_K \equiv 0 \pmod 4$, and so clearly $2 \mid d_K$.

(b) $\Leftarrow$ Suppose that $d \equiv 1 \pmod 8$. Then clearly from the previous problem $2 \nmid d_K$. We claim that $(2) = (2, (1+\sqrt{d})/2)(2, (1-\sqrt{d})/2)$. Note that

$$\left(2, \frac{1+\sqrt{d}}{2}\right)\left(2, \frac{1-\sqrt{d}}{2}\right) = (2)\left(2, \frac{1+\sqrt{d}}{2}, \frac{1-\sqrt{d}}{2}, \frac{1-d}{8}\right).$$

But the second ideal is just $\mathcal{O}_K$ since it contains $1 = (1+\sqrt{d})/2+(1-\sqrt{d})/2$.

$\Rightarrow$ Now suppose that (2) splits in $\mathcal{O}_K$. We know from part (a) that $d \equiv 1 \pmod 4$. If $(2) = \wp\wp'$, then $N(\wp) = 2$. There exists an element a which is in $\wp$ but not in $\wp\wp' = (2)$. Then $a = m + n(1+\sqrt{d})/2$ where not both m, n are even. Therefore 2 divides the norm of the ideal generated by a, and

$$N((a)) = |N(a)| = \left|\frac{(2m+n)^2}{4} - \frac{n^2 d}{4}\right|.$$

So $(2m+n)^2 \equiv n^2 d \pmod 8$. We know that $2 \nmid d$. So suppose that n is even, and further suppose that $n = 2n_1$ where n_1 is odd. Then $2 \mid (m+n_1)^2+n_1^2 d$, and since $2 \nmid n_1^2 d$, then $2 \nmid (m+n_1)^2$, which implies that m is even. But we assumed that not both m and n were even. Now suppose that $4 \mid n$. Then $4 \mid (2m+n)$ and so m is even, a contradiction. Then n must be odd, and we can find an integer n_2 such that $nn_2 \equiv 1 \pmod 8$.

Then $d \equiv n_2^2(2m+n)^2 \pmod 8$, and since $2 \nmid d$, we conclude $n_2(2m+n)$ is odd, and $d \equiv 1 \pmod 8$, as desired.

(c) Just as in Exercise 7.4.4, this follows directly from parts (a) and (b), since if $2 \nmid d_K$ and $d \not\equiv 1 \pmod 8$, then $d \equiv 5 \pmod 8$. We know that (2) cannot split or ramify in this case, so it must remain inert. □

7.5 Primes in Special Progressions

Another interesting application of quadratic reciprocity is that it can be used to show there exist infinitely many primes in certain arithmetic progressions. In the next two exercises, we imitate Euclid's proof for the existence of an infinite number of primes to show that there are infinitely many primes in the following two arithmetic progressions, $4k+1$ and $8k+7$.

Exercise 7.5.1 Show that there are infinitely many primes of the form $4k+1$.

Exercise 7.5.2 Show that there are infinitely many primes of the form $8k+7$.

The results we have just derived are just a special case of a theorem proved by Dirichlet. Dirichlet proved that if l and k are coprime integers, then there must exist an infinite number of primes p such that $p \equiv l \pmod k$. What is interesting about these two exercises, however, is the fact that we used a proof similar to Euclid's proof for the existence of an infinite number of primes. An obvious question to ask is whether questions about all arithmetic progressions can be solved in a similar fashion.

The answer, sadly, is no. However, not all is lost. It can be shown that if $l^2 \equiv 1 \pmod{k}$, then we can apply a Euclid-type proof to show there exist an infinite number of primes p such that $p \equiv l \pmod{k}$. (See Schur [S]. For instance, Exercises 1.2.6 and 5.6.10 give Euclid-type proofs for $p \equiv 1 \pmod{k}$ using cyclotomic polynomials.) Surprisingly, the converse of this statement is true as well. The proof is not difficult, but involves some Galois Theory. It is due to Murty [Mu].

We can restate our two previous exercises as follows:

(1) Are there infinitely many primes p such that $p \equiv 1 \pmod{4}$?
(2) $p \equiv 7 \pmod{8}$?

From what we have just discussed, we observe that we can indeed apply a Euclid-type proof since $1^2 \equiv 1 \pmod{4}$ and $7^2 \equiv 1 \pmod{8}$.

Exercise 7.5.3 Show that $p \equiv 4 \pmod{5}$ for infinitely many primes p.

In their paper [BL], Bateman and Low show that if l is an integer relatively prime to 24, then there are infinitely many primes p such that $p \equiv l \pmod{24}$. Their proof makes use of the interesting fact that every integer l relatively prime to 24 has the property $l^2 \equiv 1 \pmod{24}$. (All the integers relatively prime to 24 are 1, 5, 7, 11, 13, 17, 19, and 23. A quick mental calculation will show you that the statement is true.) Because of this fact, they can use a proof similar to Euclid's.

Their proof relies on the ability to "cook up" a specific polynomial $f(x) \in \mathbb{Z}[x]$. This polynomial is created in such a way so that we can use quadratic reciprocity. Notice that in our exercises there is also some polynomial sitting in the background. In Exercise 7.5.1, we used the polynomial $f(x) = 4x^2 + 1$. In Exercise 7.5.2, $f(x) = 16x^2 - 2$ was used, and finally, in the previous exercise, $f(x) = 25x^2 - 5$. Not all the polynomials used are as simple as the ones we used. The next example uses a fourth degree polynomial.

Example 7.5.4 Show there are an infinite number of primes in the arithmetic progession $15k + 4$.

Solution. Since $4^2 \equiv 1 \pmod{15}$, we can use a Euclid-type proof. We will start with a couple of observations about the polynomial

$$f(x) = x^4 - x^3 + 2x^2 + x + 1.$$

First, we note that it can be factored in the following three ways:

$$f(x) = \left(x^2 - \frac{x}{2} - 1\right)^2 + \tfrac{15}{4}x^2, \tag{7.2}$$

$$f(x) = \left(-x^2 + \frac{x}{2} - \tfrac{1}{2}\right)^2 + \tfrac{3}{4}(x+1)^2, \tag{7.3}$$

$$f(x) = \left(-x^2 + \frac{x}{2} - \tfrac{3}{2}\right)^2 - \tfrac{5}{4}(x-1)^2. \tag{7.4}$$

We note by (7.2) that if p divides $f(x)$, then -15 is a quadratic residue mod p. By quadratic reciprocity,

$$\left(\frac{p}{3}\right)\left(\frac{p}{5}\right) = 1.$$

So, there will be a solution only if $(p/3) = 1$ and $(p/5) = 1$ or if they both equal -1. The first case will happen if $p \equiv 1 \pmod 3$ and $p \equiv 1, 4 \pmod 5$. The second happens if $p \equiv 2 \pmod 3$ and $p \equiv 2, 3 \pmod 5$. So,

$$\left(\frac{p}{15}\right) = \begin{cases} 1 & \text{if } p \equiv 1, 2, 4, 8 \pmod{15}, \\ -1 & \text{if } p \equiv 7, 11, 13 \pmod{15}, \\ 0 & \text{otherwise.} \end{cases}$$

From equation (7.3), we see that $(-3/p) = 1$. Using Exercise 7.3.2,

$$\left(\frac{3}{p}\right) = \begin{cases} 1 & \text{if } p \equiv 1, 11 \pmod{12}, \\ -1 & \text{if } p \equiv 5, 7 \pmod{12}, \\ 0 & \text{otherwise.} \end{cases}$$

So, since we already know what $(-1/p)$ is, we find that

$$\left(\frac{-3}{p}\right) = \begin{cases} 1 & \text{if } p \equiv 1, 7 \pmod{12}, \\ -1 & \text{if } p \equiv 5, 11 \pmod{12}, \\ 0 & \text{otherwise.} \end{cases}$$

Finally, equation (7.4) tells us that $(5/p) = 1$. But we know this only happens when $p \equiv 1, 4 \pmod 5$.

When we combine all these results, we find that any prime divisor of $f(x)$ must be congruent to either $1 \pmod{15}$ or $4 \pmod{15}$.

We can now begin the Euclid-type proof. Suppose that there were only a finite number of primes such that $p \equiv 4 \pmod{15}$. Let $p_1, \ldots, p_n$ be these primes. We now consider the integer $d = f(15p_1p_2 \cdots p_n + 1)$. From what we have just said, d is divisible by some prime p such that $p \equiv 1, 4 \pmod 5$. Not all the prime divisors have the form $p \equiv 1 \pmod 5$. This follows from the fact that $d = f(15p_1 \cdots p_n + 1) = 15p_1 \cdots p_n g(p_1 \cdots p_n) + 4$, where $g(x)$ is some polynomial. So, there is a divisor p such that $p \equiv 4$

(mod 15). But p cannot be in $p_1, p_2, \ldots, p_n$ because when they divide d, they leave a remainder of 4. This gives us the needed contradiction.

One item that we did not discuss is how to derive a polynomial that we can use in a Euclid-style proof. One method involves a little ingenuity and some luck. By playing around with some equations, you may happen upon such a polynomial. Murty, on the other hand, describes [Mu] an explicit construction for these polynomials. Though interesting in their own right, we will refrain from going into any detail about these polynomials.

7.6 Supplementary Problems

Exercise 7.6.1 Compute $(11/p)$.

Exercise 7.6.2 Show that $(-3/p) = 1$ if and only if $p \equiv 1 \pmod 3$.

Exercise 7.6.3 If $p \equiv 1 \pmod 3$, prove that there are integers a, b such that $p = a^2 - ab + b^2$.

Exercise 7.6.4 If $p \equiv \pm 1 \pmod 8$, show that there are integers a, b such that $a^2 - 2b^2 = \pm p$.

Exercise 7.6.5 If $p \equiv \pm 1 \pmod 5$, show that there are integers a, b such that $a^2 + ab - b^2 = \pm p$.

Exercise 7.6.6 Let p be a prime greater than 3. Show that:

(a) $(-2/p) = 1$ if and only if $p \equiv 1, 3 \pmod 8$;

(b) $(3/p) = 1$ if and only if $p \equiv 1, 11 \pmod{12}$;

(c) $(-3/p) = 1$ if and only if $p \equiv 1 \pmod 6$;

(d) $(6/p) = 1$ if and only if $p \equiv 1, 5, 19, 23 \pmod{24}$; and

(e) $(-6/p) = 1$ if and only if $p \equiv 1, 5, 7, 11 \pmod{24}$.

Exercise 7.6.7 If p is a prime dividing $n^8 - n^4 + 1$, show that p is coprime to n^2, $n^3 + n$, and $n^3 - n$. Deduce that there are integers a, b, c such that

$$\begin{aligned} an^2 &\equiv 1 \pmod p, \\ b(n^3 + n) &\equiv 1 \pmod p, \\ c(n^3 - n) &\equiv 1 \pmod p. \end{aligned}$$

Exercise 7.6.8 Let the notation be as in Exercise 7.6.7 above.

(a) Observe that $x^8 - x^4 + 1 = (x^4 - 1)^2 + (x^2)^2$. Deduce that

$$(an^4 - a)^2 + 1 \equiv 0 \pmod p.$$

(b) Observe that $x^8 - x^4 + 1 = (x^4 + x^2 + 1)^2 - 2(x^3 + x)^2$. Deduce that

$$(bn^4 + bn^2 + b)^2 - 2 \equiv 0 \pmod{p}.$$

(c) Observe that $x^8 - x^4 + 1 = (x^4 - x^2 + 1)^2 + 2(x^3 - x)^2$. Deduce that

$$(cn^4 - cn^2 + c)^2 + 2 \equiv 0 \pmod{p}.$$

(d) From $x^8 - x^4 + 1 = (x^4 + 1)^2 - 3(x^2)^2$, deduce that

$$(an^4 + a)^2 \equiv 3 \pmod{p}.$$

(e) From $x^8 - x^4 + 1 = (x^4 - \frac{1}{2})^2 + 3(\frac{1}{2})^2$, deduce that

$$(2n^4 - 1)^2 \equiv -3 \pmod{p}.$$

(f) From $x^8 - x^4 + 1 = (x^4 + 3x^2 + 1)^2 - 6(x^3 + x)^2$, deduce that

$$(bn^4 + 3bn^2 + b)^2 \equiv 6 \pmod{p}.$$

(g) From $x^8 - x^4 + 1 = (x^4 - 3x^2 + 1)^2 + 6(x^3 - x)^2$, deduce that

$$(cn^4 - 3cn^2 + c)^2 \equiv -6 \pmod{p}.$$

Exercise 7.6.9 From Exercises 7.6.7 and 7.6.8, deduce that any prime divisor p of $n^8 - n^4 + 1$ satisfies

$$\left(\frac{-1}{p}\right) = \left(\frac{2}{p}\right) = \left(\frac{-2}{p}\right) = \left(\frac{3}{p}\right) = \left(\frac{6}{p}\right) = \left(\frac{-6}{p}\right) = 1.$$

Deduce that $p \equiv 1 \pmod{24}$. Prove that there are infinitely many primes $p \equiv 1 \pmod{24}$.

[Exercises 7.6.7, 7.6.8, 7.6.9 were suggested by a paper of P. Bateman and M.E. Low, Prime Numbers in Arithmetic Progressions with Difference 24, Amer. Math. Monthly, 72 (1965), 139–143.]

Exercise 7.6.10 Show that the number of solutions of the congruence

$$x^2 + y^2 \equiv 1 \pmod{p},$$

with $0 < x < p$, $0 < y < p$, (p an odd prime) is even if and only if $p \equiv \pm 3 \pmod{8}$.

Exercise 7.6.11 If p is a prime such that $p - 1 = 4q$ with q prime, show that 2 is a primitive root mod p.

Exercise 7.6.12 (The Jacobi Symbol) Let Q be a positive odd number. We can write $Q = q_1 q_2 \cdots q_s$ where the q_i are odd primes, not necessarily distinct. Define the Jacobi symbol

$$\left(\frac{a}{Q}\right) = \prod_{j=1}^{s} \left(\frac{a}{q_i}\right).$$

If Q and Q' are odd and positive, show that:

(a) $(a/Q)(a/Q') = (a/QQ')$.

(b) $(a/Q)(a'/Q) = (aa'/Q)$.

(c) $(a/Q) = (a'/Q)$ if $a \equiv a' \pmod{Q}$.

Exercise 7.6.13 If Q is odd and positive, show that

$$\left(\frac{-1}{Q}\right) = (-1)^{(Q-1)/2}.$$

Exercise 7.6.14 If Q is odd and positive, show that $(2/Q) = (-1)^{(Q^2-1)/8}$.

Exercise 7.6.15 (Reciprocity Law for the Jacobi Symbol) Let P and Q be odd, positive, and coprime. Show that

$$\left(\frac{P}{Q}\right)\left(\frac{Q}{P}\right) = (-1)^{\frac{P-1}{2}\cdot\frac{Q-1}{2}}.$$

Exercise 7.6.16 (The Kronecker Symbol) We can define (a/n) for any integer $a \equiv 0$ or $1 \pmod 4$, as follows. Define

$$\left(\frac{a}{2}\right) = \left(\frac{a}{-2}\right) = \begin{cases} 0 & \text{if } a \equiv 0 \pmod{4}, \\ 1 & \text{if } a \equiv 1 \pmod{8}, \\ -1 & \text{if } a \equiv 5 \pmod{8}. \end{cases}$$

For general n, write $n = 2^c n_1$, with n_1 odd, and define

$$\left(\frac{a}{n}\right) = \left(\frac{a}{2}\right)^c \left(\frac{a}{n_1}\right),$$

where $(a/2)$ is defined as above and (a/n_1) is the Jacobi symbol.

Show that if d is the discriminant of a quadratic field, and n, m are positive integers, then

$$\left(\frac{d}{n}\right) = \left(\frac{d}{m}\right) \quad \text{for } n \equiv m \pmod{d}$$

and

$$\left(\frac{d}{n}\right) = \left(\frac{d}{m}\right) \operatorname{sgn} d \quad \text{for } n \equiv -m \pmod{d}.$$

Exercise 7.6.17 If p is an odd prime show that the least positive quadratic nonresidue is less than $\sqrt{p}+1$.

(It is a famous conjecture of Vinogradov that the least quadratic non-residue mod p is $O(p^\varepsilon)$ for any $\varepsilon > 0$.)

Exercise 7.6.18 Show that $x^4 \equiv 25 \pmod{1013}$ has no solution.

Exercise 7.6.19 Show that $x^4 \equiv 25 \pmod{p}$ has no solution if p is a prime congruent to 13 or 17 (mod 20).

Exercise 7.6.20 If p is a prime congruent to 13 or 17 (mod 20), show that $x^4 + py^4 = 25z^4$ has no solutions in integers.

Exercise 7.6.21 Compute the class number of $\mathbb{Q}(\sqrt{33})$.

Exercise 7.6.22 Compute the class number of $\mathbb{Q}(\sqrt{21})$.

Exercise 7.6.23 Show that $\mathbb{Q}(\sqrt{-11})$ has class number 1.

Exercise 7.6.24 Show that $\mathbb{Q}(\sqrt{-15})$ has class number 2.

Exercise 7.6.25 Show that $\mathbb{Q}(\sqrt{-31})$ has class number 3.

Chapter 8

The Structure of Units

8.1 Dirichlet's Unit Theorem

Let K be a number field and $\mathcal{O}_K$ its ring of integers. An element $\alpha \in \mathcal{O}_K$ is called a *unit* if $\exists \beta \in \mathcal{O}_K$ such that $\alpha\beta = 1$. Evidently, the set of all units in $\mathcal{O}_K$ forms a multiplicative subgroup of K^*, which we will call the unit group of K.

In this chapter, we will prove the following fundamental theorem, which gives an almost complete description of the structure of the unit group of K, for any number field K.

Theorem (Dirichlet's Unit Theorem) *Let U_K be the unit group of K. Let $n = [K : \mathbb{Q}]$ and write $n = r_1 + 2r_2$, where, as usual, r_1 and $2r_2$ are, respectively, the number of real and nonreal embeddings of K in $\mathbb{C}$. Then there exist fundamental units $\varepsilon_1, \ldots, \varepsilon_r$, where $r = r_1 + r_2 - 1$, such that every unit $\varepsilon \in U_K$ can be written uniquely in the form*

$$\varepsilon = \zeta \varepsilon_1^{n_1} \cdots \varepsilon_r^{n_r},$$

where $n_1, \ldots, n_r \in \mathbb{Z}$ and ζ is a root of unity in $\mathcal{O}_K$. More precisely, if W_K is the subgroup of U_K consisting of roots of unity, then W_K is finite and cyclic and $U_K \simeq W_K \times \mathbb{Z}^r$.

Definition. $\alpha \in \mathcal{O}_K$ is called a *root of unity* if $\exists m \in \mathbb{N}$ such that $\alpha^m = 1$.

Exercise 8.1.1 (a) Let K be an algebraic number field. Show that there are only finitely many roots of unity in K.

(b) Show, similarly, that for any positive constant c, there are only finitely many $\alpha \in \mathcal{O}_K$ for which $|\alpha^{(i)}| \leq c$ for all i.

If α is an algebraic integer all of whose conjugates lie on the unit circle, then α must be a root of unity by the argument in (a). Indeed, the polynomials

$$f_{\alpha,h}(x) = \prod_{i=1}^{n}(x - \alpha^{(i)h})$$

cannot all be distinct since $|\alpha^{(i)h}| = 1$. If $f_{\alpha,h}(x)$ is identical with $f_{\alpha,k}(x)$ where $h < k$ (say), then the roots must coincide. If $\alpha^h = \alpha^k$, then α is a root of unity and we are done. If not, after a suitable relabelling we may suppose that $\alpha^{(1)h} = \alpha^{(2)k}$, $\alpha^{(2)h} = \alpha^{(3)k}$, ..., $\alpha^{(n-1)h} = \alpha^{(n)k}$, $\alpha^{(n)h} = \alpha^{(1)k}$. Therefore,

$$\alpha^{(1)h^n} = \alpha^{(2)kh^{n-1}} = \alpha^{(3)k^2h^{n-2}} = \cdots = \alpha^{(1)k^n}$$

so that again, α is a root of unity.

This is a classical result due to Kronecker.

Exercise 8.1.2 Show that W_K, the group of roots of unity in K, is cyclic, of even order.

Definition.

(i) An (additive) subgroup Γ of $\mathbb{R}^m$ is called *discrete* if any bounded subset of $\mathbb{R}^m$ contains only finitely many elements of Γ.

(ii) Let $\{\gamma_1, \dots, \gamma_r\}$ be a linearly independent set of vectors in $\mathbb{R}^m$ (so that $r \leq m$). The additive subgroup of $\mathbb{R}^m$ generated by $\gamma_1, \dots, \gamma_r$ is called a *lattice* of *dimension* r, *generated by* $\gamma_1, \dots, \gamma_r$.

Theorem 8.1.3 *Any discrete subgroup* Γ *of* $\mathbb{R}^m$ *is a lattice.*

Proof. We prove this by induction on m:

In the trivial case, where $\Gamma = (0)$, Γ is a lattice of dimension 0. We will thus, heretofore, assume that $\Gamma \neq (0)$.

Suppose first that $m = 1$, so that $\Gamma \subseteq \mathbb{R}$.

Let α be a nonzero element of Γ and let $A = \{\lambda \in \mathbb{R} : \lambda\alpha \in \Gamma\}$. By hypothesis, the set $\{\gamma \in \Gamma : |\gamma| \leq |\alpha|\}$ is finite. Then $A \cap [-1, 1]$ is finite and contains a least positive element $0 < \mu \leq 1$.

Let $\beta = \mu\alpha$ and suppose that $\lambda\beta \in \Gamma$, with $\lambda \in \mathbb{R}$. Then

$$\lambda\beta - [\lambda]\beta = (\lambda - [\lambda])\beta = (\lambda - [\lambda])\mu\alpha \in \Gamma,$$

which, by the minimality of μ, implies that $\lambda = [\lambda]$, i.e., $\lambda \in \mathbb{Z}$ which implies that $\Gamma = \mathbb{Z}\beta$ is a lattice of dimension 1.

Now, suppose that $m > 1$.

Let $\{v_1, \dots, v_k\}$ be a maximal linearly independent subset of Γ (so that $\Gamma \subseteq \mathbb{R}v_1 + \cdots + \mathbb{R}v_k$), let V be the subspace of $\mathbb{R}^m$ spanned by $\{v_1, \dots, v_{k-1}\}$ and let $\Gamma_0 = \Gamma \cap V$. Then Γ_0 is a discrete subgroup of $V \simeq \mathbb{R}^{k-1}$ (as vector spaces) so, by the induction hypothesis, is a lattice. That is, there are linearly independent vectors $w_1, \ldots, w_l \in V$ such that $\Gamma_0 = \mathbb{Z}w_1 + \cdots + \mathbb{Z}w_l$ and, since $v_1, \ldots, v_{k-1} \in \Gamma_0$, we must have $l = k - 1$.

Evidently, $\{w_1, \ldots, w_{k-1}, w_k := v_k\}$ is also a maximal linearly independent subset of Γ (since $\text{span}\{v_1, \ldots, v_{k-1}\} = \text{span}\{w_1, \ldots, w_{k-1}\} = V$). Let

$$T = \left\{ \sum_{i=1}^{k} a_i w_i \in \Gamma : 0 \le a_i < 1 \text{ for } 1 \le i \le k-1 \text{ and } 0 \le a_k \le 1 \right\}.$$

T is bounded, hence finite, by hypothesis. We may therefore choose an element $x \in T$ with smallest nonzero coefficient a_k of w_k, say $x = \sum_{i=1}^{k} b_i w_i$. Since $b_k \neq 0$, the set $\{w_1, \ldots, w_{k-1}, x\}$ is linearly independent. Moreover, for any $\gamma \in \Gamma$, writing $\gamma = c_1 w_1 + \cdots + c_{k-1} w_{k-1} + c_k x$, we see that there are integers $d_1, \ldots, d_{k-1}$ so that

$$\gamma' = \gamma - [c_k]x - \sum_{i=1}^{k-1} d_i w_i \in T.$$

Since the coefficient of w_k in γ' is $(c_k - [c_k])b_k < b_k$, by the minimality of b_k, we must have that $c_k - [c_k] = 0$ so that $c_k \in \mathbb{Z}$ and $\gamma' \in \Gamma_0 \Rightarrow \gamma \in \Gamma_0 + \mathbb{Z}x \Rightarrow \Gamma = \Gamma_0 + \mathbb{Z}x = \mathbb{Z}w_1 + \cdots + \mathbb{Z}w_{k-1} + \mathbb{Z}x$ is a lattice of dimension k. □

Below, we will develop the proof of Dirichlet's Unit Theorem.

Let $\sigma_1, \ldots, \sigma_{r_1}, \sigma_{r_1+1}, \overline{\sigma}_{r_1+1}, \ldots, \sigma_{r_1+r_2}, \overline{\sigma}_{r_1+r_2}$ be the real and complex conjugate embeddings of K in $\mathbb{C}$. Let $E = \{k \in \mathbb{Z} : 1 \le k \le r_1 + r_2\}$. For $k \in E$, set

$$\overline{k} = \begin{cases} k & \text{if } k \le r_1, \\ k + r_2 & \text{if } k > r_1. \end{cases}$$

If $A \subseteq E$, set $\overline{A} = \{\overline{k} : k \in A\}$. Note that $E \cup \overline{E} = \{k \in \mathbb{Z} : 1 \le k \le r_1 + 2r_2\}$ and that, if $E = A \cup B$ is a partition of E, then $E \cup \overline{E} = (A \cup \overline{A}) \cup (B \cup \overline{B})$ is a partition of $E \cup \overline{E}$.

Lemma 8.1.4 (a) *Let $m, n \in \mathbb{Z}$ with $0 < m \le n$ and let $\Delta = (d_{ij}) \in M_{n \times m}(\mathbb{R})$. For any integer $t > 1$, there is a nonzero $\mathbf{x} = (x_1, \ldots, x_n) \in \mathbb{Z}^n$ with each $|x_i| \le t$ such that, if $\mathbf{y} = \mathbf{x}\Delta = (y_1, \ldots, y_m) \in \mathbb{R}^m$, then each $|y_i| \le ct^{1-n/m}$, where c is a constant depending only on the matrix Δ.*

(b) *Let $E = A \cup B$ be a partition of E and let $m = |A \cup \overline{A}|$, $n = r_1 + 2r_2 = [K : \mathbb{Q}]$. Then there is a constant c, depending only on K, such that, for t sufficiently large, $\exists \alpha \in \mathcal{O}_K$ such that*

$$\begin{array}{rcccll} c^{1-m}t^{1-n/m} & \le & |\alpha^{(k)}| & \le & ct^{1-n/m} & \text{for} \quad k \in A, \\ c^{-m}t & \le & |\alpha^{(k)}| & \le & t & \text{for} \quad k \in B. \end{array}$$

Proof. (a) Let $\delta = \max_{1 \le j \le m} \sum_{i=1}^{n} |d_{ij}|$. Then, for

$$\mathbf{0} \neq \mathbf{x} = (x_1, \ldots, x_n) \in \mathbb{Z}_{\geq 0}^n$$

with each $|x_i| \le t$,

$$|y_j| = \left| \sum_{i=1}^{n} x_i d_{ij} \right| \le \delta t.$$

Consider the cube $[-\delta t, \delta t]^m \in \mathbb{R}^m$. Let h be an integer ≥ 1 and divide the given cube into h^m equal subcubes so that each will have side length $2\delta t/h$. Now, for each $\mathbf{x} = (x_1, \ldots, x_n) \in [0, t]^n \cap \mathbb{Z}^n, \mathbf{y} = (y_1, \ldots, y_m) \in [-\delta t, \delta t]^m$ which means that there are $(t+1)^n$ such points $\mathbf{y} \in [-\delta t, \delta t]^m$. Thus, if $h^m < (t+1)^n$, then two of the points must lie in the same subcube. That is, for some $\mathbf{x}' \neq \mathbf{x}''$, we have that, if $\mathbf{y} = (\mathbf{x}' - \mathbf{x}'')\Delta = (y_1, \ldots, y_m)$, then each $|y_i| \le 2\delta t/h$.

Since $t > 1$ and $n/m \geq 1$, $(t+1)^{n/m} > t^{n/m} + 1$ so there exists an integer h with $t^{n/m} < h < (t+1)^{n/m}$ (in particular, $h^m < (t+1)^n$). Then $|y_i| \le 2\delta t/h < 2\delta t^{1-n/m}$ for each i.

(b) Let $\{\omega_1, \ldots, \omega_n\} \subset \mathcal{O}_K$ be linearly independent over $\mathbb{Q}$ and suppose that $(x_1, \ldots, x_n) \in \mathbb{Z}^n$. If $\alpha = \sum_{j=1}^{n} x_j \omega_j$, we have $\alpha^{(k)} = \sum_{i=1}^{n} x_i \omega_i^{(k)}$. Let $k_1, \ldots, k_u$ be the elements of A with $\overline{k_i} = k_i$ and let $l_1, \ldots, l_v$ be the elements of A with $\overline{l_i} \neq l_i$, so that $m = u + 2v$. Let

$$d_{ij} = \begin{cases} \omega_i^{(k_j)} & \text{for } 1 \le j \le u, \\ \text{Re } \omega_i^{(l_j)} & \text{for } u < j \le u + v, \\ \text{Im } \omega_i^{(l_j)} & \text{for } u + v < j \le 2u + v = m, \end{cases}$$

and let $\Delta = (d_{ij})$. By (a), there is a nonzero $\mathbf{x} = (x_1, \ldots, x_n) \in \mathbb{Z}^n$ with each $|x_i| \le t$ such that, if $\mathbf{y} = \mathbf{x}\Delta$, then each $|y_j| = |\sum_{i=1}^{n} x_i d_{ij}| \le Ct^{1-n/m}$, for some constant C.

For $1 \le j \le u$,

$$y_j = \sum_{i=1}^{n} x_i d_{ij} = \sum_{i=1}^{n} x_i \omega_i^{(k_j)} = \alpha^{(k_j)},$$

for $u < j < u + v$, $y_j = \text{Re } \alpha^{(l_j)}$ and for $u + v < j < u + 2v$, $y_j = \text{Im } \alpha^{(l_j)} \Rightarrow \alpha^{(l_j)} = y_j + y_{\bar{j}} \Rightarrow |\alpha^{(l_j)}| \le 2Ct^{1-n/m} = ct^{1-n/m}$. Therefore, for any $k \in A \cup \overline{A}$, $|\alpha^{(k)}| \le ct^{1-n/m}$.

If $k \in B \cup \overline{B}$, then each $|x_i| \le t$, and therefore

$$|\alpha^{(k)}| = \left| \sum_{j=1}^{n} x_j \omega_j \right| \le t \sum_{j=1}^{n} |\omega_j| = \delta t.$$

Choosing $t_0 = \delta$, we see that $|\alpha^{(k)}| \le t$, for all $t \geq t_0$.

On the other hand, for $h \in B$,

$$\begin{aligned} 1 \leq |N_K(\alpha)| &= \prod_{l=1}^{n} |\alpha^{(l)}| \\ &= \prod_{k \in A \cup \overline{A}} |\alpha^{(k)}| \prod_{l \in B \cup \overline{B}} |\alpha^{(l)}| \leq c^m (t^{1-n/m})^m |\alpha^{(h)}| t^{n-m-1} \\ &\Rightarrow |\alpha^{(h)}| \geq t c^{-m}. \end{aligned}$$

Similarly, for $j \in A$,

$$\begin{aligned} 1 \leq |N_K(\alpha)| &\leq |\alpha^{(j)}| (ct^{1-n/m})^{m-1} t^{n-m} = |\alpha^{(j)}| c^{m-1} t^{m/n-1} \\ &\Rightarrow |\alpha^{(j)}| \geq c^{1-m} t^{1-n/m}. \end{aligned}$$

□

Lemma 8.1.5 *Let $E = A \cup B$ be a proper partition of E.*

(a) *There exists a sequence of nonzero integers $\{\alpha_v\} \subseteq \mathcal{O}_K$ such that*

$$\begin{aligned} |\alpha_v^{(k)}| &> |\alpha_{v+1}^{(k)}| \quad \text{for } k \in A, \\ |\alpha_v^{(k)}| &< |\alpha_{v+1}^{(k)}| \quad \text{for } k \in B, \end{aligned}$$

and $|N_K(\alpha_v)| \leq c^m$, where c is a positive constant depending only on K and $m = |A \cup \overline{A}|$.

(b) *There exists a unit ε with $|\varepsilon^{(k)}| < 1$, for $k \in A$ and $|\varepsilon^{(k)}| > 1$, for $k \in B$.*

Proof. (a) Let t_1 be an integer greater than 1 and let $\{t_v\}$ be the sequence defined recursively by the relation $t_{v+1} = Mt_v$ for all $v \geq 1$, where M is a positive constant that will be suitably chosen. By Lemma 8.1.4, for each v, $\exists \alpha_v \in \mathcal{O}_K$ such that

$$\begin{aligned} c^{1-m} t_v^{1-n/m} &\leq |\alpha_v^{(k)}| \leq c t_v^{1-n/m} \quad \text{for } k \in A, \\ c^{-m} t_v &\leq |\alpha_v^{(k)}| \leq t_v \quad \text{for } k \in B. \end{aligned}$$

Now, let $\kappa = \min\{1, n/m - 1\}$ and choose M such that $M^\kappa > c^m$ so that both $M > c^m$ and $M^{n/m-1} > c^m$. Then, if $k \in A$,

$$|\alpha_v^{(k)}| \geq c^{-m+1} t_v^{1-n/m} = c^{-m+1} \left(\frac{t_{v+1}}{M}\right)^{1-n/m} > c t_{v+1}^{1-n/m} \geq |\alpha_{v+1}^{(k)}|$$

and if $k \in B$, then

$$|\alpha_v^{(k)}| \leq t_v = \frac{t_{v+1}}{M} < c^{-m} t_{v+1} \leq |\alpha_{v+1}^{(k)}|.$$

Also,

$$|N_K(\alpha_v)| = \prod_{i \in A \cup \overline{A}} |\alpha_v^{(i)}| \prod_{j \in B \cup \overline{B}} |\alpha_v^{(j)}| \leq (c t_v^{1-n/m})^m t_v^{n-m} = c^m.$$

(b) Let $\{\alpha_v\} \subseteq \mathcal{O}_K$ be the sequence of algebraic integers in (a). Define the sequence of principal ideals $\mathcal{A}_v = (\alpha_v)$. Then $N(\mathcal{A}_v) = N_K(\alpha_v) \leq c^m$ (where $N(\mathcal{A}_v) := \#\mathcal{O}_K/\mathcal{A}_v$).

Since there are only finitely many integral ideals $\mathcal{A}_v$ of bounded norm, $\exists \mu \in \mathbb{N}$ such that, for some $v > \mu$, $\mathcal{A}_v = \mathcal{A}_\mu$ which means that $\alpha_\mu = \varepsilon\alpha_v$, for some unit ε. We conclude that

$$|\varepsilon^{(k)}| = \left|\frac{\alpha_\mu^{(k)}}{\alpha_v^{(k)}}\right| \quad \begin{cases} < 1 & \text{for } k \in A, \\ > 1 & \text{for } k \in B. \end{cases}$$

□

Theorem 8.1.6 (Dirichlet's Unit Theorem) *Let U_K be the unit group of K. Let $n = [K : \mathbb{Q}]$ and write $n = r_1 + 2r_2$, where, as usual, r_1 and $2r_2$ are, respectively, the number of real and nonreal embeddings of K in $\mathbb{C}$. Then there exist fundamental units $\varepsilon_1, \ldots, \varepsilon_r$, where $r = r_1 + r_2 - 1$, such that every unit $\varepsilon \in U_K$ can be written uniquely in the form*

$$\varepsilon = \zeta\varepsilon_1^{n_1} \cdots \varepsilon_r^{n_r},$$

where $n_1, \ldots, n_r \in \mathbb{Z}$ and ζ is a root of unity in $\mathcal{O}_K$. More precisely, if W_K is the subgroup of U_K consisting of roots of unity, then W_K is finite and cyclic and $U_K \simeq W_K \times \mathbb{Z}^r$.

Proof. Let U_K be the unit group of K and consider the homomorphism

$$\begin{aligned} f : U_K &\rightarrow \mathbb{R}^r \\ \varepsilon &\mapsto (\log|\varepsilon^{(1)}|, \ldots, \log|\varepsilon^{(r)}|). \end{aligned}$$

We will show that:

(a) $\ker f = W_K$; and
(b) $\operatorname{Im} f = \Gamma$ is a lattice of dim r in $\mathbb{R}^r$.

a) Suppose that $\varepsilon \in \ker f$, i.e.,

$$\begin{aligned} |\varepsilon^{(1)}| &= \cdots = |\varepsilon^{(r_1+r_2-1)}| = 1, \\ \Rightarrow \quad |\varepsilon^{(r_1+r_2+1)}| &= \cdots = |\varepsilon^{(r_1+2r_2-1)}| = 1. \end{aligned}$$

But, since $\varepsilon \in U_K$,

$$\begin{aligned} 1 = |N_K(\varepsilon)| = \prod_{i=1}^{n} |\varepsilon^{(i)}| &= |\varepsilon^{(r_1+r_2)}||\varepsilon^{(r_1+2r_2)}| = |\varepsilon^{(r_1+r_2)}|^2, \\ \Rightarrow \quad |\varepsilon^{(r_1+r_2)}| &= |\varepsilon^{(r_1+2r_2)}| = 1. \end{aligned}$$

By Exercise 8.1.1, the number of $\varepsilon \in \mathcal{O}_K$ such that $|\varepsilon^{(i)}| \leq 1$ for all i is finite. ε must, therefore, have finite order in U_K, i.e, $\varepsilon^k = 1$, for some positive integer k and so $\varepsilon \in W_K$.

(b) If $-M < \log|\varepsilon^{(i)}| < M$, for $i = 1, \ldots, r$, then $e^{-M} < |\varepsilon^{(i)}| < e^{M}$ for all $i \notin S = \{r_1 + r_2, r_1 + 2r_2\}$. But

$$|\varepsilon^{(r_1+r_2)}|^2 = \frac{|N_K(\varepsilon)|}{\prod_{i \notin S} |\varepsilon^i|} < e^{Mr_1+2(r_2-1)} < e^{Mn}.$$

Thus, we have that each $|\varepsilon^{(i)}| < e^{Mn/2}$. By Exercise 8.1.1, there are only finitely many ε for which this inequality holds. Therefore, any bounded region in $\mathbb{R}^m$ contains only finitely many points of Γ so, by Theorem 8.1.3, Γ is a lattice of dimension $t \leq r$.

By Lemma 8.1.5 (b), we can find for each $1 \leq i \leq r$, a unit ϵ_i such that $|\epsilon_i^{(i)}| > 1$ and $|\epsilon_i^{(j)}| < 1$ for $j \neq i$ and $1 \leq j \leq r$. Let x_i be the image of ϵ_i under the map f. We claim that $x_1, ..., x_r$ are linearly independent. For suppose that

$$c_1 x_1 + \cdots + c_r x_r = 0,$$

with the c_i's not all zero. We may suppose without loss of generality that $c_1 > 0$ and $c_1 \geq c_j$ for $1 \leq j \leq r$. Then,

$$0 = \sum_{i=1}^{r} c_i \log|\epsilon_1^{(i)}| \geq c_1 \sum_{i=1}^{r} \log|\epsilon_1^{(i)}|,$$

so that

$$\sum_{i=1}^{r} \log|\epsilon_1^{(i)}| \leq 0.$$

Now the product of the conjugates of ϵ_1 has absolute value 1. By our choicc of ϵ_1, we see that $|\epsilon_1^{(i)}| < 1$ and we deduce that

$$\sum_{i=1}^{r} \log|\epsilon_1^{(i)}| > 0,$$

which is a contradiction. Thus, $U_K \simeq W_K \times \Gamma$, and as shown above, $\Gamma \simeq \mathbb{Z}^r$. □

Exercise 8.1.7 (a) Let Γ be a lattice of dimension n in $\mathbb{R}^n$ and suppose that $\{v_1, \ldots, v_n\}$ and $\{w_1, \ldots, w_n\}$ are two bases for Γ over $\mathbb{Z}$. Let V and W be the $n \times n$ matrices with rows consisting of the v_i's and w_i's, respectively. Show that $|\det V| = |\det W|$. Thus, we can unambiguously define the *volume of the lattice* Γ, $\text{vol}(\Gamma) =$ the absolute value of the determinant of the matrix formed by taking, as its rows, any basis for Γ over $\mathbb{Z}$.

(b) Let $\varepsilon_1, \ldots, \varepsilon_r$ be a fundamental system of units for a number field K. Show that the *regulator of* K, $R_K = |\det(\log|\varepsilon_j^{(i)}|)|$, is independent of the choice of $\varepsilon_1, \ldots, \varepsilon_r$.

If ε is a fundamental unit in $\mathbb{Q}(\sqrt{d})$, then so are $-\varepsilon, \varepsilon^{-1}, -\varepsilon^{-1}$. Subject to the constraint $\varepsilon > 1$, ε is uniquely determined and is called *the* fundamental unit of $\mathbb{Q}(\sqrt{d})$.

Exercise 8.1.8 (a) Show that, for any real quadratic field $K = \mathbb{Q}(\sqrt{d})$, where d is a positive squarefree integer, $U_K \simeq \mathbb{Z}/2\mathbb{Z} \times \mathbb{Z}$. That is, there is a fundamental unit $\varepsilon \in U_K$ such that $U_K = \{\pm\varepsilon^k : k \in \mathbb{Z}\}$. Conclude that the equation $x^2 - dy^2 = 1$ (erroneously dubbed *Pell's equation*) has infinitely many integer solutions for $d \equiv 2, 3 \ mod \ 4$ and that the equation $x^2 - dy^2 = 4$ has infinitely many integer solutions for $d \equiv 1 \ mod \ 4$.

(b) Let $d \equiv 2, 3 \pmod 4$. Let b be the smallest positive integer such that one of $db^2 \pm 1$ is a square, say a^2, $a > 0$. Then $a + b\sqrt{d}$ is a unit. Show that it is the fundamental unit. Using this algorithm, determine the fundamental units of $\mathbb{Q}(\sqrt{2})$, $\mathbb{Q}(\sqrt{3})$.

(c) Devise a similar algorithm to compute the fundamental unit in $\mathbb{Q}(\sqrt{d})$, for $d \equiv 1 \pmod 4$. Determine the fundamental unit of $\mathbb{Q}(\sqrt{5})$.

Exercise 8.1.9 (a) For an imaginary quadratic field $K = \mathbb{Q}(\sqrt{-d})$ (d a positive, squarefree integer), show that

$$U_K \simeq \begin{cases} \mathbb{Z}/4\mathbb{Z} & \text{for } d = 1, \\ \mathbb{Z}/6\mathbb{Z} & \text{for } d = 3, \\ \mathbb{Z}/2\mathbb{Z} & \text{otherwise.} \end{cases}$$

(b) Show that U_K is finite $\Leftrightarrow K = \mathbb{Q}$ or K is an imaginary quadratic field.

(c) Show that, if there exists an embedding of K in $\mathbb{R}$, then $W_K \simeq \{\pm 1\} \simeq \mathbb{Z}/2\mathbb{Z}$. Conclude that, in particular, this is the case if $[K : \mathbb{Q}]$ is odd.

Theorem 8.1.10 (a) *Let* $\zeta_m = e^{2\pi i/m}$, $K = \mathbb{Q}(\zeta_m)$. *If* m *is even, the only roots of unity in* K *are the* m*th roots of unity, so that* $W_K \simeq \mathbb{Z}/m\mathbb{Z}$. *If* m *is odd, the only ones are the* $2m$*th roots of unity, so that* $W_K \simeq \mathbb{Z}/2m\mathbb{Z}$.

(b) *Suppose that* $[K : \mathbb{Q}] = 4$. *Then* W_K *is one of the six groups* $\mathbb{Z}/2l\mathbb{Z}$, $1 \leq l \leq 6$. *If, furthermore,* K *has no real embedding, then* $U_K \simeq W_K \times \mathbb{Z}$.

(c) *Let* $K = \mathbb{Q}(\zeta_p)$, p *an odd prime. For any unit* $\varepsilon \in U_K$, $\varepsilon = \zeta_p^k u$, *for some real unit* $u \in U_K \cap \mathbb{R}$, $k \in \mathbb{Z}/p\mathbb{Z}$.

(d) *Let* K *be as in* (c) *and let* $L = \mathbb{Q}(\zeta_p + \zeta_p^{-1})$. *Then* $L = K \cap \mathbb{R}$ *and conclude that* $U_K = \langle \zeta_p \rangle \times U_L$.

(e) *For* $K = \mathbb{Q}(\zeta_5)$,

$$U_K = \{\pm \zeta_5^k \varepsilon^l : k \in \mathbb{Z}/5\mathbb{Z}, l \in \mathbb{Z}\},$$

where $\varepsilon = (1 + \sqrt{5})/2$ *is the fundamental unit of* $\mathbb{Q}(\sqrt{5})$.

Proof. (a) If m is odd, then $\zeta_{2m} = -\zeta_{2m}^{m+1} = -\zeta_m^{(m+1)/2}$ which implies that $\mathbb{Q}(\zeta_m) = \mathbb{Q}(\zeta_{2m})$. It will, therefore, suffice to establish the statement for m even. Suppose that $\theta \in \mathbb{Q}(\zeta_m)$ is a primitive kth root of unity, $k \nmid m$. Then

$\mathbb{Q}(\zeta_m)$ contains a primitive rth root of unity, where $r = \text{lcm}(k, m) > m$. Then $\mathbb{Q}(\zeta_r) \subseteq \mathbb{Q}(\zeta_m)$

$$\Rightarrow \qquad \varphi(r) = [\mathbb{Q}(\zeta_r) : \mathbb{Q}] \leq [\mathbb{Q}(\zeta_m) : \mathbb{Q}] = \varphi(m)$$

(where φ denotes the Euler phi-function). But m is even and m properly divides r implies that $\varphi(m)$ properly divides $\varphi(r)$, so that, in particular, $\varphi(m) < \varphi(r)$, a contradiction. Thus, the mth roots of unity are the only roots of unity in $\mathbb{Q}(\zeta_m)$.

(b) W_K is cyclic, generated by an rth root of unity, for some even $r \geq 2$. $\mathbb{Q}(\zeta_r) \subseteq K$ means that $4 = [K : \mathbb{Q}] \geq \varphi(r)$. A straightforward computation shows that

$$\varphi(r) \leq 4 \qquad \Rightarrow \qquad r \in \{2, 4, 6, 8, 10, 12\}.$$

If K has no real embedding, then $r_1 = 0, r_2 = 2 \Rightarrow r = r_1 + r_2 - 1 = 1$. By Dirichlet's theorem, $U_K \simeq W_K \times \mathbb{Z}$.

(c) Let $\varepsilon \in U_K$. Then $1 = |(\varepsilon/\overline{\varepsilon})^{(i)}| = |\varepsilon^{(i)}/\overline{\varepsilon^{(i)}}|$, for $i = 1, \ldots, n = [K : \mathbb{Q}]$. By the remark in the solution to Exercise 8.1.1, $\varepsilon/\overline{\varepsilon}$ is a root of unity in K and so $\varepsilon/\overline{\varepsilon} = \pm\zeta_p^k$, for some k.

Since $\mathcal{O}_K = \mathbb{Z}[\zeta_p]$, we may write $\varepsilon = \sum_{i=0}^{p-2} a_i\zeta_p^i$, each $a_i \in \mathbb{Z}$. Then

$$\varepsilon^p = \left(\sum_{i=0}^{p-2} a_i\zeta_p^i\right)^p \equiv \sum_{i=0}^{p-2} a_i \pmod p.$$

Since $\overline{\varepsilon} = \sum_{i=0}^{p-2} a_i\zeta^{-i}$,

$$\overline{\varepsilon}^p \equiv \sum_{i=0}^{p-2} a_i \equiv \varepsilon^p \pmod p.$$

If $\varepsilon = -\zeta_p^k\overline{\varepsilon}$, then $\varepsilon^p \equiv -\overline{\varepsilon}^p \pmod p$. This implies that $2\varepsilon^p \equiv 0 \pmod p$ and so $\varepsilon^p \equiv 0 \pmod p$. In other words, $\varepsilon^p \in (p)$, a contradiction, since ε is a unit.

Thus $\varepsilon/\overline{\varepsilon} = \zeta_p^k$, for some k. Let $r \in \mathbb{Z}$ such that $2r \equiv k \pmod p$ and let $\varepsilon_1 = \zeta_p^{-r}\varepsilon$. Then $\overline{\varepsilon_1} = \zeta_p^r\overline{\varepsilon} = \zeta_p^{r-k}\varepsilon = \zeta_p^{-r}\varepsilon = \varepsilon_1 \Rightarrow \varepsilon_1 \in \mathbb{R}$ and $\varepsilon = \zeta_p^r\varepsilon_1$.

(d) Let $\alpha = \zeta_p + \zeta_p^{-1}$. Then $\alpha = 2\cos(2\pi/p) \in \mathbb{R}$ so $L = \mathbb{Q}(\alpha) \subseteq K \cap \mathbb{R}$. But since $[K : K \cap \mathbb{R}] = 2$ and $\zeta_p^2 - \alpha\zeta_p + 1 = 0$ so that $[K : \mathbb{Q}(\alpha)] \leq 2$, we have that $L = K \cap \mathbb{R}$. Thus, $\varepsilon \in U_K$, meaning $\varepsilon = \pm\zeta_p^k\varepsilon_1$, for some $\varepsilon_1 \in L$, so that

$$U_K = \langle\zeta_p\rangle \times U_L.$$

(e) It remains only to show that $\mathbb{Q}(\zeta_5 + \zeta_5^{-1}) = \mathbb{Q}(\sqrt{5})$. Let $\alpha = \zeta_5 + \zeta_5^{-1}$.

$$\begin{aligned} \zeta_5^4 + \zeta_5^3 + \zeta_5^2 + \zeta_5 + 1 &= 0, \\ \Rightarrow \qquad \zeta_5^2 + \zeta_5 + 1 + \zeta_5^{-1} + \zeta_5^{-2} &= 0, \end{aligned}$$

or $\alpha^2 + \alpha - 1 = 0$. Since $\alpha = 2\cos(2\pi/5) > 0$, this implies that $\alpha = (-1 + \sqrt{5})/2$ and we conclude that $\mathbb{Q}(\alpha) = \mathbb{Q}(\sqrt{5})$. □

Exercise 8.1.11 Let $[K : \mathbb{Q}] = 3$ and suppose that K has only one real embedding. Then, by Exercise 8.1.9 (c), $W_K = \{\pm 1\}$ implies that $U_K = \{\pm u^k : k \in \mathbb{Z}\}$, where $u > 1$ is the fundamental unit in K.

(a) Let $u, \rho e^{i\theta}, \rho e^{-i\theta}$ be the $\mathbb{Q}$-conjugates of u. Show that $u = \rho^{-2}$ and that $d_{K/\mathbb{Q}}(u) = -4\sin^2\theta(\rho^3 + \rho^{-3} - 2\cos\theta)^2$.

(b) Show that $|d_{K/\mathbb{Q}}(u)| < 4(u^3 + u^{-3} + 6)$.

(c) Conclude that $u^3 > d/4 - 6 - u^{-3} > d/4 - 7$, where $d = |d_K|$.

Exercise 8.1.12 Let $\alpha = \sqrt[3]{2}, K = \mathbb{Q}(\alpha)$. Given that $d_K = -108$:

(a) Show that, if u is the fundamental unit in K, $u^3 > 20$.

(b) Show that $\beta = (\alpha - 1)^{-1} = \alpha^2 + \alpha + 1$ is a unit, $1 < \beta < u^2$. Conclude that $\beta = u$.

Exercise 8.1.13 (a) Show that, if $\alpha \in K$ is a root of a monic polynomial $f \in \mathbb{Z}[x]$ and $f(r) = \pm 1$, for some $r \in \mathbb{Z}$, then $\alpha - r$ is a unit in K.

(b) Using the fact that if $K = \mathbb{Q}(\sqrt[3]{m})$, then $d_K = -27m^2$, for any cubefree integer m, determine the fundamental unit in $K = \mathbb{Q}(\sqrt[3]{7})$.

(c) Determine the fundamental unit in $K = \mathbb{Q}(\sqrt[3]{3})$.

8.2 Units in Real Quadratic Fields

In Exercise 8.1.8, we developed a simple algorithm with which we can determine the fundamental unit of a real quadratic field. However, this algorithm is extremely inefficient and, moreover, there is no way of determining the number of steps it will take to terminate. In this section, we develop a more efficient and more enlightening algorithm, using continued fractions.

Definition.

(i) A *finite continued fraction* is an expression of the form

$$a_0 + \cfrac{1}{a_1 + \cfrac{1}{a_2 + \cdots + \cfrac{1}{a_{n-1} + \cfrac{1}{a_n}}}},$$

where each $a_i \in \mathbb{R}$ and $a_i \geq 0$ for $1 \leq i \leq n$. We use the notation $[a_0, \ldots, a_n]$ to denote the above expression.

(ii) $[a_0, \ldots, a_n]$ is called a *simple* continued fraction if $a_1, \ldots, a_n \in \mathbb{Z}$.

(iii) The continued fraction $C_k = [a_0, \ldots, a_k]$, $0 \leq k \leq n$, is called the kth convergent of $[a_0, \ldots, a_n]$.

Evidently, a finite simple continued fraction represents a rational number. Conversely, using the Euclidean algorithm, one can show that every rational number can be expressed as a finite simple continued fraction.

Exercise 8.2.1 (a) Consider the continued fraction $[a_0, \ldots, a_n]$. Define the sequences $p_0, \ldots, p_n$ and $q_0, \ldots, q_n$ recursively as follows:

$$\begin{aligned} p_0 &= a_0, & q_0 &= 1, \\ p_1 &= a_0 a_1 + 1, & q_1 &= a_1, \\ p_k &= a_k p_{k-1} + p_{k-2}, & q_k &= a_k q_{k-1} + q_{k-2}, \end{aligned}$$

for $k \geq 2$. Show that the kth convergent $C_k = p_k / q_k$.

(b) Show that $p_k q_{k-1} - p_{k-1} q_k = (-1)^{k-1}$, for $k \geq 1$.

(c) Derive the identities

$$C_k - C_{k-1} = \frac{(-1)^{k-1}}{q_k q_{k-1}},$$

for $1 \leq k \leq n$, and

$$C_k - C_{k-2} = \frac{a_k (-1)^k}{q_k q_{k-2}},$$

for $2 \leq k \leq n$.

(d) Show that

$$C_1 > C_3 > C_5 > \cdots,$$

$$C_0 < C_2 < C_4 < \cdots,$$

and that every odd-numbered convergent C_{2j+1}, $j \geq 0$, is greater than every even-numbered convergent C_{2k}, $k \geq 0$.

Remark. By (b), we can conclude that if $[a_0, \ldots, a_n]$ is a *simple* continued fraction, then the integers p_k, q_k are relatively prime.

It is also useful to note for $k \geq 1$,

$$\begin{pmatrix} a_0 & 1 \\ 1 & 0 \end{pmatrix} \begin{pmatrix} a_1 & 1 \\ 1 & 0 \end{pmatrix} \cdots \begin{pmatrix} a_k & 1 \\ 1 & 0 \end{pmatrix} = \begin{pmatrix} p_k & p_{k-1} \\ q_k & q_{k-1} \end{pmatrix},$$

which is easily proved by induction. Thus, the convergents can be easily retrieved by matrix multiplication.

Exercise 8.2.2 Let $\{a_i\}_{i \geq 0}$ be an infinite sequence of integers with $a_i \geq 0$ for $i \geq 1$ and let $C_k = [a_0, \ldots, a_k]$. Show that the sequence $\{C_k\}$ converges.

Definition. We define the *continued fraction* $[a_0, a_1, \ldots]$ to be the limit as $k \to \infty$ of its *kth convergent* C_k.

$$[a_0, a_1, \ldots] = \lim_{k \to \infty} C_k.$$

Exercise 8.2.3 Let $\alpha = \alpha_0$ be an irrational real number greater than 0. Define the sequence $\{a_i\}_{i\geq 0}$ recursively as follows:

$$a_k = [\alpha_k], \qquad \alpha_{k+1} = \frac{1}{\alpha_k - a_k}.$$

Show that $\alpha = [a_0, a_1, \ldots]$ is a representation of α as a simple continued fraction.

By Exercise 8.2.3, it is evident that every real number α has an expression as a simple continued fraction. We can also show that the representation of an *irrational* number as a simple continued fraction is *unique*. From now on, we will call the representation of α as a simple continued fraction simply the *continued fraction of* α.

Theorem 8.2.4 (a) *Let α be an irrational number and let $C_j = p_j/q_j$, for $j \in \mathbb{N}$, be the convergents of the simple continued fraction of α. If $r, s \in \mathbb{Z}$ with $s > 0$ and k is a positive integer such that*

$$|s\alpha - r| < |q_k\alpha - p_k|,$$

then $s \geq q_{k+1}$.

(b) *If α is an irrational number and r/s is a rational number in lowest terms, $s > 0$, such that*

$$|\alpha - r/s| < \frac{1}{2s^2},$$

then r/s is a convergent of the continued fraction of α.

Proof. (a) Suppose, on the contrary, that $1 \leq s < q_{k+1}$. For each $k \geq 0$, consider the system of linear equations

$$\begin{aligned} p_k x + p_{k+1} y &= r, \\ q_k x + q_{k+1} y &= s. \end{aligned}$$

Using Gaussian elimination, we easily find that

$$\begin{aligned} (p_{k+1}q_k - p_k q_{k+1})y &= rq_k - sp_k, \\ (p_k q_{k+1} - p_{k+1}q_k)x &= rq_{k+1} - sp_{k+1}. \end{aligned}$$

By Exercise 8.2.1 (b), $p_{k+1}q_k - p_k q_{k+1} = (-1)^k$ so $p_k q_{k+1} - p_{k+1}q_k = (-1)^{k+1}$. Thus, the unique solution to this system is given by

$$\begin{aligned} x &= (-1)^k(sp_{k+1} - rq_{k+1}), \\ y &= (-1)^k(rq_k - sp_k). \end{aligned}$$

We will show that x and y are nonzero and have opposite signs.

If $x = 0$, then

$$\frac{r}{s} = \frac{p_{k+1}}{q_{k+1}}$$

and since $(p_{k+1}, q_{k+1}) = 1$, this implies that $q_{k+1}|s$, and so $q_{k+1} \leq s$, contradicting our hypothesis.

If $y = 0$, then $r = p_k x, s = q_k x$, so that

$$|s\alpha - r| = |x| \cdot |q_k\alpha - p_k| \geq |q_k\alpha - p_k|,$$

again contradicting our hypothesis.

Suppose now that $y < 0$. Then since $q_k x = s - q_{k+1}y$ and each $q_j \geq 0, x > 0$. On the other hand, if $y > 0$, then $q_{k+1}y \geq q_{k+1} > s$ so $q_k x = s - q_{k+1}y < 0$ and $x < 0$.

By Exercise 8.2.1 (d), if k is even,

$$\frac{p_k}{q_k} < \alpha < \frac{p_{k+1}}{q_{k+1}},$$

while, if k is odd,

$$\frac{p_{k+1}}{q_{k+1}} < \alpha < \frac{p_k}{q_k}.$$

Thus, in either case, $q_k\alpha - p_k$ and $q_{k+1}\alpha - p_{k+1}$ have opposite signs so that $x(q_k\alpha - p_k)$ and $y(q_{k+1}\alpha - p_{k+1})$ have the same sign.

$$\begin{aligned}
\Rightarrow \qquad |s\alpha - r| &= |(q_k x + q_{k+1}y)\alpha - (p_k x + p_{k+1}y)| \\
&= |x(q_k\alpha - p_k) + y(q_{k+1}\alpha - p_{k+1})| \\
&\geq |x| \cdot |q_k\alpha - p_k| + |y| \cdot |q_{k+1}\alpha - p_{k+1}| \\
&\geq |x| \cdot |q_k\alpha - p_k| \geq |q_k\alpha - p_k|,
\end{aligned}$$

a contradiction, thus establishing our assertion.

(b) Suppose that r/s is not a convergent of the continued fraction of α, i.e., $r/s \neq p_n/q_n$ for all n. Let k be the largest nonnegative integer such that $s \geq q_k$. (Since $s \geq q_0 = 1$ and $q_k \to \infty$ as $k \to \infty$, we know that such a k exists.) Then $q_k \leq s \leq q_{k+1}$ and by (a),

$$\begin{aligned}
|q_k\alpha - p_k| \leq |s\alpha - r| &= s|\alpha - r/s| < \frac{1}{2s}, \\
\Rightarrow \qquad \left|\alpha - \frac{p_k}{q_k}\right| &< \frac{1}{2sq_k}.
\end{aligned}$$

Since $r/s \neq p_k/q_k$, $|sp_k - rq_k| \geq 1$,

$$\begin{aligned}
\frac{1}{sq_k} \leq \frac{|sp_k - rq_k|}{sq_k} &= \left|\frac{p_k}{q_k} - \frac{r}{s}\right| \\
&= \left|\frac{p_k}{q_k} - \frac{r}{s} + \alpha - \alpha\right| \\
&\leq \left|\alpha - \frac{p_k}{q_k}\right| + \left|\alpha - \frac{r}{s}\right| \\
&< \frac{1}{2sq_k} + \frac{1}{2s^2}.
\end{aligned}$$

This would imply that

$$\frac{1}{2sq_k} < \frac{1}{2s^2},$$

so $q_k > s$, a contradiction. □

Exercise 8.2.5 Let d be a positive integer, not a perfect square. Show that, if $|x^2 - dy^2| < \sqrt{d}$ for positive integers x, y, then x/y is a convergent of the continued fraction of $\sqrt{d}$.

Definition. A simple continued fraction is called *periodic* with *period* k if there exist positive integers N, k such that $a_n = a_{n+k}$ for all $n \geq N$. We denote such a continued fraction by $[a_0, \dots, a_{N-1}, \overline{a_N, a_{N+1}, \dots, a_{N+k-1}}]$.

Exercise 8.2.6 Let α be a quadratic irrational (i.e, the minimal polynomial of the real number α over $\mathbb{Q}$ has degree 2). Show that there are integers P_0, Q_0, d such that

$$\alpha = \frac{P_0 + \sqrt{d}}{Q_0} \qquad \text{with} \qquad Q_0 | (d - P_0^2).$$

Recursively define

$$\begin{aligned}
\alpha_k &= \frac{P_k + \sqrt{d}}{Q_k}, \\
a_k &= [\alpha_k], \\
P_{k+1} &= a_k Q_k - P_k, \\
Q_{k+1} &= \frac{d - P_{k+1}^2}{Q_k},
\end{aligned}$$

for $k = 0, 1, 2, \dots$. Show that $[a_0, a_1, a_2, \dots]$ is the simple continued fraction of α.

Exercise 8.2.7 Show that the simple continued fraction expansion of a quadratic irrational α is periodic.

Exercise 8.2.8 Show that, if d is a positive integer but not a perfect square, and $\alpha = \alpha_0 = \sqrt{d}$, then

$$p_{k-1}^2 - dq_{k-1}^2 = (-1)^k Q_k,$$

for all $k \geq 1$, where p_k/q_k is the kth convergent of the continued fraction of α and Q_k is as defined in Exercise 8.2.6.

Let n be the period of the continued fraction of $\sqrt{d}$. We can show, using properties of *purely periodic* continued fractions, that n is the smallest positive integer such that $Q_n = 1$ (so that $Q_j \neq 0$, for $0 < j < n$) and that $Q_n \neq -1$ for all n. In particular, this implies that $(-1)^k Q_k = \pm 1$ if and only if $n|k$. For the sake of brevity, we omit the proof of this fact.

Theorem 8.2.9 *Let n be the period of the continued fraction of $\sqrt{d}$.*

(a) *All integer solutions to the equation $x^2 - dy^2 = \pm 1$ are given by*

$$x + y\sqrt{d} = \pm(p_{n-1} + q_{n-1}\sqrt{d})^l \ : l \in \mathbb{Z},$$

where p_{n-1}/q_{n-1} is the $(n-1)$th convergent of the continued fraction of $\sqrt{d}$.

(b) *If d is squarefree, $d \equiv 2, 3 \pmod 4$, then $p_{n-1} + q_{n-1}\sqrt{d}$ is the fundamental unit of $\mathbb{Q}(\sqrt{d})$.*

(c) *The equation $x^2 - dy^2 = -1$ has an integer solution if and only if n is odd.*

(d) *If d has a prime divisor $p \equiv 3 \pmod 4$, then the equation $x^2 - dy^2 = -1$ has no integer solution.*

Proof. (a) For any solution (x, y) to the given equation,

$$(x + \sqrt{d}y)^{-1} = \pm(x - \sqrt{d}y).$$

Therefore, one of $\pm(a, \pm b)$ is a solution to $x^2 - dy^2 = \pm 1$ if and only if each of the four pairs is a solution. It will, thus, suffice to show that all positive solutions are given by

$$x + y\sqrt{d} = (p_{n-1} + q_{n-1}\sqrt{d})^m : m > 0.$$

By Exercise 8.2.5, if $x^2 - dy^2 = \pm 1$, then $x = p_{k-1}, y = q_{k-1}$, for some k. By Exercise 8.2.8, $p_{k-1}^2 - dq_{k-1}^2 = (-1)^k Q_k = \pm 1 \Rightarrow Q_k = \pm 1$ and, by our Remark, this implies that $n|k$. Since

$$p_{n-1} < p_{2n-1} < \cdots \qquad \text{and} \qquad q_{n-1} < q_{2n-1} < \cdots,$$

we have that, in particular, the least positive solution to the given equation is $x_1 = p_{n-1}, y_1 = q_{n-1}$. We will now show that all positive solutions (x_m, y_m) are given by $x_m + y_m\sqrt{d} = (x_1 + y_1\sqrt{d})^m$, $m > 0$.

Taking $\mathbb{Q}$-conjugates, $x_m - y_m\sqrt{d} = (x_1 - y_1\sqrt{d})^m$

$$(x_m + y_m\sqrt{d})(x_m - y_m\sqrt{d}) = (x_1^2 - dy_1^2)^m = (\pm 1)^m = \pm 1,$$

so that (x_m, y_m) is indeed a solution. Evidently, $x_1 < x_m, y_1 < y_m$, so that (x_m, y_m) is a positive solution.

Now, suppose that (X, Y) is a positive solution that is *not* one of the (x_m, y_m). Then $\exists$ an integer $\kappa \geq 0$ such that

$$(x_1 + y_1\sqrt{d})^\kappa < X + Y\sqrt{d} < (x_1 + y_1\sqrt{d})^{\kappa+1},$$

or

$$1 < (x_1 + y_1\sqrt{d})^{-\kappa}(X + Y\sqrt{d}) < x_1 + y_1\sqrt{d}.$$

But $x_1^2 - dy_1^2 = \pm 1$ which implies that $(x_1 + y_1\sqrt{d})^{-\kappa} = [\pm(x_1 - y_1\sqrt{d})]^\kappa$. Define the integers s, t such that

$$s + t\sqrt{d} = (x_1 + y_1\sqrt{d})^{-\kappa}(X + Y\sqrt{d}) = \pm(x_1 - y_1\sqrt{d})^\kappa(X + Y\sqrt{d}).$$

Then

$$\begin{aligned} s^2 - dt^2 &= [\pm(x_1 - y_1\sqrt{d})^\kappa(X + Y\sqrt{d})][\pm(x_1 + y_1\sqrt{d})^\kappa(X - Y\sqrt{d})] \\ &= X^2 - dY^2 = \pm 1. \end{aligned}$$

Thus, (s, t) is a solution to the given equation with

$$1 < s + t\sqrt{d} < x_1 + y_1\sqrt{d}.$$

Also,

$$0 < (x_1 + y_1\sqrt{d})^{-1} < (s + t\sqrt{d})^{-1} < 1 < s + t\sqrt{d}.$$

But this implies that

$$\begin{aligned} 2s &= s + t\sqrt{d} \pm [\pm(s - t\sqrt{d})] = s + t\sqrt{d} \pm (s + t\sqrt{d})^{-1} > 0, \\ 2t\sqrt{d} &= s + t\sqrt{d} \mp [\pm(s - t\sqrt{d})] > 0, \end{aligned}$$

and so (s, t) is a positive solution. By hypothesis, then, $s \geq x_1, t \geq y_1$ and, since $s + t\sqrt{d} < x_1 + y_1\sqrt{d}$, we have a contradiction.

(b) Since $p_{n-1} + q_{n-1}\sqrt{d} > 1$, this follows immediately from part (a).

(c) $x^2 - dy^2 = -1 \Rightarrow x = p_{k-1}, y = q_{k-1}$ for some k, by Exercise 8.2.5. But $p_{k-1}^2 - dq_{k-1}^2 = (-1)^k Q_k = -1$ if and only if $n|k$ and k is odd. Clearly then, a solution exists if and only if n is odd.

(d) $x^2 - dy^2 = -1$ implies that $x^2 \equiv -1 (\bmod p)$ for all $p|d$. But, for $p \equiv 3 \bmod 4$, this congruence has no solutions. □

Exercise 8.2.10 (a) Find the simple continued fractions of $\sqrt{6}$, $\sqrt{23}$.

(b) Using Theorem 8.2.9 (c), compute the fundamental unit in both $\mathbb{Q}(\sqrt{6})$ and $\mathbb{Q}(\sqrt{23})$.

Exercise 8.2.11 (a) Show that $[d, \overline{2d}]$ is the continued fraction of $\sqrt{d^2+1}$.

(b) Conclude that, if d^2+1 is squarefree, $d \equiv 1, 3 \pmod 4$, then the fundamental unit of $\mathbb{Q}(\sqrt{d^2+1})$ is $d+\sqrt{d^2+1}$. Compute the fundamental unit of $\mathbb{Q}(\sqrt{2}), \mathbb{Q}(\sqrt{10}), \mathbb{Q}(\sqrt{26})$.

(c) Show that the continued fraction of $\sqrt{d^2+2}$ is $[d, \overline{d, 2d}]$.

(d) Conclude that, if d^2+2 is squarefree, then the fundamental unit of $\mathbb{Q}(\sqrt{d^2+2})$ is $d^2+1+d\sqrt{d^2+2}$. Compute the fundamental unit in $\mathbb{Q}(\sqrt{3})$, $\mathbb{Q}(\sqrt{11})$, $\mathbb{Q}(\sqrt{51})$, $\mathbb{Q}(\sqrt{66})$.

8.3 Supplementary Problems

Exercise 8.3.1 If n^2-1 is squarefree, show that $n+\sqrt{n^2-1}$ is the fundamental unit of $\mathbb{Q}(\sqrt{n^2-1})$.

Exercise 8.3.2 Determine the units of an imaginary quadratic field from first principles.

Exercise 8.3.3 Suppose that $2^{2n}+1 = dy^2$ with d squarefree. Show that $2^n + y\sqrt{d}$ is the fundamental unit of $\mathbb{Q}(\sqrt{d})$, whenever $\mathbb{Q}(\sqrt{d}) \neq \mathbb{Q}(\sqrt{5})$.

Exercise 8.3.4 (a) Determine the continued fraction expansion of $\sqrt{51}$ and use it to obtain the fundamental unit ε of $\mathbb{Q}(\sqrt{51})$.

(b) Prove from first principles that all units of $\mathbb{Q}(\sqrt{51})$ are given by $\varepsilon^n, n \in \mathbb{Z}$.

Exercise 8.3.5 Determine a unit $\neq \pm 1$ in the ring of integers of $\mathbb{Q}(\theta)$ where $\theta^3+6\theta+8=0$.

Exercise 8.3.6 Let p be an odd prime > 3 and supose that it does not divide the class number of $\mathbb{Q}(\zeta_p)$. Show that

$$x^p + y^p + z^p = 0$$

is impossible for integers x, y, z such that $p \nmid xyz$.

Exercise 8.3.7 Let K be a quadratic field of discriminant d. Let P_0 denote the group of principal fractional ideals $\alpha \mathcal{O}_K$ with $\alpha \in K$ satisfying $N_K(\alpha) > 0$. The quotient group H_0 of all nonzero fractional ideals modulo P_0 is called the *restricted class group* of K. Show that H_0 is a subgroup of the ideal class group H of K and $[H : H_0] \leq 2$.

Exercise 8.3.8 Given an ideal $\mathfrak{a}$ of a quadratic field K, let $\mathfrak{a}'$ denote the conjugate ideal. If K has discriminant d, write

$$|d| = p_1^{\alpha_1} p_2 \cdots p_t$$

where $p_1 = 2$, $\alpha_1 = 0, 2$, or 3 and $p_2, \ldots, p_t$ are distinct odd primes. If we write $p_i \mathcal{O}_K = \wp_i^2$ show that for any ideal $\mathfrak{a}$ of $\mathcal{O}_K$ satisfying $\mathfrak{a} = \mathfrak{a}'$ we can write

$$\mathfrak{a} = r\wp_1^{a_1} \cdots \wp_t^{a_t},$$

$r > 0$, $a_i = 0, 1$ uniquely.

Exercise 8.3.9 An ideal class C of H_0 is said to be *ambiguous* if $C^2 = 1$ in H_0. Show that any ambiguous ideal class is equivalent (in the restricted sense) to one of the at most 2^t ideal classes

$$\wp_1^{a_1} \cdots \wp_t^{a_t}, \qquad a_i = 0, 1.$$

Exercise 8.3.10 With the notation as in the previous two questions, show that there is exactly one relation of the form

$$\wp_1^{a_1} \cdots \wp_t^{a_t} = \rho \mathcal{O}_K, \qquad N_K(\rho) > 0,$$

with $a_i = 0$ or 1, $\sum_{i=1}^t a_i > 0$.

Exercise 8.3.11 Let K be a quadratic field of discriminant d. Show that the number of ambiguous ideal classes is 2^{t-1} where t is the number of distinct primes dividing d. Deduce that 2^{t-1} divides the order of the class group.

Exercise 8.3.12 If K is a quadratic field of discriminant d and class number 1, show that d is prime or $d = 4$ or 8.

Exercise 8.3.13 If a real quadratic field K has odd class number, show that K has a unit of norm -1.

Exercise 8.3.14 Show that $15 + 4\sqrt{14}$ is the fundamental unit of $\mathbb{Q}(\sqrt{14})$.

Exercise 8.3.15 In Chapter 6 we showed that $\mathbb{Z}[\sqrt{14}]$ is a PID (principal ideal domain). Assume the following hypothesis: given $\alpha, \beta \in \mathbb{Z}[\sqrt{14}]$, such that $\gcd(\alpha, \beta) = 1$, there is a prime $\pi \equiv \alpha \pmod{\beta}$ for which the fundamental unit $\varepsilon = 15 + 4\sqrt{14}$ generates the coprime residue classes $\pmod{\pi}$. Show that $\mathbb{Z}[\sqrt{14}]$ is Euclidean.

It is now known that $\mathbb{Z}[\sqrt{14}]$ is Euclidean and is the main theorem of the doctoral thesis of Harper [Ha]. The hypothesis of the previous exercise is still unknown however and is true if the Riemann hypothesis holds for Dedekind zeta functions of number fields (see Chapter 10). The hypothesis in the question should be viewed as a number field version of a classical conjecture of Artin on primitive roots. Previously the classification of Euclidean rings of algebraic integers relied on some number field generalization of the Artin primitive root conjecture. But recently, Harper and Murty [HM] have found new techniques which circumvent the need of such a hypothesis in such questions. No doubt, these techniques will have further applications.

Exercise 8.3.16 Let $d = a^2 + 1$. Show that if $|u^2 - dv^2| \neq 0, 1$ for integers u, v, then

$$|u^2 - dv^2| > \sqrt{d}.$$

Exercise 8.3.17 Suppose that n is odd, $n \geq 5$, and that $n^{2g} + 1 = d$ is squarefree. Show that the class group of $\mathbb{Q}(\sqrt{d})$ has an element of order $2g$.

Chapter 9

Higher Reciprocity Laws

9.1 Cubic Reciprocity

Let $\rho = (-1 + \sqrt{-3})/2$ be as in Chapter 2, and let $\mathbb{Z}[\rho]$ be the ring of Eisenstein integers. Recall that $\mathbb{Z}[\rho]$ is a Euclidean domain and hence a PID. We set $N(a + b\rho) = a^2 - ab + b^2$ which is the Euclidean norm as proved in Section 2.3.

Exercise 9.1.1 If π is a prime of $\mathbb{Z}[\rho]$, show that $N(\pi)$ is a rational prime or the square of a rational prime.

Exercise 9.1.2 If $\pi \in \mathbb{Z}[\rho]$ is such that $N(\pi) = p$, a rational prime, show that π is a prime of $\mathbb{Z}[\rho]$.

Exercise 9.1.3 If p is a rational prime congruent to 2 (mod 3), show that p is prime in $\mathbb{Z}[\rho]$. If $p \equiv 1 \pmod 3$, show that $p = \pi\overline{\pi}$ where π is prime in $\mathbb{Z}[\rho]$.

Exercise 9.1.4 Let π be a prime of $\mathbb{Z}[\rho]$. Show that $\alpha^{N(\pi)-1} \equiv 1 \pmod{\pi}$ for all $\alpha \in \mathbb{Z}[\rho]$ which are coprime to π.

Exercise 9.1.5 Let π be a prime not associated to $(1 - \rho)$. First show that $3 \mid N(\pi) - 1$. If $(\alpha, \pi) = 1$, show that there is a unique integer $m = 0, 1,$ or 2 such that

$$\alpha^{(N(\pi)-1)/3} \equiv \rho^m \pmod{\pi}.$$

Let $N(\pi) \neq 3$. We define the *cubic residue character* of $\alpha \pmod{\pi}$ by the symbol $(\alpha/\pi)_3$ as follows:

(i) $(\alpha/\pi)_3 = 0$ if $\pi \mid \alpha$;

(ii) $\alpha^{(N(\pi)-1)/3} \equiv (\alpha/\pi)_3 \pmod{\pi}$ where $(\alpha/\pi)_3$ is the unique cube root of unity determined by the previous exercise.

Exercise 9.1.6 Show that:

(a) $(\alpha/\pi)_3 = 1$ if and only if $x^3 \equiv \alpha \pmod{\pi}$ is solvable in $\mathbb{Z}[\rho]$;

(b) $(\alpha\beta/\pi)_3 = (\alpha/\pi)_3(\beta/\pi)_3$; and

(c) if $\alpha \equiv \beta \pmod{\pi}$, then $(\alpha/\pi)_3 = (\beta/\pi)_3$.

Let us now define the *cubic character* $\chi_\pi(\alpha) = (\alpha/\pi)_3$.

Exercise 9.1.7 Show that:

(a) $\overline{\chi_\pi(\alpha)} = \chi_\pi(\alpha)^2 = \chi_\pi(\alpha^2)$; and

(b) $\overline{\chi_\pi(\alpha)} = \chi_{\overline{\pi}}(\alpha)$.

Exercise 9.1.8 If $q \equiv 2 \pmod 3$, show that $\chi_q(\overline{\alpha}) = \chi_q(\alpha^2)$ and $\chi_q(n) = 1$ if n is a rational integer coprime to q.

This exercise shows that any rational integer is a cubic residue mod q. If π is prime in $\mathbb{Z}[\rho]$, we say π is *primary* if $\pi \equiv 2 \pmod 3$. Therefore if $q \equiv 2 \pmod 3$, then q is primary in $\mathbb{Z}[\rho]$. If $\pi = a + b\rho$, then this means $a \equiv 2 \pmod 3$ and $b \equiv 0 \pmod 3$.

Exercise 9.1.9 Let $N(\pi) = p \equiv 1 \pmod 3$. Among the associates of π, show there is a unique one which is primary.

We can now state the law of cubic reciprocity: let π_1, π_2 be primary. Suppose $N(\pi_1), N(\pi_2) \neq 3$ and $N(\pi_1) \neq N(\pi_2)$. Then

$$\chi_{\pi_1}(\pi_2) = \chi_{\pi_2}(\pi_1).$$

To prove the law of cubic reciprocity, we will introduce Jacobi sums and more general Gauss sums than the ones used in Chapter 7. Let $\mathbb{F}_p$ denote the finite field of p elements. A multiplicative character on $\mathbb{F}_p$ is a homomorphism $\chi : \mathbb{F}_p^\times \to \mathbb{C}^\times$. The Legendre symbol (a/p) is an example of such a character. Another example is the trivial character χ_0 defined by $\chi_0(a) = 1$ for all $a \in \mathbb{F}_p^\times$. It is useful to extend the definition of χ to all of $\mathbb{F}_p$. We set $\chi(0) = 0$ for $\chi \neq \chi_0$ and $\chi_0(0) = 1$.

For $a \in \mathbb{F}_p^\times$, define the Gauss sum

$$g_a(\chi) = \sum_{t \in \mathbb{F}_p} \chi(t)\zeta^{at},$$

where $\zeta = e^{2\pi i/p}$ is a primitive pth root of unity. We also write $g(\chi)$ for $g_1(\chi)$.

Theorem 9.1.10 *If* $\chi \neq \chi_0$, *then* $|g(\chi)| = \sqrt{p}$.

Proof. We first observe that $a \neq 0$ and $\chi \neq \chi_0$ together imply that $g_a(\chi) = \chi(a^{-1})g(\chi)$ because

$$\begin{aligned} \chi(a)g_a(\chi) &= \chi(a)\sum_{t\in\mathbb{F}_p}\chi(t)\zeta^{at} \\ &= \sum_{t\in\mathbb{F}_p}\chi(at)\zeta^{at} \\ &= g(\chi). \end{aligned}$$

With our conventions that $\chi_0(0) = 1$, we see for $a \neq 0$,

$$g_a(\chi_0) = \sum_{t\in\mathbb{F}_p}\zeta^{at} = 0,$$

since this is just the sum of the pth roots of unity. Finally, $g_0(\chi) = 0$ if $\chi \neq \chi_0$ and $g_0(\chi_0) = p$.

Now, by our first observation,

$$\sum_{a\in\mathbb{F}_p} g_a(\chi)\overline{g_a(\chi)} = |g(\chi)|^2(p-1).$$

On the other hand,

$$\sum_{a\in\mathbb{F}_p} g_a(\chi)\overline{g_a(\chi)} = \sum_{s\in\mathbb{F}_p}\sum_{t\in\mathbb{F}_p}\chi(s)\chi(t)\sum_{a\in\mathbb{F}_p}\zeta^{as-at}.$$

If $s \neq t$, the innermost sum is zero, being the sum of all the pth roots of unity. If $s = t$, the sum is p. Hence $|g(\chi)|^2 = p$. □

Let $\chi_1, \chi_2, \ldots, \chi_r$ be characters of $\mathbb{F}_p$. A *Jacobi sum* is defined by

$$J(\chi_1, \ldots, \chi_r) = \sum_{t_1+\cdots+t_r=1} \chi_1(t_1)\cdots\chi_r(t_r),$$

where the summation is over all solutions of $t_1 + \cdots + t_r = 1$ in $\mathbb{F}_p$. The relationship between Gauss sums and Jacobi sums is given by the following exercise.

Exercise 9.1.11 If $\chi_1, \ldots, \chi_r$ are nontrivial and the product $\chi_1\cdots\chi_r$ is also nontrivial, prove that $g(\chi_1)\cdots g(\chi_r) = J(\chi_1, \ldots, \chi_r)g(\chi_1\cdots\chi_r)$.

Exercise 9.1.12 If $\chi_1, \ldots, \chi_r$ are nontrivial, and $\chi_1\cdots\chi_r$ is trivial, show that

$$g(\chi_1)\cdots g(\chi_r) = \chi_r(-1)pJ(\chi_1, \ldots, \chi_{r-1}).$$

We are now ready to prove the cubic reciprocity law. It will be convenient to work in the ring Ω of all algebraic integers.

Lemma 9.1.13 *Let π be a prime of $\mathbb{Z}[\rho]$ such that $N(\pi) = p \equiv 1 \pmod 3$. The character χ_π introduced above can be viewed as a character of the finite field $\mathbb{Z}[\rho]/(\pi)$ of p elements. $J(\chi_\pi, \chi_\pi) = \pi$.*

Proof. If χ is any cubic character, Exercise 9.1.12 shows that $g(\chi)^3 = pJ(\chi,\chi)$ since $\chi(-1) = 1$. We can write $J(\chi,\chi) = a + b\rho$ for some $a, b \in \mathbb{Z}$. But

$$\begin{aligned} g(\chi)^3 &= \left(\sum_t \chi(t)\zeta^t\right)^3 \\ &\equiv \sum_t \chi^3(t)\zeta^{3t} \pmod{3\Omega} \\ &\equiv \sum_{t\neq 0} \zeta^{3t} \pmod{3\Omega} \\ &\equiv -1 \pmod{3\Omega}. \end{aligned}$$

Therefore, $a + b\rho \equiv -1 \pmod{3\Omega}$. In a similar way,

$$g(\overline{\chi})^3 \equiv a + b\overline{\rho} \equiv -1 \pmod{3\Omega}.$$

Thus, $b\sqrt{-3} \equiv 0 \pmod{3\Omega}$ which means $-3b^2/9$ is an algebraic integer and by Exercise 3.1.2, it is an ordinary integer. Thus, $b \equiv 0 \pmod 3$ and $a \equiv -1 \pmod 3$. Also, from Exercise 9.1.12 and Theorem 9.1.10, it is clear that $|J(\chi,\chi)|^2 = p = J(\chi,\chi)\overline{J(\chi,\chi)}$. Therefore, $J(\chi,\chi)$ is a primary prime of norm p. Set $J(\chi_\pi,\chi_\pi) = \pi'$. Since $\pi\overline{\pi} = p = \pi'\overline{\pi'}$, we have $\pi \mid \pi'$ or $\pi \mid \overline{\pi'}$. We want to eliminate the latter possibility.

By definition,

$$\begin{aligned} J(\chi_\pi,\chi_\pi) &= \sum_t \chi_\pi(t)\chi_\pi(1-t) \\ &\equiv \sum_t t^{(p-1)/3}(1-t)^{(p-1)/3} \pmod{\pi}. \end{aligned}$$

The polynomial $x^{(p-1)/3}(1-x)^{(p-1)/3}$ has degree $\frac{2}{3}(p-1) < p-1$. Let g be a primitive root $\pmod{\pi}$. Then

$$\sum_t t^j = \sum_{a=0}^{p-2} g^{aj} \equiv 0 \pmod{\pi}$$

if $g^j \not\equiv 1 \pmod{\pi}$, which is the case since $j < p-1$. Thus, $J(\chi_\pi,\chi_\pi) \equiv 0 \pmod{\pi}$. Therefore $\pi \mid \pi'$, as desired. □

Exercise 9.1.14 Show that $g(\chi)^3 = p\pi$.

Lemma 9.1.15 *Let $\pi_1 = q \equiv 2 \pmod 3$ and $\pi_2 = \pi$ be a prime of $\mathbb{Z}[\rho]$ of norm p. Then $\chi_\pi(q) = \chi_q(\pi)$. In other words,*

$$\left(\frac{q}{\pi}\right)_3 = \left(\frac{\pi}{q}\right)_3.$$

Proof. Let $\chi_\pi = \chi$, and consider the Jacobi sum $J(\chi, \dots, \chi)$ with q terms. Since $3 \mid q+1$, we have, by Exercise 9.1.12, $g(\chi)^{q+1} = pJ(\chi, \dots, \chi)$. By Exercise 9.1.14, $g(\chi)^3 = p\pi$ so that

$$g(\chi)^{q+1} = (p\pi)^{(q+1)/3}.$$

Recall that

$$J(\chi, \dots, \chi) = \sum \chi(t_1) \cdots \chi(t_q),$$

where the sum is over all $t_1, \dots, t_q \in \mathbb{Z}/p\mathbb{Z}$ such that $t_1 + \cdots + t_q = 1$. The term in which $t_1 = \cdots = t_q$ satisfies $qt_1 = 1$ and so $\chi(q)\chi(t_1) = 1$. Raising both sides to the qth power and noting that $q \equiv 2 \pmod 3$ gives

$$\chi(q)^2 \chi(t_1)^q = 1$$

and so $\chi(t_1)^q = \chi(q)$. Therefore, the "diagonal" term which corresponds to $t_1 = \cdots = t_q$ has the value $\chi(q)$. If not all the t_i are equal, then we get q different q-tuples from a cyclic permutation of $(t_1, \dots, t_q)$. Thus $J(\chi, \dots, \chi) \equiv \chi(q) \pmod q$.

Hence $(p\pi)^{(q+1)/3} \equiv p\chi(q) \pmod q$ so that

$$p^{(q-2)/3}\pi^{(q+1)/3} \equiv \chi(q) \pmod q.$$

We raise both sides of this congruence to the $(q-1)$st power (recalling that $q - 1 \equiv 1 \pmod 3$):

$$p^{(q-2)(q-1)/3}\pi^{(q^2-1)/3} \equiv \chi(q)^{q-1} \equiv \chi(q) \pmod q.$$

By Fermat's little Theorem, $p^{q-1} \equiv 1 \pmod q$. Therefore,

$$\chi_\pi(q) \equiv \pi^{(q^2-1)/3} \equiv \chi_q(\pi) \pmod q$$

so that $\chi_\pi(q) = \chi_q(\pi)$ as desired. □

Theorem 9.1.16 *Let π_1 and π_2 be two primary primes of $\mathbb{Z}[\rho]$, of norms p_1, p_2, respectively. Then $\chi_{\pi_1}(\pi_2) = \chi_{\pi_2}(\pi_1)$. In other words,*

$$\left(\frac{\pi_2}{\pi_1}\right)_3 = \left(\frac{\pi_1}{\pi_2}\right)_3.$$

Proof. To begin, let $\gamma_1 = \overline{\pi_1}, \gamma_2 = \overline{\pi_2}$. Then $p_1 = \pi_1\gamma_1, p_2 = \pi_2\gamma_2$, and $p_1, p_2 \equiv 1 \pmod 3$. Now,

$$g(\chi_{\gamma_1})^{p_2} = J(\chi_{\gamma_1}, \ldots, \chi_{\gamma_1})g(\chi_{\gamma_1}^{p_2})$$

by Exercise 9.1.11. (There are p_2 characters in the Jacobi sum.) Since $p_2 \equiv 1 \pmod 3), \chi_{\gamma_1}^{p_2} = \chi_{\gamma_1}$. Thus,

$$\left[g(\chi_{\gamma_1})^3\right]^{(p_2-1)/3} = J(\chi_{\gamma_1}, \ldots, \chi_{\gamma_1}).$$

As before, isolating the "diagonal" term in the Jacobi sum and observing that the contribution from the other terms is congruent to 0 (mod p_2), we find

$$J(\chi_{\gamma_1}, \ldots, \chi_{\gamma_1}) \equiv \chi_{\gamma_1}(p_2^{-1}) \equiv \chi_{\gamma_1}(p_2^2) \pmod{p_2}.$$

By Exercise 9.1.14, $g(\chi_{\gamma_1})^3 = p_1\gamma_1$ so that

$$(p_1\gamma_1)^{(p_2-1)/3} \equiv \chi_{\gamma_1}(p_2^2) \pmod{p_2}.$$

Hence $\chi_{\pi_2}(p_1\gamma_1) = \chi_{\gamma_1}(p_2^2)$. Similarly, $\chi_{\pi_1}(p_2\pi_2) = \chi_{\pi_2}(p_1^2)$. Now, by Exercise 9.1.7, $\chi_{\gamma_1}(p_2^2) = \chi_{\pi_1}(p_2)$. Thus

$$\begin{aligned} \chi_{\pi_2}(p_1\gamma_1) &= \chi_{\pi_1}(p_2), \\ \chi_{\pi_1}(p_2\gamma_2) &= \chi_{\pi_2}(p_1). \end{aligned}$$

Therefore

$$\begin{aligned} \chi_{\pi_1}(\pi_2)\chi_{\pi_2}(p_1\gamma_1) &= \chi_{\pi_1}(\pi_2 p_2) \\ &= \chi_{\pi_2}(p_1^2) \\ &= \chi_{\pi_2}(p_1\pi_1\gamma_1) \\ &= \chi_{\pi_2}(\pi_1)\chi_{\pi_2}(p_1\gamma_1), \end{aligned}$$

which gives $\chi_{\pi_1}(\pi_2) = \chi_{\pi_2}(\pi_1)$, as desired. □

Exercise 9.1.17 Let π be a prime of $\mathbb{Z}[\rho]$. Show that $x^3 \equiv 2 \pmod \pi$ has a solution if and only if $\pi \equiv 1 \pmod 2$.

9.2 Eisenstein Reciprocity

The Eisenstein reciprocity law generalizes both the laws of quadratic and cubic reciprocity. In 1850, Eisenstein published the proof of this generalization by using the (then) new language of ideal numbers due to Kummer. We do not prove this law here but content ourselves with understanding its formulation and applying it to Fermat's Last Theorem, in a particular instance. We begin with the definition of the power residue symbol.

Let m be a positive integer. Then, it is known that $\mathbb{Z}[\zeta_m]$ is the ring of integers of $\mathbb{Q}(\zeta_m)$. In the case m is prime, this was proved in Chapter 4 (Exercise 4.3.7). The general case is deduced from Exercises 4.5.9 and 4.5.13. Let $\wp$ be a prime ideal of $\mathbb{Z}[\zeta_m]$ not containing m. Let q be its norm.

Exercise 9.2.1 Show that $q \equiv 1 \pmod{m}$ and that $1, \zeta_m, \zeta_m^2, \ldots, \zeta_m^{m-1}$ are distinct coset representatives mod $\wp$.

Exercise 9.2.2 Let $\alpha \in \mathbb{Z}[\zeta_m]$, $\alpha \notin \wp$. Show that there is a unique integer i (modulo m) such that

$$\alpha^{(q-1)/m} \equiv \zeta_m^i \pmod{\wp}.$$

We can now define the power residue symbol. For $\alpha \in \mathbb{Z}[\zeta_m]$, and $\wp$ a prime ideal not containing m, define $(\alpha/\wp)_m$ as:

(i) $(\alpha/\wp)_m = 0$ if $\alpha \in \wp$; and

(ii) if $\alpha \notin \wp$, $(\alpha/\wp)_m$ is the unique mth root of unity such that

$$\alpha^{(N(\wp)-1)/m} \equiv \left(\frac{\alpha}{\wp}\right)_m \pmod{m}$$

as determined by Exercise 9.2.2.

Exercise 9.2.3 Show that:

(a) $(\alpha/\wp)_m = 1$ if and only if $x^m \equiv \alpha \pmod{\wp}$ is solvable in $\mathbb{Z}[\zeta_m]$.

(b) for all $\alpha \in \mathbb{Z}[\zeta_m]$, $\alpha^{(N(\wp)-1)/m} \equiv (\alpha/\wp)_m \pmod{\wp}$.

(c) $(\alpha\beta/\wp)_m = (\alpha/\wp)_m(\beta/\wp)_m$.

(d) if $\alpha \equiv \beta \pmod{\wp}$, then $(\alpha/\wp)_m = (\beta/\wp)_m$.

Exercise 9.2.4 If $\wp$ is a prime ideal of $\mathbb{Z}[\zeta_m]$ not containing m, show that

$$\left(\frac{\zeta_m}{\wp}\right)_m = \zeta_m^{(N(\wp)-1)/m}.$$

We will now extend the definition of $(\alpha/\wp)_m$. Let $\mathfrak{a} = \wp_1\wp_2\cdots\wp_r$ be the prime ideal decomposition of $\mathfrak{a}$. Suppose $\mathfrak{a}$ is coprime to m. For $\alpha \in \mathbb{Z}[\zeta_m]$, define

$$\left(\frac{\alpha}{\mathfrak{a}}\right)_m = \prod_{i=1}^{r}\left(\frac{\alpha}{\wp_i}\right)_m.$$

If $\beta \in \mathbb{Z}[\zeta_m]$ is coprime to m, define

$$\left(\frac{\alpha}{\beta}\right)_m = \left(\frac{\alpha}{(\beta)}\right)_m.$$

Exercise 9.2.5 Suppose $\mathfrak{a}$ and $\mathfrak{b}$ are ideals coprime to (m). Show that:

(a) $(\alpha\beta/\mathfrak{a})_m = (\alpha/\mathfrak{a})_m(\beta/\mathfrak{a})_m$;

(b) $(\alpha/\mathfrak{a}\mathfrak{b})_m = (\alpha/\mathfrak{a})_m(\beta/\mathfrak{b})_m$; and

(c) if α is prime to $\mathfrak{a}$ and $x^m \equiv \alpha \pmod{\mathfrak{a}}$ is solvable in $\mathbb{Z}[\zeta_m]$, then $(\alpha/\mathfrak{a})_m = 1$.

Exercise 9.2.6 Show that the converse of (c) in the previous exercise is not necessarily true.

Now let ℓ be an odd prime number. Recall that by Exercise 5.5.4, we have $(\ell) = (1-\zeta_\ell)^{\ell-1}$ in $\mathbb{Z}[\zeta_\ell]$ and $(1-\zeta_\ell)$ is a prime ideal of degree 1. We will say that $\alpha \in \mathbb{Z}[\zeta_\ell]$ is *primary* if it is prime to ℓ and congruent to a rational integer modulo $(1-\zeta_\ell)^2$. In the case $\ell = 3$, we required $\alpha \equiv 2 \pmod 3$ which is the same as $\alpha \equiv 2 \pmod{(1-\zeta_3)^2}$. So this new definition is weaker but will suffice for our purpose.

Exercise 9.2.7 If $\alpha \in \mathbb{Z}[\zeta_\ell]$ is coprime to ℓ, show that there is an integer $c \in \mathbb{Z}$ (unique mod ℓ) such that $\zeta_\ell^c \alpha$ is primary.

We can now state the Eisenstein reciprocity law: let ℓ be an odd prime, $a \in \mathbb{Z}$ prime to ℓ and let $\alpha \in \mathbb{Z}[\zeta_\ell]$ be primary. If α and a are coprime, then

$$\left(\frac{\alpha}{a}\right)_\ell = \left(\frac{a}{\alpha}\right)_\ell.$$

We will now apply this to establish the theorems of Wieferich and Furtwangler on Fermat's Last Theorem: let ℓ be an odd prime and suppose $x^\ell + y^\ell + z^\ell = 0$ for three mutually coprime integers x, y, z with $\ell \nmid xyz$. (This is the so-called first case.) We let $\zeta = \zeta_\ell$ be a primitive ℓth root of unity and factor the above equation as

$$(x+y)(x+\zeta y)\cdots(x+\zeta^{\ell-1}y) = (-z)^\ell.$$

Exercise 9.2.8 With notation as above, show that $(x+\zeta^i y)$ and $(x+\zeta^j y)$ are coprime in $\mathbb{Z}[\zeta_\ell]$ whenever $i \neq j$, $0 \leq i, j < \ell$.

Exercise 9.2.9 Show that the ideals $(x+\zeta^i y)$ are perfect ℓth powers.

Exercise 9.2.10 Consider the element

$$\alpha = (x+y)^{\ell-2}(x+\zeta y).$$

Show that:

(a) the ideal (α) is a perfect ℓth power.

(b) $\alpha \equiv 1 - u\lambda \pmod{\lambda^2}$ where $u = (x+y)^{\ell-2}y$.

Exercise 9.2.11 Show that $\zeta^{-u}\alpha$ is primary.

Exercise 9.2.12 Use Eisenstein reciprocity to show that if $x^\ell + y^\ell + z^\ell = 0$ has a solution in integers, $\ell \nmid xyz$, then for any $p \mid y$, $(\zeta/p)_\ell^{-u} = 1$. (Hint: Evaluate $(p/\zeta^{-u}\alpha)_\ell$.)

Exercise 9.2.13 Show that if

$$x^\ell + y^\ell + z^\ell = 0$$

has a solution in integers, $l \nmid xyz$, then for any $p \mid xyz$, $(\zeta/p)_\ell^{-u} = 1$.

Exercise 9.2.14 Show that $(\zeta/p)_\ell^{-u} = 1$ implies that $p^{\ell-1} \equiv 1 \pmod{\ell^2}$.

Exercise 9.2.15 If ℓ is an odd prime and

$$x^\ell + y^\ell + z^\ell = 0$$

for some integers x, y, z coprime to ℓ, then show that $p^{\ell-1} \equiv 1 \pmod{\ell^2}$ for every $p \mid xyz$. Deduce that $2^{\ell-1} \equiv 1 \pmod{\ell^2}$.

The congruence $2^{\ell-1} \equiv 1 \pmod{\ell^2}$ was first established by Wieferich in 1909 as a necessary condition in the first case of Fermat's Last Theorem. The only primes less than 3×10^9 satisfying this congruence are 1093 and 3511 as a quick computer calculation shows. It is not known if there are infinitely many such primes. (See also Exercise 1.3.4.)

9.3 Supplementary Problems

Exercise 9.3.1 Show that there are infinitely many primes p such that $(2/p) = -1$.

Exercise 9.3.2 Let a be a nonsquare integer greater than 1. Show that there are infinitely many primes p such that $(a/p) = -1$.

Exercise 9.3.3 Suppose that $x^2 \equiv a \pmod p$ has a solution for all but finitely many primes. Show that a is a perfect square.

Exercise 9.3.4 Let K be a quadratic extension of $\mathbb{Q}$. Show that there are infinitely many primes which do not split completely in K.

Exercise 9.3.5 Suppose that a is an integer coprime to the odd prime q. If $x^q \equiv a \pmod p$ has a solution for all but finitely many primes, show that a is a perfect qth power. (This generalizes the penultimate exercise.)

Exercise 9.3.6 Let $p \equiv 1 \pmod 3$. Show that there are integers A and B such that

$$4p = A^2 + 27B^2.$$

A and B are unique up to sign.

Exercise 9.3.7 Let $p \equiv 1 \pmod 3$. Show that $x^3 \equiv 2 \pmod p$ has a solution if and only if $p = C^2 + 27D^2$ for some integers C, D.

Exercise 9.3.8 Show that the equation

$$x^3 - 2y^3 = 23z^m$$

has no integer solutions with $\gcd(x, y, z) = 1$.

Chapter 10

Analytic Methods

10.1 The Riemann and Dedekind Zeta Functions

The Riemann zeta function $\zeta(s)$ is defined initially for $\mathrm{Re}(s) > 1$ as the infinite series

$$\zeta(s) = \sum_{n=1}^{\infty} \frac{1}{n^s}.$$

Exercise 10.1.1 Show that for $\mathrm{Re}(s) > 1$,

$$\zeta(s) = \prod_p \left(1 - \frac{1}{p^s}\right)^{-1},$$

where the product is over prime numbers p.

Exercise 10.1.2 Let K be an algebraic number field and $\mathcal{O}_K$ its ring of integers. The Dedekind zeta function $\zeta_K(s)$ is defined for $\mathrm{Re}(s) > 1$ as the infinite series

$$\zeta_K(s) = \sum_{\mathfrak{a}} \frac{1}{(N\mathfrak{a})^s},$$

where the sum is over all ideals of $\mathcal{O}_K$. Show that the infinite series is absolutely convergent for $\mathrm{Re}(s) > 1$.

Exercise 10.1.3 Prove that for $\mathrm{Re}(s) > 1$,

$$\zeta_K(s) = \prod_{\wp} \left(1 - \frac{1}{(N\wp)^s}\right)^{-1}.$$

Theorem 10.1.4 *Let $\{a_m\}_{m=1}^{\infty}$ be a sequence of complex numbers, and let $A(x) = \sum_{m \le x} a_m = O(x^\delta)$, for some $\delta \ge 0$. Then*

$$\sum_{m=1}^{\infty} \frac{a_m}{m^s}$$

converges for $\text{Re}(s) > \delta$ *and in this half-plane we have*

$$\sum_{m=1}^{\infty} \frac{a_m}{m^s} = s \int_1^{\infty} \frac{A(x)\, dx}{x^{s+1}}$$

for $\text{Re}(s) > 1$.

Proof. We write

$$\begin{aligned} \sum_{m=1}^{M} \frac{a_m}{m^s} &= \sum_{m=1}^{M} (A(m) - A(m-1))m^{-s} \\ &= A(M)M^{-s} + \sum_{m=1}^{M-1} A(m)\{m^{-s} - (m+1)^{-s}\}. \end{aligned}$$

Since

$$m^{-s} - (m+1)^{-s} = s \int_m^{m+1} \frac{dx}{x^{s+1}},$$

we get

$$\sum_{m=1}^{M} \frac{a_m}{m^s} = \frac{A(M)}{M^s} + s \int_1^{M} \frac{A(x)\, dx}{x^{s+1}}.$$

For $\text{Re}(s) > \delta$, we find

$$\lim_{M \to \infty} \frac{A(M)}{M^s} = 0,$$

since $A(x) = O(x^\delta)$. Hence, the partial sums converge for $\text{Re}(s) > \delta$ and we have

$$\sum_{m=1}^{\infty} \frac{a_m}{m^s} = s \int_1^{\infty} \frac{A(x)\, dx}{x^{s+1}}$$

in this half-plane. □

Exercise 10.1.5 Show that $(s-1)\zeta(s)$ can be extended analytically for $\text{Re}(s) > 0$.

Exercise 10.1.6 Evaluate

$$\lim_{s \to 1} (s-1)\zeta(s).$$

Example 10.1.7 Let $K = \mathbb{Q}(i)$. Show that $(s-1)\zeta_K(s)$ extends to an analytic function for $\text{Re}(s) > \frac{1}{2}$.

Solution. Since every ideal $\mathfrak{a}$ of $\mathcal{O}_K$ is principal, we can write $\mathfrak{a} = (a + ib)$ for some integers a, b. Moreover, since

$$\mathfrak{a} = (a + ib) = (-a - ib) = (-a + ib) = (a - ib)$$

we can choose a, b to be both positive. In this way, we can associate with each ideal $\mathfrak{a}$ a unique lattice point $(a, b), a \geq 0, b \geq 0$. Conversely, to each such lattice point (a, b) we can associate the ideal $\mathfrak{a} = (a + ib)$. Moreover, $N\mathfrak{a} = a^2 + b^2$. Thus, if we write

$$\zeta_K(s) = \sum_{\mathfrak{a}} \frac{1}{N\mathfrak{a}^s} = \sum_{n=1}^{\infty} \frac{a_n}{n^s}$$

we find that

$$A(x) = \sum_{n \leq x} a_n$$

is equal to the number of lattice points lying in the positive quadrant defined by the circle $a^2 + b^2 \leq x$. We will call such a lattice point (a, b) internal if $(a + 1)^2 + (b + 1)^2 \leq x$. Otherwise, we will call it a boundary lattice point. Let I be the number of internal lattice points, and B the number of boundary lattice points. Then

$$I \leq \frac{\pi}{4} x \leq I + B.$$

Any boundary point (a, b) is contained in the annulus

$$(\sqrt{x} - \sqrt{2})^2 \leq a^2 + b^2 \leq (\sqrt{x} + \sqrt{2})^2$$

and an upper bound for B is provided by the area of the annulus. This is easily seen to be

$$\pi(\sqrt{x} + \sqrt{2})^2 - \pi(\sqrt{x} - \sqrt{2})^2 = O(\sqrt{x}).$$

Thus $A(x) = \pi x/4 + O(\sqrt{x})$. By Theorem 10.1.4, we deduce that

$$\zeta_K(s) = \frac{\pi}{4} s \int_1^{\infty} \frac{dx}{x^s} + s \int_1^{\infty} \frac{E(x)}{x^{s+1}}\, dx,$$

where $E(x) = O(\sqrt{x})$, so that the latter integral converges for $\mathrm{Re}(s) > \frac{1}{2}$. Thus

$$(s - 1)\zeta_K(s) = \frac{\pi}{4} s + s(s - 1) \int_1^{\infty} \frac{E(x)}{x^{s+1}}\, dx$$

is analytic for $\mathrm{Re}(s) > \frac{1}{2}$.

Exercise 10.1.8 For $K = \mathbb{Q}(i)$, evaluate

$$\lim_{s \to 1^+} (s - 1)\zeta_K(s).$$

Exercise 10.1.9 Show that the number of integers (a, b) with $a > 0$ satisfying $a^2 + Db^2 \leq x$ is

$$\frac{\pi x}{2\sqrt{D}} + O(\sqrt{x}).$$

Exercise 10.1.10 Suppose $K = \mathbb{Q}(\sqrt{-D})$ where $D > 0$ and $-D \not\equiv 1 \pmod 4$ and $\mathcal{O}_K$ has class number 1. Show that $(s-1)\zeta_K(s)$ extends analytically to $\text{Re}(s) > \frac{1}{2}$ and find

$$\lim_{s\to 1}(s-1)\zeta_K(s).$$

(Note that there are only finitely many such fields.)

10.2 Zeta Functions of Quadratic Fields

In this section, we will derive the analytic continuation of zeta functions of quadratic fields to the region $\text{Re}(s) > \frac{1}{2}$.

Exercise 10.2.1 Let $K = \mathbb{Q}(\sqrt{d})$ with d squarefree, and denote by a_n the number of ideals in $\mathcal{O}_K$ of norm n. Show that a_n is multiplicative. (That is, prove that if $(n, m) = 1$, then $a_{nm} = a_n a_m$.)

Exercise 10.2.2 Show that for an odd prime p, $a_p = 1 + \left(\frac{d}{p}\right)$.

Exercise 10.2.3 Let d_K be the discriminant of $K = \mathbb{Q}(\sqrt{d})$. Show that for all primes p, $a_p = 1 + \left(\frac{d_K}{p}\right)$.

Exercise 10.2.4 Show that for all primes p,

$$a_{p^\alpha} = \sum_{j=1}^{\alpha}\left(\frac{d_K}{p^j}\right) = \sum_{\delta | p^\alpha}\left(\frac{d_K}{\delta}\right).$$

Exercise 10.2.5 Prove that

$$a_n = \sum_{\delta | n}\left(\frac{d_K}{\delta}\right).$$

Exercise 10.2.6 Let d_K be the discriminant of the quadratic field K. Show that there is an $n > 0$ such that $\left(\frac{d_K}{n}\right) = -1$.

Exercise 10.2.7 Show that

$$\left|\sum_{n\leq x}\left(\frac{d_K}{n}\right)\right| \leq |d_K|.$$

Theorem 10.2.8 (Dirichlet's Hyperbola Method) *Let*

$$f(n) = \sum_{\delta | n} g(\delta) h\left(\frac{n}{\delta}\right)$$

and define

$$\begin{aligned} G(x) &= \sum_{n \leq x} g(n), \\ H(x) &= \sum_{n \leq x} h(n). \end{aligned}$$

Then for any $y > 0$,

$$\sum_{n \leq x} f(n) = \sum_{\delta \leq y} g(\delta) H\left(\frac{x}{\delta}\right) + \sum_{\delta < \frac{x}{y}} h(\delta) G\left(\frac{x}{\delta}\right) - G(y) H\left(\frac{x}{y}\right).$$

Proof. We have

$$\begin{aligned} \sum_{n \leq x} f(n) &= \sum_{\delta e \leq x} g(\delta) h(e) \\ &= \sum_{\substack{\delta e \leq x \\ \delta \leq y}} g(\delta) h(e) + \sum_{\substack{\delta e \leq x \\ \delta > y}} g(\delta) h(e) \\ &= \sum_{\delta \leq y} g(\delta) H\left(\frac{x}{\delta}\right) + \sum_{e \leq \frac{x}{y}} h(e) \left\{ G\left(\frac{x}{e}\right) - G(y) \right\} \\ &= \sum_{\delta \leq y} g(\delta) H\left(\frac{x}{\delta}\right) + \sum_{e \leq \frac{x}{y}} h(e) G\left(\frac{x}{e}\right) - G(y) H\left(\frac{x}{y}\right) \end{aligned}$$

as desired. □

Example 10.2.9 Let K be a quadratic field, and a_n the number of ideals of norm n in $\mathcal{O}_K$. Show that

$$\sum_{n \leq x} a_n = cx + O(\sqrt{x}),$$

where

$$c = \sum_{\delta=1}^{\infty} \left(\frac{d_K}{\delta}\right) \frac{1}{\delta}.$$

Solution. By Exercise 10.2.5,

$$a_n = \sum_{\delta | n} \left(\frac{d_K}{\delta}\right)$$

so that we can apply Theorem 10.2.8 with $g(\delta) = \left(\frac{d_K}{\delta}\right)$ and $h(\delta) = 1$, $y = \sqrt{x}$. We get

$$\sum_{n \le x} a_n = \sum_{\delta \le \sqrt{x}} \left(\frac{d_K}{\delta}\right)\left[\frac{x}{\delta}\right] + \sum_{\delta < \sqrt{x}} G\left(\frac{x}{\delta}\right) - G(\sqrt{x})[\sqrt{x}].$$

By Exercise 10.2.7, $|G(x)| \le |d_K|$. Hence

$$\sum_{n \le x} a_n = \sum_{\delta \le \sqrt{x}} \left(\frac{d_K}{\delta}\right)\left[\frac{x}{\delta}\right] + O(\sqrt{x}).$$

Now $[x/\delta] = x/\delta + O(1)$ so that

$$\sum_{n \le x} a_n = \sum_{\delta \le \sqrt{x}} \left(\frac{d_K}{\delta}\right)\frac{x}{\delta} + O(\sqrt{x}).$$

Finally,

$$\sum_{\delta \le \sqrt{x}} \left(\frac{d_K}{\delta}\right)\frac{1}{\delta} = \sum_{\delta=1}^{\infty} \left(\frac{d_K}{\delta}\right)\frac{1}{\delta} - \sum_{\delta > \sqrt{x}} \left(\frac{d_K}{\delta}\right)\frac{1}{\delta}$$

and by Theorem 10.1.4 we see that

$$c = \sum_{\delta=1}^{\infty} \left(\frac{d_K}{\delta}\right)\frac{1}{\delta}$$

converges and

$$\sum_{\delta > \sqrt{x}} \left(\frac{d_K}{\delta}\right)\frac{1}{\delta} = O\left(\frac{1}{\sqrt{x}}\right).$$

Therefore

$$\sum_{n \le x} a_n = cx + O(\sqrt{x}).$$

Exercise 10.2.10 If K is a quadratic field, show that $(s-1)\zeta_K(s)$ extends to an analytic function for $\mathrm{Re}(s) > \frac{1}{2}$.

Dedekind conjectured in 1877 that $(s-1)\zeta_K(s)$ extends to an entire function for all $s \in \mathbb{C}$ and this was proved by Hecke in 1917 for *all* algebraic number fields K. Moreover, Hecke established a functional equation for $\zeta_K(s)$ and proved that

$$\lim_{s \to 1}(s-1)\zeta_K(s) = \frac{2^{r_1}(2\pi)^{r_2} h_K R_K}{w\sqrt{|d_K|}},$$

where r_1 is the number of real embeddings of K, r_2 is the number of complex embeddings, h_K is the class number of K, d_K is the discriminant of K, w is the number of roots of unity in K, and R_K is the regulator defined as the determinant of the $r \times r$ matrix $(\log|\sigma_i(\varepsilon_j)|)$, and $\varepsilon_1, \dots, \varepsilon_r$ are a system of fundamental units, $\sigma_1, \dots, \sigma_r, \sigma_{r+1}, \overline{\sigma_{r_1+1}}, \dots, \overline{\sigma_{r_1+r_2}}$ are the n embeddings of K into $\mathbb{C}$.

10.3 Dirichlet's L-Functions

Let m be a natural number and χ a Dirichlet character mod m. That is, χ is a homomorphism

$$\chi : (\mathbb{Z}/m\mathbb{Z})^* \to \mathbb{C}^*.$$

We extend the definition of χ to all natural numbers by setting

$$\chi(a) = \begin{cases} \chi\ (a \bmod m) & \text{if } (a,m) = 1, \\ 0 & \text{otherwise.} \end{cases}$$

Now define the Dirichlet L-function:

$$L(s,\chi) = \sum_{n=1}^{\infty} \frac{\chi(n)}{n^s}.$$

Exercise 10.3.1 Show that $L(s,\chi)$ converges absolutely for $\mathrm{Re}(s) > 1$.

Exercise 10.3.2 Prove that

$$\left| \sum_{n \le x} \chi(n) \right| \le m.$$

Exercise 10.3.3 If χ is nontrivial, show that $L(s,\chi)$ extends to an analytic function for $\mathrm{Re}(s) > 0$.

Exercise 10.3.4 For $\mathrm{Re}(s) > 1$, show that

$$L(s,\chi) = \prod_p \left(1 - \frac{\chi(p)}{p^s}\right)^{-1}.$$

Exercise 10.3.5 Show that

$$\sum_{\chi \bmod m} \overline{\chi}(a)\chi(b) = \begin{cases} \varphi(m) & \text{if } a \equiv b \pmod m, \\ 0 & \text{otherwise.} \end{cases}$$

Exercise 10.3.6 For $\mathrm{Re}(s) > 1$, show that

$$\sum_{\chi \bmod m} \log L(s,\chi) = \varphi(m) \sum_{p^n \equiv 1 \bmod m} \frac{1}{np^{ns}}.$$

Exercise 10.3.7 For $\mathrm{Re}(s) > 1$, show that

$$\sum_{\chi \bmod m} \overline{\chi}(a) \log L(s,\chi) = \varphi(m) \sum_{p^n \equiv a \bmod m} \frac{1}{np^{ns}}.$$

Exercise 10.3.8 Let $K = \mathbb{Q}(\zeta_m)$. Set

$$f(s) = \prod_{\chi} L(s,\chi).$$

Show that $\zeta_K(s)/f(s)$ is analytic for $\mathrm{Re}(s) > \frac{1}{2}$.

10.4 Primes in Arithmetic Progressions

In this section we will establish the infinitude of primes $p \equiv a \pmod{m}$ for any a coprime to m.

Lemma 10.4.1 *Let $\{a_n\}$ be a sequence of nonnegative numbers. There exists a $\sigma_0 \in \mathbb{R}$ (possibly infinite) such that*

$$f(s) = \sum_{n=1}^{\infty} \frac{a_n}{n^s}$$

converges for $\sigma > \sigma_0$ and diverges for $\sigma < \sigma_0$. Moreover, if $s \in \mathbb{C}$, with $\mathrm{Re}(s) > \sigma_0$*, then the series converges uniformly in* $\mathrm{Re}(s) \geq \sigma_0 + \delta$ *for any $\delta > 0$ and*

$$f^{(k)}(s) = (-1)^k \sum_{n=1}^{\infty} \frac{a_n (\log n)^k}{n^s}$$

for $\mathrm{Re}(s) > \sigma_0$. *($\sigma_0$ is called the* abscissa of convergence *of the (Dirichlet) series $\sum_{n=1}^{\infty} a_n n^{-s}$.)*

Proof. If there is no real value of s for which the series converges, we take $\sigma_0 = \infty$. Therefore, suppose there is some real s_0 for which the series converges. Clearly by the comparison test, the series converges for $\mathrm{Re}(s) > s_0$ since the coefficients are nonnegative. Now let σ_0 be the infimum of all real s_0 for which the series converges. The uniform convergence in $\mathrm{Re}(s) \geq \sigma_0 + \delta$ for any $\delta > 0$ is now immediate. Because of this, we can differentiate term by term to calculate $f^{(k)}(s)$ for $\mathrm{Re}(s) > \sigma_0$. □

Theorem 10.4.2 *Let $a_n \geq 0$ be a sequence of nonnegative numbers. Then*

$$f(s) = \sum_{n=1}^{\infty} \frac{a_n}{n^s}$$

defines a holomorphic function in $\mathrm{Re}(s) > \sigma_0$ *and $s = \sigma_0$ is a singular point of $f(s)$. (Here σ_0 is the abscissa of convergence of the Dirichlet series.)*

Proof. By the previous lemma, it is clear that $f(s)$ is holomorphic in $\mathrm{Re}(s) > \sigma_0$. If f is not singular at $s = \sigma_0$, then there is a disk

$$D = \{s : |s - \sigma_1| < \delta\},$$

where $\sigma_1 > \sigma_0$ such that $|\sigma_0 - \sigma_1| < \delta$ and a holomorphic function g in D such that $g(s) = f(s)$ for $\mathrm{Re}(s) > \sigma_0$, $s \in D$. By Taylor's formula

$$g(s) = \sum_{k=0}^{\infty} \frac{g^{(k)}(\sigma_1)}{k!}(s - \sigma_1)^k = \sum_{k=0}^{\infty} \frac{f^{(k)}(\sigma_1)}{k!}(s - \sigma_1)^k$$

since $g(s) = f(s)$ for s in a neighborhood of σ_1. Thus, the series

$$\sum_{k=0}^{\infty} \frac{(-1)^k f^{(k)}(\sigma_1)}{k!}(\sigma_1 - s)^k$$

converges absolutely for any $s \in D$. By the lemma, we can write this series as the double series

$$\sum_{k=0}^{\infty} \frac{(\sigma_1 - s)^k}{k!} \sum_{n=1}^{\infty} \frac{a_n (\log n)^k}{n^{\sigma_1}}.$$

If $\sigma_1 - \delta < s < \sigma_1$, this convergent double series consists of nonnegative terms and we may interchange the summations to find

$$\sum_{n=1}^{\infty} \frac{a_n}{n^{\sigma_1}} \sum_{k=0}^{\infty} \frac{(\sigma_1 - s)^k (\log n)^k}{k!} = \sum_{n=1}^{\infty} \frac{a_n}{n^s} < \infty.$$

Since $\sigma_1 - \delta < \sigma_0 < \sigma_1$, this is a contradiction for $s = \sigma_0$. Therefore, the abscissa of convergence is a singular point of $f(s)$. □

Exercise 10.4.3 With the notation as in Section 10.3, write

$$f(s) = \prod_{\chi} L(s, \chi) = \sum_{n=1}^{\infty} \frac{c_n}{n^s}.$$

Show that $c_n \geq 0$.

Exercise 10.4.4 With notation as in the previous exercise, show that

$$\sum_{n=1}^{\infty} \frac{c_n}{n^s}$$

diverges for $s = 1/\varphi(m)$.

Theorem 10.4.5 *Let $L(s, \chi)$ be defined as above. Then $L(1, \chi) \neq 0$ for $\chi \neq \chi_0$.*

Proof. By the previous exercise, the abscissa of convergence of

$$f(s) = \prod_{\chi} L(s, \chi) = \sum_{n=1}^{\infty} \frac{c_n}{n^s}$$

is greater than or equal to $1/\varphi(m)$. If for some $\chi \neq \chi_0$ we have $L(1, \chi) = 0$, then $f(s)$ is holomorphic at $s = 1$ since the zero of $L(s, \chi)$ cancels the simple pole at $s = 1$ of

$$L(s, \chi_0) = \zeta(s) \prod_{p|m} \left(1 - \frac{1}{p^s}\right).$$

By Exercise 10.3.3, each $L(s,\chi)$ extends to an analytic function for $\text{Re}(s) > 0$. By Exercise 10.1.5, $\zeta(s)$ (and hence $L(s,\chi_0)$) is analytic for $\text{Re}(s) > 0$, $s \neq 1$. Thus, $f(s)$ is analytic for $\text{Re}(s) > 0$. By Theorem 10.4.2, the abscissa of convergence of the Dirichlet series

$$\sum_{n=1}^{\infty} \frac{c_n}{n^s}$$

is not in $\text{Re}(s) > 0$ which contradicts the divergence of the series at $s = 1/\varphi(m)$. $\square$

Exercise 10.4.6 Show that

$$\sum_{p \equiv 1 \pmod{m}} \frac{1}{p} = +\infty.$$

Exercise 10.4.7 Show that if $\gcd(a,m) = 1$, then

$$\sum_{p \equiv a \pmod{m}} \frac{1}{p} = +\infty.$$

10.5 Supplementary Problems

Exercise 10.5.1 Define for each character $\chi \pmod{m}$ the Gauss sum

$$g(\chi) = \sum_{a \pmod{m}} \chi(a) e^{2\pi i a/m}.$$

If $(n,m) = 1$, show that

$$\chi(n) g(\overline{\chi}) = \sum_{b \pmod{m}} \overline{\chi}(b) e^{2\pi i b n/m}.$$

Exercise 10.5.2 Show that $|g(\chi)| = \sqrt{m}$.

Exercise 10.5.3 Establish the Pólya–Vinogradov inequality:

$$\left| \sum_{n \leq x} \chi(n) \right| \leq \tfrac{1}{2} m^{1/2} (1 + \log m)$$

for any nontrivial character $\chi \pmod{m}$.

Exercise 10.5.4 Let p be prime. Let χ be a character mod p. Show that there is an $a \leq p^{1/2}(1 + \log p)$ such that $\chi(a) \neq 1$.

Exercise 10.5.5 Show that if χ is a nontrivial character mod m, then

$$L(1,\chi) = \sum_{n \leq u} \frac{\chi(n)}{n} + O\left(\frac{\sqrt{m} \log m}{u} \right).$$

Exercise 10.5.6 Let D be a bounded open set in $\mathbb{R}^2$ and let $N(x)$ denote the number of lattice points in xD. Show that

$$\lim_{x\to\infty} \frac{N(x)}{x^2} = \text{vol}(D).$$

Exercise 10.5.7 Let K be an algebraic number field, and C an ideal class of K. Let $N(x, C)$ be the number of nonzero ideals of $\mathcal{O}_K$ belonging to C with norm $\leq x$. Fix an integral ideal $\mathfrak{b}$ in C^{-1}. Show that $N(x, C)$ is the number of nonzero principal ideals (α) with $\alpha \in \mathfrak{b}$ with $|N_K(\alpha)| \leq xN(\mathfrak{b})$.

Exercise 10.5.8 Let K be an imaginary quadratic field, C an ideal class of $\mathcal{O}_K$, and d_K the discriminant of K. Prove that

$$\lim_{x\to\infty} \frac{N(x,C)}{x} = \frac{2\pi}{w\sqrt{|d_K|}},$$

where w is the number of roots of unity in K.

Exercise 10.5.9 Let K be a real quadratic field with discriminant d_K, and fundamental unit ε. Let C be an ideal class of $\mathcal{O}_K$. Show that

$$\lim_{x\to\infty} \frac{N(x,C)}{x} = \frac{2\log\varepsilon}{\sqrt{d_K}},$$

where $N(x, C)$ denotes the number of integral ideals of norm $\leq x$ lying in the class C.

Exercise 10.5.10 Let K be an imaginary quadratic field. Let $N(x; K)$ denote the number of integral ideals of norm $\leq x$. Show that

$$\lim_{x\to\infty} \frac{N(x;K)}{x} = \frac{2\pi h}{w\sqrt{|d_K|}},$$

where h denotes the class number of K.

Exercise 10.5.11 Let K be a real quadratic field. Let $N(x; K)$ denote the number of integral ideals of norm $\leq x$. Show that

$$\lim_{x\to\infty} \frac{N(x;K)}{x} = \frac{2h\log\varepsilon}{\sqrt{|d_K|}},$$

where h is the class number of K.

Exercise 10.5.12 (Dirichlet's Class Number Formula) Suppose that K is a quadratic field with discriminant d_K. Show that

$$\sum_{n=1}^{\infty} \left(\frac{d_K}{n}\right)\frac{1}{n} = \begin{cases} \frac{2\pi h}{w\sqrt{|d_K|}} & \text{if } d_K < 0, \\ \\ \frac{2h\log\varepsilon}{\sqrt{|d_K|}} & \text{if } d_K > 0, \end{cases}$$

where h denotes the class number of K.

Exercise 10.5.13 Let d be squarefree and positive. Using Dirichlet's class number formula, prove that the class number of $\mathbb{Q}(\sqrt{-d})$ is $O(\sqrt{d}\log d)$.

Exercise 10.5.14 Let d be squarefree and positive. Using Dirichlet's class number formula, prove that the class number h of $\mathbb{Q}(\sqrt{d})$ is $O(\sqrt{d})$.

Exercise 10.5.15 With $\psi(x)$ defined (as in Chapter 1) by

$$\psi(x) = \sum_{p^\alpha \leq x} \log p,$$

prove that for $\mathrm{Re}(s) > 1$,

$$-\frac{\zeta'}{\zeta}(s) = s\int_1^\infty \frac{\psi(x)}{x^{s+1}}\,dx.$$

Exercise 10.5.16 If for any $\varepsilon > 0$,

$$\psi(x) = x + O(x^{1/2+\varepsilon}),$$

show that $\zeta(s) \neq 0$ for $\mathrm{Re}(s) > \frac{1}{2}$.

A famous hypothesis of Riemann asserts that $\zeta(s) \neq 0$ for $\mathrm{Re}(s) > \frac{1}{2}$ and this is still (as of 2004) unresolved.

Chapter 11

Density Theorems

Given an algebraic number field K, we may ask how the ideals are distributed in the ideal classes. We may ask the same of the distribution of prime ideals. It turns out that in both cases, they are equidistributed in the sense of probability. For many reasons, it has been customary to view the latter set of results as generalizations of the celebrated theorem of Dirichlet about primes in arithmetic progressions.

11.1 Counting Ideals in a Fixed Ideal Class

As usual, let K be a fixed algebraic number field of degree n over $\mathbb{Q}$, and denote by $N(x; K)$ the number of ideals of $\mathcal{O}_K$ with norm $\leq x$. For an ideal class C, let $N(x, C)$ be the number of ideals in C with norm $\leq x$. Clearly,

$$N(x; K) = \sum_C N(x, C)$$

where the summation is over all ideal classes of $\mathcal{O}_K$. Let us fix an ideal $\mathfrak{b}$ in C^{-1} and note that if $\mathfrak{a}$ is an ideal in C with norm $\leq x$, then $\mathfrak{ab} = (\alpha)$ with $\alpha \in \mathfrak{b}$ and $|N(\alpha)| \leq xN(\mathfrak{b})$. Conversely, if $\alpha \in \mathfrak{b}$ and $|N(\alpha)| \leq xN(\mathfrak{b})$, then $\mathfrak{a} = (\alpha)\mathfrak{b}^{-1}$ is an integral ideal in C with norm $\leq x$. Thus, $N(x, C)$ is the number of principal ideals (α) contained in $\mathfrak{b}$ with norm less than or equal to $xN(\mathfrak{b})$.

If $\beta_1, ..., \beta_n$ is an integral basis of $\mathfrak{b}$, then we may write

$$\alpha = x_1\beta_1 + \cdots + x_n\beta_n$$

for some integers $x_1, ..., x_n$. Thus, $N(x, C)$ is the number of such α's (up to associates), with $|N(\alpha)| \leq xN(\mathfrak{b})$. We will now try to extract a single element from the set of such associates by means of inequalities. Let $\epsilon_1, ..., \epsilon_r$ be a system of fundamental units (with $r = r_1 + r_2 - 1$ as in Theorem 8.1.6). Recall that it is customary (as we did in Chapter 8) to order our embeddings $K \to K^{(i)}$ in such a way that for $1 \le i \le r_1$, $K^{(i)}$ are real, and $K^{(i)} = \overline{K^{(i+r_2)}}$ for $r_1 + 1 \le i \le r_1 + r_2$. We keep this convention throughout this discussion. By Exercise 8.1.7, the regulator R_K is non-zero and we may find real numbers $c_1, ..., c_r$ such that

$$\sum_{j=1}^{r} c_j \log|\epsilon_j^{(i)}| = \log\left(|\alpha^{(i)}||N(\alpha)|^{-1/n}\right), \qquad 1 \le i \le r.$$

Following Hecke [He], we will call the c_j's the exponents of α. We now want to show that this equation also holds for $i = r+1$. Setting $e_i = 1$ if $K^{(i)}$ is real, and $e_i = 2$ if $K^{(i)}$ is non-real, we see that

$$\sum_{i=1}^{r+1} e_i \log\left(|\alpha^{(i)}||N(\alpha)|^{-1/n}\right) = 0,$$

because

$$|\alpha^{(1)} \cdots \alpha^{(n)}| = |N(\alpha)|.$$

Also,

$$\sum_{i=1}^{r+1} e_i \log|\epsilon_j^{(i)}| = 0.$$

Consequently,

$$\sum_{j=1}^{r} c_j \log|\epsilon_j^{(r+1)}| = \log\left(|\alpha^{(r+1)}||N(\alpha)|^{-1/n}\right),$$

as desired. Thus, this equation holds for all $\alpha^{(i)}, 1 \le i \le n$. (Why?) By Theorem 8.1.6, every unit u of $\mathcal{O}_K$ has the form

$$\zeta\epsilon_1^{n_1} \cdots \epsilon_r^{n_r}$$

where ζ is a root of unity in K and the n_i's are rational integers. Thus, any associate $u\alpha$ of α has exponents

$$c_1 + n_1, ..., c_r + n_r.$$

Therefore, each α has an associate with a set of exponents satisfying the inequalities

$$0 \le c_j < 1, \qquad j = 1, 2, ..., r.$$

If w denotes the number of roots of unity in K, we see then that $wN(x, C)$ is equal to the number of rational integers $(x_1, ..., x_n)$ not all zero, satisfying the following conditions:

$$\alpha = x_1\beta_1 + \cdots + x_n\beta_n;$$

$$|N(\alpha)| = |\alpha^{(1)} \cdots \alpha^{(n)}| \le xN(\mathfrak{b})$$

and for $1 \le i \le n$,

$$\log\left(|\alpha^{(i)}||N(\alpha)|^{-1/n}\right) = \sum_{j=1}^{r} c_j \log|\epsilon_j^{(i)}|, \qquad 0 \le c_j < 1, \quad j = 1, ..., r.$$

In this way, we reduce the problem to counting lattice points in a region of the Euclidean space $\mathbb{R}^n$. This region can be described as follows. We choose arbitrary real values for the x_j and "define" the set of numbers

$$\alpha^{(i)} = \sum_{j=1}^{n} x_j \beta_j^{(i)}.$$

Corresponding to this set of numbers there is a uniquely determined set of "exponents" $c_1, ..., c_r$ provided the x_j's do not lie on the subspace defined by $\alpha^{(i)} = 0$ for some i satisfying $1 \le i \le n$. Thus, if we include $\alpha^{(i)} \neq 0$ to the above set of inequalities, these inequalities describe a well-defined region B_x (say) in $\mathbb{R}^n$. That is, if we put

$$\alpha^{(i)} = x_1\beta_1^{(i)} + \cdots + x_n\beta_n^{(i)}, \qquad 1 \le i \le n,$$

and

$$N(\alpha) = \alpha^{(1)} \cdots \alpha^{(n)},$$

then B_x is the set of n-tuples $(x_1, ..., x_n) \in \mathbb{R}^n$ satisfying

$$|\alpha^{(1)} \cdots \alpha^{(n)}| \le xN(\mathfrak{b})$$

and either $\alpha^{(i)} = 0$ for some i or for all $1 \le i \le n$, we have

$$\log\left(|\alpha^{(i)}||N(\alpha)|^{-1/n}\right) = \sum_{j=1}^{r} c_j \log|\epsilon_j^{(i)}|,$$

with $0 \le c_j < 1$, $j = 1, ..., r$.

Exercise 11.1.1 Show that B_x is a bounded region in $\mathbb{R}^n$.

Exercise 11.1.2 Show that $tB_1 = B_{t^n}$ for any $t > 0$.

Exercise 11.1.3 Show that $N(x, C) = O(x)$. Deduce that $N(x; K) = O(x)$.

The idea now is to approximate the number of lattice points satisfying the above inequalities by the volume of B_x. This we do below. As will be seen, the calculation is an exercise in multivariable calculus.

Before we begin, it is important to note that for some $\delta > 0$ and $t = x^{1/n}$, we have

$$\text{vol}(B_{(t-\delta)^n}) \le wN(x, C) \le \text{vol}(B_{(t+\delta)^n}).$$

To see this, we may associate each lattice point lying inside B_x with an appropriate translate of the standard unit cube, namely $[0,1]^n$. Each translate lying entirely within B_x contributes 1 to the volume of B_x. The error term arises from the cubes intersecting with the boundary. In view of Exercise 11.1.2, it is intuitively clear (see also Exercise 11.1.12 below) that by enlarging the region by some fixed quantity and reducing the region by a fixed quantity δ in the way indicated, the above inequalities are assured. Thus,

$$(x^{1/n}-\delta)^n\text{vol}(B_1) \le wN(x,C) \le (x^{1/n}+\delta)^n\text{vol}(B_1)$$

so that

$$wN(x,C) = \text{vol}(B_1)x + O(x^{1-\frac{1}{n}}).$$

The essential feature of the theorem below is that this volume is independent of the ideal class under consideration.

Theorem 11.1.4 (Dedekind)

$$\text{vol}(B_1) = \frac{2^{r_1}(2\pi)^{r_2}R_K}{\sqrt{|d_K|}}.$$

Proof. Let M be the maximal value of $|\log|\epsilon_j^{(i)}||$ for $j=1,...,r$. We first complete the domain B_x by adding the points of the space lying in the subspace $\alpha^{(i)}=0$ for some i and that also satisfy the inequalities

$$|\alpha^{(j)}| \le e^{rM}(xN(\mathfrak{b}))^{1/n}, \qquad j=1,2,...,n.$$

Since these subspaces are of lower dimension, their contribution to the volume is negligible. We denote the completed space by B_x^*. If we now change variables and put $x_j = y_j x^{1/n}$, we see that the volume is equal to

$$x\int\cdots\int_{B_1^*} dy_1\cdots dy_n = J \qquad (say).$$

Now B_1^* is the domain described by

$$\alpha^{(i)} = \sum_{j=1}^{n} y_j\beta_j^{(i)}, \qquad 1\le i\le n$$

and

$$0 < |\alpha^{(1)}\cdots\alpha^{(n)}| \le N(\mathfrak{b}),$$

so that there exist c_j's for $1\le j\le r$ satisfying $0\le c_j<1$ and

$$\log\left(|\alpha^{(i)}||N(\alpha)|^{-1/n}\right) = \sum_{j=1}^{r} c_j\log|\epsilon_j^{(i)}|, \quad 1\le i\le n$$

or

$$|\alpha^{(i)}| \leq e^{rM}(N(\mathfrak{b}))^{1/n}, \qquad 1 \leq i \leq n$$

and at least one $\alpha^{(i)} = 0$. As noted, the region defined by these latter conditions are manifolds of lower dimension and thus make no contribution to the n-fold integral and thus, these conditions may be omitted in the evaluation of J. To evaluate the integral, we change variables:

$$u_i := \alpha^{(i)} = \sum_{j=1}^{n} y_j \beta_j^{(i)}, \qquad 1 \leq i \leq r_1,$$

$$u_i + u_{i+r_2}\sqrt{-1} := \sum_{j=1}^{n} y_j \beta_j^{(i)}, \qquad r_1 + 1 \leq i \leq r_1 + r_2.$$

Thus, with our convention concerning the ordering of embeddings,

$$u_i = \sum_{j=1}^{n} y_j \left(\frac{\beta_j^{(i)} + \beta_j^{(i+r_2)}}{2} \right),$$

$$u_{i+r_2} = \sum_{j=1}^{n} y_j \left(\frac{\beta_j^{(i)} - \beta_j^{(i+r_2)}}{2\sqrt{-1}} \right),$$

for $r_1 + 1 \leq i \leq r_1 + r_2$. The absolute value of the Jacobian for this change of variables is easily computed to be

$$2^{-r_2} N(\mathfrak{b}) \sqrt{|d_K|}.$$

Hence,

$$\mathrm{vol}(B_1^*) = \frac{2^{r_2}}{N(\mathfrak{b})\sqrt{|d_K|}} \int \cdots \int_{\tilde{B}_1^*} du_1 \cdots du_n,$$

where $\tilde{B}_1^*$ is the image of B_1^* under the change of variables. The variables $u_1, ..., u_{r_1}$ may take one of two signs and so if we insist $u_i \geq 0$ for $i = 1, ..., r_1$, we must multiply our volume (with this additional constraint) by a factor of 2^{r_1}. We now shift to polar co-ordinates. Put

$$\rho_j = u_j \qquad 1 \leq j \leq r_1$$

and

$$\rho_j \cos\theta_j = u_j, \quad \rho_j \sin\theta_j = u_{j+r_2}, \quad r_1 + 1 \leq j \leq r_1 + r_2,$$

with $0 \leq \theta_j < 2\pi$ and $\rho_j \geq 0$. The Jacobian of this transformation is easily seen to be

$$\rho_{r_1+1} \cdots \rho_{r_1+r_2}.$$

Thus,

$$\mathrm{vol}(B_1^*) = \frac{2^{r_1+r_2}(2\pi)^{r_2}}{N(\mathfrak{b})\sqrt{|d_K|}} \int \cdots \int_{C_1^*} \rho_{r_1+1} \cdots \rho_{r_1+r_2} d\rho_1 \cdots d\rho_{r_1+r_2}$$

where C_1^* is the domain described by:

$$0 \leq \prod_{j=1}^{r_1+r_2} \rho_j^{e_j} \leq N(\mathfrak{b})$$

$$\log \rho_i \left(\prod_{j=1}^{r} \rho_j^{e_j} \right)^{-1/n} = \sum_{j=1}^{r} c_j \log |\epsilon_j^{(i)}|$$

for $1 \leq i \leq r_1 + r_2$. (Recall that $e_i = 1$ for $1 \leq i \leq r_1$ and 2 otherwise.) We make (yet) another change of variables. Put

$$\tau_j = \rho_j^{e_j}, \qquad 1 \leq j \leq r_1 + r_2.$$

The Jacobian of this transformation is easily seen to be

$$2^{-r_2} \rho_{r_1+1}^{-1} \cdots \rho_{r_1+r_2}^{-1}$$

so that the integral becomes

$$\frac{2^{r_1}(2\pi)^{r_2}}{N(\mathfrak{b})\sqrt{|d_K|}} \int \cdots \int_{D_1^*} d\tau_1 \cdots d\tau_{r_1+r_2}$$

where the region D_1^* is described by the conditions

$$\tau_1 \cdots \tau_{r_1+r_2} \leq N(\mathfrak{b}), \quad \tau_i > 0,$$

$$\log \tau_i = \frac{e_i}{n} \log(\tau_1 \cdots \tau_{r_1+r_2}) + e_i \sum_{j=1}^{r} c_j \log |\epsilon_j^{(i)}|.$$

We make one final change of variables: write the c_i's in terms of the τ_i's and put

$$u = \tau_1 \cdots \tau_{r+1}.$$

The Jacobian of this transformation is now seen to be R_K and the final result is

$$\mathrm{vol}(B_1) = \frac{2^{r_1}(2\pi)^{r_2} R_K}{N(\mathfrak{b})\sqrt{|d_K|}} \int_0^{N(\mathfrak{b})} du \int_0^1 \cdots \int_0^1 dc_1 \cdots dc_r = \frac{2^{r_1}(2\pi)^{r_2} R_K}{\sqrt{|d_K|}}$$

which completes the proof. □

By our remarks preceding the statement of Theorem 11.1.4, we immediately deduce:

Theorem 11.1.5 (Weber)

$$N(x, C) = \frac{2^{r_1}(2\pi)^{r_2} R_K}{w\sqrt{|d_K|}} x + O(x^{1-\frac{1}{n}}).$$

If $N(x;K)$ is the number of integral ideals of K with norm $\leq x$, then

$$N(x;K) = \frac{2^{r_1}(2\pi)^{r_2} h_K R_K}{w\sqrt{|d_K|}} x + O(x^{1-\frac{1}{n}}),$$

where h_K denotes the class number of K.

Following Hecke, we define the ideal class zeta function as

$$\zeta(s,C) = \sum_{\mathfrak{a}\in C} \frac{1}{N(\mathfrak{a})^s}.$$

Note that the Dedekind zeta function defined in the previous chapter may now be written as

$$\zeta_K(s) = \sum_C \zeta(s,C)$$

where the summation is over all ideal classes.

Exercise 11.1.6 Prove that $\zeta(s,C)$ extends to the region $\Re(s) > 1 - \frac{1}{n}$ except for a simple pole at $s = 1$ with residue

$$\frac{2^{r_1}(2\pi)^{r_2} R_K}{w\sqrt{|d_K|}}.$$

Deduce that $\zeta_K(s)$ extends to $\Re(s) > 1 - \frac{1}{n}$ except for a simple pole at $s = 1$ with residue

$$\rho_K := \frac{2^{r_1}(2\pi)^{r_2} h_K R_K}{w\sqrt{|d_K|}},$$

where h_K denotes the class number of K. (This is usually called the *analytic class number formula.*)

Exercise 11.1.7 Prove that there are infinitely many prime ideals $\wp$ in $\mathcal{O}_K$ which are of degree 1.

Exercise 11.1.8 Prove that the number of prime ideals $\wp$ of degree ≥ 2 and with norm $\leq x$ is $O(x^{1/2}\log x)$.

Exercise 11.1.9 Let μ be defined on integral ideals $\mathfrak{a}$ of $\mathcal{O}_K$ as follows. $\mu(\mathcal{O}_K) = 1$, and if $\mathfrak{a}$ is divisible by the square of a prime ideal, we set $\mu(\mathfrak{a}) = 0$. Otherwise, we let $\mu(\mathfrak{a}) = (-1)^k$ when $\mathfrak{a}$ is the product of k distinct prime ideals. Show that

$$\sum_{\mathfrak{b}|\mathfrak{a}} \mu(\mathfrak{b}) = 0$$

unless $\mathfrak{a} = \mathcal{O}_K$.

Exercise 11.1.10 Prove that the number of ideals of $\mathcal{O}_K$ of odd norm $\leq x$ is

$$\rho_K x \prod_{\wp|2}\left(1-\frac{1}{N(\wp)}\right)+O(x^{1-\frac{1}{n}}),$$

where the product is over non-zero prime ideals $\wp$ of $\mathcal{O}_K$ dividing $2\mathcal{O}_K$.

Exercise 11.1.11 Let $A(x)$ be the number of ideals of $\mathcal{O}_K$ of even norm $\leq x$ and $B(x)$ of odd norm $\leq x$. Show that

$$\lim_{x\to\infty}\frac{A(x)}{B(x)}=1$$

if and only if $K=\mathbb{Q}$ or K is a quadratic field in which 2 ramifies.

Exercise 11.1.12 With notation as in the discussion preceding Theorem 11.1.4, let V_x denote the set of n-tuples $(x_1,...,x_n)$ satisfying

$$|\alpha^{(1)}\cdots\alpha^{(n)}|\leq xN(\mathfrak{b}).$$

Let $t=x^{1/n}$. Show that there is a $\delta>0$ such that for each lattice point P contained in $V_{(t-\delta)^n}$, all the points contained in the translate of the standard cube by P belong to V_x.

11.2 Distribution of Prime Ideals

Let $\mathcal{H}$ be the ideal class group of K. Following Hecke, we define for each character

$$\chi:\mathcal{H}\to\mathbb{C}^*$$

the Hecke L-function

$$L(s,\chi):=\sum_{\mathfrak{a}}\frac{\chi(\mathfrak{a})}{N(\mathfrak{a})^s},$$

where $\chi(\mathfrak{a})$ is simply $\chi(C)$ if $\mathfrak{a}$ belongs to the ideal class C. If χ is the trivial character χ_0, note that $L(s,\chi_0)=\zeta_K(s)$, the Dedekind zeta function. Since $\mathcal{H}$ is a finite abelian group of order h_K, its character group is also finite of order h_K and so, in this way, we have attached h_K L-functions to K.

Exercise 11.2.1 Show that $L(s,\chi)$ converges absolutely for $\Re(s)>1$ and that

$$L(s,\chi)=\prod_{\wp}\left(1-\frac{\chi(\wp)}{N(\wp)^s}\right)^{-1},$$

in this region. Deduce that $L(s,\chi)\neq 0$ for $\Re(s)>1$.

Exercise 11.2.2 If χ is not the trivial character, show that

$$\sum_C \chi(C) = 0$$

where the summation is over the ideal classes C of $\mathcal{H}$.

Exercise 11.2.3 If C_1 and C_2 are distinct ideal classes, show that

$$\sum_\chi \overline{\chi(C_1)}\chi(C_2) = 0.$$

If $C_1 = C_2$, show that the sum is h_K. (This is analogous to Exercise 10.3.5.)

From Theorem 11.1.5, we obtain:

Theorem 11.2.4 *Let n be the degree of $K/\mathbb{Q}$. If $\chi \neq \chi_0$, then $L(s,\chi)$ extends analytically to $\Re(s) > 1 - \frac{1}{n}$.*

Proof. By Theorem 11.1.5, we have

$$\sum_C \sum_{\mathfrak{a}\in C, N(\mathfrak{a})\leq x} \chi(\mathfrak{a}) = \sum_C \chi(C)N(x,C) = O(x^{1-\frac{1}{n}}),$$

since (by Exercise 11.2.2)

$$\sum_C \chi(C) = 0.$$

□

In 1917, Hecke proved that $L(s,\chi)$ extends to an entire function if $\chi \neq \chi_0$, and satisfies a suitable functional equation relating $L(1-s,\overline{\chi})$ with $L(s,\chi)$. In the case $\chi = \chi_0$, he showed that $\zeta_K(s)$ extends meromorphically to the entire complex plane, with only a simple pole at $s = 1$. Moreover, it satisfies the functional equation

$$\xi_K(s) = \xi_K(1-s),$$

where

$$\xi_K(s) := \left(\frac{\sqrt{|d_K|}}{2^{r_2}\pi^{n/2}}\right)^s \Gamma(s/2)^{r_1}\Gamma(s)^{r_2}\zeta_K(s)$$

with $\Gamma(s)$ denoting the Γ-function. Recall that this is defined by

$$\Gamma(s) = \int_0^\infty e^{-t}t^{s-1}dt$$

for $\Re(s) > 0$ and can be extended meromorphically to the entire complex plane via the functional equation

$$\Gamma(s+1) = s\Gamma(s).$$

Our goal is to show that each ideal class contains infinitely many prime ideals. This is analogous to Dirichlet's theorem about primes in arithmetic progressions. Indeed, as we will indicate later, the result is more than an analogue. It is a generalization that includes the celebrated theorem of Dirichlet.

Exercise 11.2.5 Let C be an ideal class of $\mathcal{O}_K$. For $\Re(s) > 1$, show that

$$\sum_{\chi} \overline{\chi}(C) \log L(s,\chi) = h_K \sum_{\wp^m \in C} \frac{1}{mN(\wp)^{ms}}$$

where the first summation is over the characters of the ideal class group and the second summation is over all prime ideals $\wp$ of $\mathcal{O}_K$ and natural numbers m such that $\wp^m \in C$.

We now proceed as we did in Chapter 10, Section 4. For the sum on the right hand side in the previous exercise, we separate the contribution from $n = 1$ and $n \geq 2$. The latter part is shown to converge for $\Re(s) > 1/2$ (see 11.2.6 below). Thus, if we can show that $L(1,\chi) \neq 0$ for $\chi \neq \chi_0$, we may conclude as in Exercise 10.4.7 that

$$\sum_{\wp \in C} \frac{1}{N(\wp)} = +\infty.$$

Exercise 11.2.6 Show that

$$\sum_{n \geq 2, \wp^m \in C} \frac{1}{mN(\wp)^{ms}}$$

converges for $\Re(s) > 1/2$.

Exercise 11.2.7 If $\chi^2 \neq \chi_0$ show that $L(1,\chi) \neq 0$.

This gives us a fairly self-contained proof of the infinitude of prime ideals in a fixed ideal class in the case h_K is odd, for in that case, there are no characters of order 2 in the character group. A genuine difficulty arises in trying to show $L(1,\chi) \neq 0$ for χ real. Historically, this was first circumvented using class field theory (and in most treatments, many authors still follow this route). A somewhat easier argument allows us to deduce the non-vanishing from the relatively simpler assertion that for real valued characters χ, $L(s,\chi)$ admits an analytic continuation to $\Re(s) \geq 1/2$. Then, we may consider the function

$$f(s) = \sum_{\mathfrak{a}} \frac{r(\mathfrak{a})}{N(\mathfrak{a})^s} := \zeta_K(s) L(s,\chi)$$

which is easily seen to be a Dirichlet series with non-negative coefficients. Moreover, it is easily verified that

$$r(\mathfrak{b}^2) \geq 1.$$

If $L(1, \chi) = 0$, then $f(s)$ is regular at $s = 1$ and by Theorem 10.4.2, is analytic for $\Re(s) > \sigma_0$, where σ_0 is the abscissa of convergence of $f(s)$. Now

$$\lim_{s \to \frac{1}{2}^+} f(s) \geq \lim_{s \to \frac{1}{2}^+} \sum_{\mathfrak{b}} \frac{1}{N(\mathfrak{b})^{2s}} = +\infty.$$

Thus, $\sigma_0 \geq 1/2$. However, the assumption that $L(s, \chi)$ admits an analytic continuation to $\Re(s) \geq 1/2$ implies that $f(s)$ is analytic for $\Re(s) \geq 1/2$, a contradiction. Thus, $L(1, \chi) \neq 0$.

In the above argument, we only used the fact that we can continue $L(s, \chi)$ to the real line segment $[1/2, 1]$. It seems unrealistic to expect any refinement of this argument unless we use the fact that we are dealing with a quadratic character in some fundamental way. Indeed, if we consider the series

$$g(s) = \frac{\zeta(2s)}{\zeta(s)},$$

then it is easy to see that the Dirichlet coefficients of $g(s)$ are ± 1. Moreover, $g(s)$ has a zero at $s = 1$ and $\zeta(s)g(s) = \zeta(2s)$ has non-negative coefficients. However, it does not admit an analytic continuation to the line segment $[1/2, 1]$ as it has a simple pole at $s = 1/2$.

So far, we have been able to extend the results of Chapter 10 to show the infinitude of prime ideals in a fixed ideal class. It is possible to refine these results by introducing the notion of Dirichlet density. We say that a set of prime ideals S of prime ideals of $\mathcal{O}_K$ has *Dirichlet density* δ if

$$\lim_{s \to 1^+} \frac{\sum_{\wp \in S} 1/N(\wp)^s}{\log \zeta_K(s)} = \delta.$$

Clearly, any set of prime ideals with a positive Dirichlet density is infinite.

Exercise 11.2.8 Let C be a fixed ideal class in $\mathcal{O}_K$. Show that the set of prime ideals $\wp \in C$ has Dirichlet density $1/h_K$.

Exercise 11.2.9 Let m be a natural number and $(a, m) = 1$. Show that the set of primes $p \equiv a(\operatorname{mod} m)$ has Dirichlet density $1/\phi(m)$.

Exercise 11.2.10 Show that the set of primes p which can be written as $a^2 + 5b^2$ is $1/4$.

By using more sophisticated methods, it is possible to obtain asymptotic formulas for the number of prime ideals lying in a given ideal class. Indeed, using standard Tauberian theory, one can show that the number of prime ideals $\wp$ in a given ideal class with norm $\leq x$ is asymptotic to

$$\frac{1}{h_K} \frac{x}{\log x},$$

as x tends to infinity.

It is possible to go further. Let $\mathfrak{f}_0$ be an ideal of $\mathcal{O}_K$ and $\mathfrak{f}_\infty$ a subset of real embeddings of K. We write formally $\mathfrak{f} = \mathfrak{f}_0\mathfrak{f}_\infty$ and define the $\mathfrak{f}$-ideal class group as follows. We define an equivalence relation on the set of ideals coprime to $\mathfrak{f}_0$ by declaring that two ideals $\mathfrak{a}$ and $\mathfrak{b}$ are equivalent if

$$(\alpha)\mathfrak{a} = (\beta)\mathfrak{b}$$

for some $\alpha, \beta \in \mathcal{O}_K$ coprime to $\mathfrak{f}_0$, $\alpha - \beta \in \mathfrak{f}_0$ and $\sigma(\alpha/\beta) > 0$ for all $\sigma \in \mathfrak{f}_\infty$. The set of equivalence classes turns out to be finite and can be given the structure of an abelian group, which we denote by $\mathcal{H}(\mathfrak{f})$ and call the $\mathfrak{f}$-ideal class group. In the case that $\mathfrak{f}_0 = \mathcal{O}_K$ and $\mathfrak{f}_\infty$ is the empty set, this group is the usual ideal class group. If $\mathfrak{f}_\infty$ consists of all the real embeddings of the given field, we call $\mathcal{H}(\mathfrak{f})$ the *ray class group* (mod $\mathfrak{f}_0$). One may now define L-functions (following Hecke) attached to characters of these groups. Proceeding as above, the theory can be developed to deduce the expected density theorems. Indeed, for a fixed $\mathfrak{f}$-ideal class C, the set of prime ideals $\wp$ lying in C has Dirichlet density $1/|\mathcal{H}(\mathfrak{f})|$. It is possible to derive even an asymptotic formula for the number of such prime ideals $\wp \in C$ with norm $\leq x$ of the form

$$\sim \frac{1}{|\mathcal{H}(\mathfrak{f})|} \frac{x}{\log x}$$

as x tends to infinity.

Exercise 11.2.11 Show that if $K = \mathbb{Q}$, the principal ray class group mod m is isomorphic to $(\mathbb{Z}/m\mathbb{Z})^*$.

The previous exercise realizes the coprime residue classes mod m as a ray class group. In this way, the theorem of Hecke generalizes the classical theorem of Dirichlet about the uniform distribution of prime numbers in arithmetic progressions.

11.3 The Chebotarev density theorem

Let K/k be a Galois extension of algebraic number fields with $\mathrm{Gal}(K/k) = G$. Recall that if $\wp$ is a prime ideal then so is $\sigma(\wp)$ for any $\sigma \in G$. For each prime ideal $\mathfrak{p}$ of k, we have (analogous to the situation in Chapter 5) a factorization

$$\mathfrak{p}\mathcal{O}_K = \wp_1^{e_1} \cdots \wp_r^{e_r}.$$

If we apply a Galois automorphism σ to both sides of this equality, we get

$$\mathfrak{p}\mathcal{O}_K = \sigma(\wp_1)^{e_1} \cdots \sigma(\wp_r)^{e_r}.$$

By uniqueness of factorization, we deduce that G acts on the set of prime ideals $\wp_1, \ldots, \wp_r$ that lie above a fixed prime ideal $\mathfrak{p}$.

Exercise 11.3.1 Show that the action of the Galois group on the set of prime ideals lying above a fixed prime of k is a transitive action.

The *decomposition group* of $\wp$, denoted $D_\wp$, is the subgroup of G consisting of elements σ satisfying $\sigma(\wp) = \wp$. The *inertia group* of $\wp$, denoted $I_\wp$, is the subgroup of elements σ satisfying

$$\sigma(x) \equiv x (\operatorname{mod} \wp) \qquad \forall x \in \mathcal{O}_K.$$

It is easily seen that $I_\wp$ is a normal subgroup of $D_\wp$. The quotient $D_\wp/I_\wp$ can be shown to be a cyclic group canonically isomorphic to the Galois group of the finite field $\mathcal{O}_K/\wp$ viewed as an extension of $\mathcal{O}_k/\mathfrak{p}$. Thus, there is an element (well-defined modulo $I_\wp$), denoted $\sigma_\wp$, whose image in $\operatorname{Gal}((\mathcal{O}_K/\wp)/(\mathcal{O}_k/\mathfrak{p}))$ is the mapping

$$x \mapsto x^{N(\mathfrak{p})}.$$

We call $\sigma_\wp$ the *Frobenius automorphism* of $\wp$. For $\mathfrak{p}$ unramified in K, one can show easily that as $\wp$ ranges over the prime ideals above $\mathfrak{p}$, the $\sigma_\wp$ comprise a conjugacy class of G. This conjugacy class is denoted $\sigma_\mathfrak{p}$ and is called the *Artin symbol* of $\mathfrak{p}$.

Now fix a conjugacy class C of G. The *Chebotarev density theorem* states the following.

Theorem 11.3.2 (Chebotarev) *Let K/k be a finite Galois extension of algebraic number fields with Galois group G. If C is a conjugacy class of G, the prime ideals $\mathfrak{p}$ of $\mathcal{O}_k$ with $\sigma_\mathfrak{p} \in C$ has Dirichlet density $|C|/|G|$. Thus, the Artin symbols are equidistributed in the conjugacy classes with the expected probability.*

A prime ideal $\mathfrak{p}$ of k is said to *split completely* in K if

$$\mathfrak{p}\mathcal{O}_K = \wp_1 \cdots \wp_n$$

where $n = [K : k]$. This is equivalent to the assertion that the Artin symbol $\sigma_\mathfrak{p}$ is equal to 1. Thus, from Chebotarev's density theorem, we immediately deduce:

Theorem 11.3.3 *The set of prime ideals $\mathfrak{p}$ which split completely in K has Dirichlet density $1/[K : k]$.*

Exercise 11.3.4 By taking $k = \mathbb{Q}$ and $K = \mathbb{Q}(\zeta_m)$, deduce from Chebotarev's theorem the infinitude of primes in a given arithmetic progression $a \pmod{m}$ with $(a, m) = 1$.

Exercise 11.3.5 If $k = \mathbb{Q}$ and $K = \mathbb{Q}(\sqrt{D})$, deduce from Chebotarev's theorem that the set of primes p with Legendre symbol $(D/p) = 1$ is of Dirichlet density $1/2$.

Exercise 11.3.6 If $f(x) \in \mathbb{Z}[x]$ is an irreducible normal polynomial of degree n (that is, its splitting field has degree n over $\mathbb{Q}$), then show that the set of primes p for which $f(x) \equiv 0 \pmod{p}$ has a solution is of Dirichlet density $1/n$.

Exercise 11.3.7 If $f(x) \in \mathbb{Z}[x]$ is an irreducible polynomial of degree $n > 1$, show that the set of primes p for which $f(x) \equiv 0 \pmod{p}$ has a solution has Dirichlet density < 1.

Exercise 11.3.8 Let q be prime. Show that the set of primes p for which $p \equiv 1 \pmod{q}$ and

$$2^{\frac{p-1}{q}} \equiv 1 (\operatorname{mod} p),$$

has Dirichlet density $1/q(q-1)$.

Exercise 11.3.9 If a natural number n is a square mod p for a set of primes p which has Dirichlet density 1, show that n must be a square.

Let K/k be a finite Galois extension of algebraic number fields with Galois group G as above. Let V be a finite dimensional vector space over $\mathbb{C}$. Suppose we have a representation

$$\rho : G \to GL(V)$$

where $GL(V)$ denotes the group of invertible linear transformations of V into itself. E. Artin defined an L-function attached to ρ by setting:

$$L(s, \rho; K/k) = \prod_{\mathfrak{p}} \det\left(1 - \rho(\sigma_{\wp}) N(\mathfrak{p})^{-s} | V^{I_{\wp}}\right)^{-1}$$

where the product is over all prime ideals $\mathfrak{p}$ of k and $\wp$ is any prime ideal of K lying above $\mathfrak{p}$, which is well-defined modulo the inertia group $I_{\wp}$. Thus taking the characteristic polynomial of $\rho(\sigma_{\wp})$ acting on the subspace $V^{I_{\wp}}$, which is the subspace of V fixed by $I_{\wp}$, we get a well-defined factor for each prime ideal $\mathfrak{p}$. The product over all prime ideals $\mathfrak{p}$ is easily seen to converge absolutely for $\Re(s) > 1$ (why?). As these L-functions play a central role in number theory, we will briefly give a description of results pertaining to them and indicate some of the open problems of the area. The reader may find it useful to have some basic knowledge of the character theory of finite groups as explained for instance in [Se].

The celebrated *Artin's conjecture* predicts that if ρ is a non-trivial irreducible representation, then $L(s, \rho; K/k)$ extends to an entire function of s. If ρ is one-dimensional, then Artin proved his famous *reciprocity law* by showing that in this case, his L-function coincides with Hecke's L-function attached to a suitable generalized ideal class group of k. This theorem is so-called since it entails all of the classical reciprocity laws including the law of quadratic reciprocity. Subsequently, R. Brauer proved that for any ρ $L(s, \rho; K/k)$ extends to a meromorphic function for all $s \in \mathbb{C}$. He did this by proving an *induction theorem* which is really a statement about irreducible characters of finite groups. More precisely, if χ is an irreducible character of G, then Brauer's theorem states that there are nilpotent subgroups H_i of G and one dimensional characters ψ_i of H so that for some set of integers n_i, we have

$$\chi = \sum_i n_i \mathrm{Ind}_{H_i}^G \psi_i$$

where $\mathrm{Ind}_H^G \psi$ indicates the character induced from H to G by ψ.

To see how Brauer's theorem implies the meromorphy of Artin L-series, it is convenient to modify our notation slightly by writing $L(s, \chi; K/k)$ for $L(s, \rho; K/k)$ with $\chi(g) = \mathrm{tr}\,\rho(g)$. As is evident, the definition of an Artin L-series attached to ρ depends only on the character χ of ρ. With this convention, it is easy to verify that

$$L(s, \chi_1 + \chi_2; K/k) = L(s, \chi_1; K/k)L(s, \chi_2; K/k)$$

and that

$$L(s, \mathrm{Ind}_H^G \psi, K/k) = L(s, \psi; K/K^H)$$

where K^H indicates the subfield of K fixed by H. Thus, by Brauer's theorem, we may write

$$L(s, \chi; K/k) = \prod_i L(s, \mathrm{Ind}_{H_i}^G \psi_i; K/k)^{n_i}.$$

By the invariance of Artin L-series under induction, we obtain

$$L(s, \chi; K/k) = \prod_i L(s, \psi_i; K/K^{H_i})^{n_i}.$$

Now, by Artin's reciprocity law, each of the L-functions appearing in the product is a Hecke L-function, which by Hecke's theorem is known to be entire. In this way, we get the meromorphic continuation of $L(s, \chi; K/k)$. It is one of the aims of the *Langlands program* to prove Artin's conjecture. The celebrated *Langlands-Tunnell theorem* says that when ρ is 2-dimensional with solvable image, then Artin's conjecture is true. This theorem played a pivotal role in the work of Wiles resolving Fermat's last theorem.

11.4 Supplementary Problems

Following a suggestion of Kumar Murty (see [FM]), we indicate in the supplementary problems (11.4.1 to 11.4.10) below how Artin L-series may be used to give a proof of Chebotarev's theorem using the techniques developed in this chapter and Chapter 10. A modest background in the representation theory of finite groups would be useful. For instance, the first three chapters of [Se] should be sufficient background. The reader may also consult [La] for an alternative (and more classical) approach.

Exercise 11.4.1 Let G be a finite group and for each subgroup H of G and each irreducible character ψ of H define $a_H(\psi, \chi)$ by

$$\mathrm{Ind}_H^G \psi = \sum_{\chi} a_H(\psi, \chi)\chi$$

where the summation is over irreducible characters χ of G. For each χ, let A_χ be the vector $(a_H(\psi, \chi))$ as H varies over all cyclic subgroups of G and ψ varies over all irreducible characters of H. Show that the A_χ's are linearly independent over $\mathbb{Q}$.

Exercise 11.4.2 Let G be a finite group with t irreducible characters. By the previous exercise, choose a set of cyclic subgroups H_i and characters ψ_i of H_i so that the $t \times t$ matrix $(a_{H_i}(\psi_i, \chi))$ is non-singular. By inverting this matrix, show that any character χ of G can be written as a rational linear combination of characters of the form $\mathrm{Ind}_{H_i}^G \psi_i$ with H_i cyclic and ψ_i one-dimensional. (This result is usually called *Artin's character theorem* and is weaker than Brauer's induction theorem.)

Exercise 11.4.3 Deduce from the previous exercise that some positive integer power of the Artin L-function $L(s, \chi; K/k)$ attached to an irreducible character χ admits a meromorphic continuation to $\Re(s) = 1$.

Exercise 11.4.4 If K/k is a finite Galois extension of algebraic number fields with group G, show that

$$\zeta_K(s) = \prod_{\chi} L(s, \chi; K/k)^{\chi(1)},$$

where the product is over all irreducible characters χ of G.

Exercise 11.4.5 Fix a complex number $s_0 \in \mathbb{C}$ with $\Re(s_0) \geq 1$ and any finite Galois extension K/k with Galois group G. For each subgroup H of G define the *Heilbronn character* θ_H by

$$\theta_H(g) = \sum_{\chi} n(H, \chi)\chi(g)$$

where the summation is over all irreducible characters χ of H and $n(H, \chi)$ is the order of the pole of $L(s, \chi; K/K^H)$ at $s = s_0$. By Exercise 11.4.3, the order is a rational number. Show that $\theta_G|_H = \theta_H$.

Exercise 11.4.6 Show that $\theta_G(1)$ equals the order at $s = s_0$ of the Dedekind zeta function $\zeta_K(s)$.

Exercise 11.4.7 Show that

$$\sum_{\chi} n(G, \chi)^2 \leq (\mathrm{ord}_{s=s_0} \zeta_K(s))^2 .$$

Exercise 11.4.8 For any irreducible non-trivial character χ, deduce that

$$L(s, \chi; K/k)$$

admits an analytic continuation to $s = 1$ and that $L(1, \chi; K/k) \neq 0$.

Exercise 11.4.9 Fix a conjugacy class C in $G = \mathrm{Gal}(K/k)$ and choose $g_C \in C$. Show that

$$\sum_{n, \mathfrak{p}, \sigma_{\mathfrak{p}}^m \in C} \frac{1}{N(\mathfrak{p})^{ms}} = \frac{|C|}{|G|} \sum_{\chi} \overline{\chi(g_C)} \log L(s, \chi; K/k).$$

Exercise 11.4.10 Show that

$$\lim_{s \to 1+} \frac{\sum_{\mathfrak{p}, \sigma_{\mathfrak{p}} \in C} 1/N(\mathfrak{p})^s}{\log \zeta_k(s)} = \frac{|C|}{|G|}$$

which is Chebotarev's theorem.

Exercise 11.4.11 Show that $\zeta_K(s)/\zeta_k(s)$ is entire. (This is called the *Brauer-Aramata theorem.*)

Exercise 11.4.12 (Stark) Let K/k be a finite Galois extension of algebraic number fields. If $\zeta_K(s)$ has a simple zero at $s = s_0$, then $L(s, \chi; K/k)$ is analytic at $s = s_0$ for every irreducible character χ of $\mathrm{Gal}(K/k)$.

Exercise 11.4.13 (Foote-K. Murty) For any irreducible character χ of $\mathrm{Gal}(K/k)$, show that

$$L(s, \chi; K/k)\zeta_K(s)$$

is analytic for $s \neq 1$.

Exercise 11.4.14 If K/k is solvable, show that

$$\sum_{\chi \neq 1} n(G, \chi)^2 \leq \left(\mathrm{ord}_{s=s_0} \zeta_K(s)/\zeta_k(s)\right)^2.$$

Part II

Solutions

Chapter 1

Elementary Number Theory

1.1 Integers

Exercise 1.1.7 Show that

$$1+\frac{1}{3}+\frac{1}{5}+\cdots+\frac{1}{2n-1}$$

is not an integer for $n>1$.

Solution. Let $S=1+\frac{1}{3}+\frac{1}{5}+\cdots+\frac{1}{2n-1}$. We can find an integer k such that $3^k \leq 2n-1 < 3^{k+1}$. Define m to be the least common multiple of all the numbers $3, 5, \ldots, 2n-1$ except for 3^k. Then

$$mS = m+\frac{m}{3}+\frac{m}{5}+\cdots+\frac{m}{2n-1}.$$

Each of the numbers on the right side of this equation is an integer, except for $m/3^k$. If $m/3^k$ were an integer, then there would be some integer b such that $m=3^k b$, but 3^k does not divide $3, 5, \ldots, 3^k-2, 3^k+2, \ldots, 2n-1$ so it cannot divide their least common multiple. Therefore mS is not an integer, and clearly neither is S.

Exercise 1.1.8 Let $a_1, \ldots, a_n$ for $n \geq 2$ be nonzero integers. Suppose there is a prime p and positive integer h such that $p^h \mid a_i$ for some i and p^h does not divide a_j for all $j \neq i$.

Then show that

$$S=\frac{1}{a_1}+\cdots+\frac{1}{a_n}$$

is not an integer.

Solution. Let h be the maximum power of p dividing a_i. We use the notation $p^h \| a_i$ to mean that $p^h \mid a_i$ but $p^{h+1} \nmid a_i$. Let m be the least common multiple of $a_1, \ldots, a_{i-1}, a_i/p^h, a_{i+1}, \ldots, a_n$. Then

$$mS = \frac{m}{a_1} + \cdots + \frac{m}{a_{i-1}} + \frac{m}{a_i} + \cdots + \frac{m}{a_n}.$$

We see that m/a_j is an integer for $j = 1, 2, \ldots, i-1, i+1, \ldots, n$. However, a_i does not divide m, since if it did then p^h would clearly have to divide m, which means we can find a $b \in \mathbb{Z}$ such that $m = p^h b$. Since p^h does not divide a_j for $j = 1, \ldots, i-1, i+1, \ldots, n$, it does not divide their least common multiple. Hence m/a_i is not an integer, which implies that mS is not an integer. Therefore

$$S = \frac{1}{a_1} + \cdots + \frac{1}{a_n}$$

is not an integer.

Exercise 1.1.9 Prove that if n is a composite integer, then n has a prime factor not exceeding $\sqrt{n}$.

Solution. Since n is composite, we can write $n = ab$ where a and b are integers with $1 < a \leq b < n$. We have $a \leq \sqrt{n}$ since otherwise $\sqrt{n} < a \leq b$ and $ab > \sqrt{n}\sqrt{n} = n$. Now, a certainly has a prime divisor, and any prime divisor of a is also a prime divisor of n. Hence n has a prime factor which is less than or equal to $\sqrt{n}$.

Exercise 1.1.10 Show that if the smallest prime factor p of the positive integer n exceeds $\sqrt[3]{n}$, then n/p must be prime or 1.

Solution. Suppose that the smallest prime factor p of the positive integer n exceeds $\sqrt[3]{n}$. Then $p > n^{1/3}$. Hence $n/p < n^{2/3}$. If n/p is composite, n/p has a prime factor not exceeding $\sqrt{n/p}$ by Exercise 1.1.9. We see that $\sqrt{n/p} < n^{1/3}$. A prime factor of n/p is also that of n, and so we have found a prime factor which is smaller than $n^{1/3}$, which contradicts our assumption. Therefore n/p is a prime or 1.

Exercise 1.1.11 Let p be prime. Show that each of the binomial coefficients $\binom{p}{k}$, $1 \leq k \leq p-1$, is divisible by p.

Solution. Since

$$\binom{p}{k} = \frac{p!}{k!\,(p-k)!}$$

we see that the numerator is divisible by p, and the denominator is not, for $1 \leq k \leq p-1$. The result is now evident.

Exercise 1.1.12 Prove that if p is an odd prime, then $2^{p-1} \equiv 1 \pmod{p}$.

Solution.

$$\begin{aligned} 2^p = (1+1)^p &= 1 + \sum_{k=1}^{p-1} \binom{p}{k} + 1 \\ &\equiv 1 + 1 \pmod{p} \end{aligned}$$

by the previous exercise.

Exercise 1.1.13 Prove Fermat's little Theorem: If $a, p \in \mathbb{Z}$ with p a prime, and $p \nmid a$, prove that $a^{p-1} \equiv 1 \pmod{p}$.

Solution. We can apply induction. For instance,

$$\begin{aligned} 3^p = (1+2)^p &= 1 + \sum_{k=1}^{p-1} \binom{p}{k} 2^k + 2^p \\ &\equiv 1 + 2^p \pmod{p} \end{aligned}$$

since the binomial coefficients are divisible by p. By the previous exercise $2^p \equiv 2 \pmod{p}$ and so we find that

$$3^p \equiv 3 \pmod{p}.$$

Alternate Solution. We consider the field $\mathbb{Z}/p\mathbb{Z}$, obtained by taking congruences mod p. Let $\overline{a}$ denote the class of $a \pmod{p}$. If $p \nmid a$, then $a \not\equiv 0 \pmod{p}$, and so $\overline{a}$ is a unit in the field $\mathbb{Z}/p\mathbb{Z}$. The units of this field form a multiplicative group G of order $p-1$. By elementary group theory, $\overline{a}^{|G|} = \overline{a^{p-1}} = \overline{1}$, which means that $a^{p-1} \equiv 1 \pmod{p}$.

Exercise 1.1.15 Show that $n \mid \phi(a^n - 1)$ for any $a > n$.

Solution. $a^n \equiv 1 \pmod{a^n - 1}$ and n is the smallest power of a with this property. Thus, a has order $n \pmod{a^n - 1}$. Therefore, $n \mid \phi(a^n - 1)$.

Exercise 1.1.16 Show that $n \nmid 2^n - 1$ for any natural number $n > 1$.

Solution. Let us suppose the set of $n > 1$ such that $2^n \equiv 1 \pmod{n}$ is nonempty. By the well-ordering principle, there is a least such number, call it n_0. Then

$$2^{n_0} \equiv 1 \pmod{n_0}.$$

By Euler's theorem,

$$2^{\phi(n_0)} \equiv 1 \pmod{n_0}.$$

Let $d = (n_0, \phi(n_0))$. By the Euclidean algorithm, we can find integers x and y so that

$$n_0 x + \phi(n_0) y = d.$$

Thus, $2^d \equiv 1 \pmod{n_0}$. If $d > 1$, this gives $2^d \equiv 1 \pmod{d}$ contradicting the minimality of n_0. Thus, $d = 1$ and we get

$$2 \equiv 1 \pmod{n_0}$$

which is also a contradiction.

Exercise 1.1.17 Show that

$$\frac{\phi(n)}{n} = \prod_{p|n} \left(1 - \frac{1}{p}\right)$$

by interpreting the left-hand side as the probability that a random number chosen from $1 \leq a \leq n$ is coprime to n.

Solution. The probability that a number chosen from $1 \leq a \leq n$ is coprime to n is clearly $\phi(n)/n$. On the other hand, this is tantamount to insisting that our number is not divisible by any prime divisors of n, which is represented by the right-hand side of the formula.

Exercise 1.1.18 Show that ϕ is multiplicative (i.e., $\phi(mn) = \phi(m)\phi(n)$ when $\gcd(m, n) = 1$) and $\phi(p^\alpha) = p^{\alpha-1}(p-1)$ for p prime.

Solution. By the previous exercise, it is clear that ϕ is multiplicative. When $n = p^\alpha$, we find

$$\phi(p^\alpha) = p^\alpha \left(1 - \frac{1}{p}\right) = p^{\alpha-1}(p-1).$$

Exercise 1.1.19 Find the last two digits of 3^{1000}.

Solution. We find the residue class that 3^{1000} belongs to in $\mathbb{Z}/100\mathbb{Z}$. This is the same as finding the last two digits. By Euler's theorem, $3^{40} \equiv 1 \pmod{100}$, since

$$\phi(100) = \phi(4)\phi(25) = 2(20) = 40.$$

Therefore,

$$3^{1000} = (3^{40})^{25} \equiv 1 \pmod{100}.$$

The last two digits are 01.

Exercise 1.1.20 Find the last two digits of 2^{1000}.

Solution. We need to find the residue class of 2^{1000} in $\mathbb{Z}/100\mathbb{Z}$. Since 2 is not coprime to 100, we cannot apply Euler's theorem as in the previous exercise. However, we have

$$x = 2^{1000} \equiv 1 \pmod{25},$$

$$x = 2^{1000} \equiv 0 \pmod 4.$$

We determine which residue classes $z \pmod{100}$ satisfy $z \equiv 1 \pmod{25}$ and $z \equiv 0 \pmod 4$.

The last condition means $z = 4k$. We solve $4k \equiv 1 \pmod{25}$. Thus, $6(4k) \equiv 6 \pmod{25}$ so that $k \equiv -6 \pmod{25}$. That is, $k \equiv 19 \pmod{25}$. Hence, $z = (19)4 = 76$. This means

$$2^{1000} \equiv 76 \pmod{100}.$$

The last two digits are 76.

Exercise 1.1.21 Let p_k denote the kth prime. Prove that

$$p_{k+1} \leq p_1 p_2 \cdots p_k + 1.$$

Solution. We see that the number $p_1 p_2 \cdots p_k + 1$ is coprime to $p_1, p_2, \ldots, p_k$ and either must be prime, or divisible by a prime different from $p_1, \ldots, p_k$. Thus,

$$p_{k+1} \leq p_1 p_2 \cdots p_k + 1.$$

Exercise 1.1.22 Show that

$$p_k < 2^{2^k},$$

where p_k denotes the kth prime.

Solution. From the preceding exercise, we know that $p_{k+1} \leq p_1 \cdots p_k + 1$. Now we have $p_1 < 2^{2^1}$ and $p_2 < 2^{2^2}$. Suppose that $p_k < 2^{2^k}$ is true for $2 < k \leq n$.

Then

$$\begin{aligned} p_{n+1} &\leq p_1 p_2 \cdots p_n + 1 \\ &< 2^{2^1} 2^{2^2} \cdots 2^{2^n} + 1 \\ &= 2^{2^{n+1}-2} + 1 \\ &< 2^{2^{n+1}}. \end{aligned}$$

Hence $p_n < 2^{2^n}$ is true for $n \geq 1$.

Exercise 1.1.23 Prove that $\pi(x) \geq \log(\log x)$.

Solution. From the previous exercise, we have that $p_n < 2^{2^n}$ for $n \geq 1$. Hence we can see that $\pi(2^{2^n}) \geq n$.

For $x > 2$, choose an integer n so that $e^{e^{n-1}} < x \leq e^{e^n}$. Then

$$\log x \leq e^n \log e = e^n$$

and

$$\log(\log x) \leq n \log e = n.$$

If $n > 2$, then $e^{n-1} > 2^n$.

$$\begin{aligned}\pi(x) &\geq \pi(e^{e^{n-1}}) \\ &\geq \pi(e^{2^n}) \\ &\geq \pi(2^{2^n}) \\ &\geq n \\ &\geq \log(\log x).\end{aligned}$$

This proves the result.

Exercise 1.1.24 By observing that any natural number can be written as sr^2 with s squarefree, show that

$$\sqrt{x} \leq 2^{\pi(x)}.$$

Deduce that

$$\pi(x) \geq \frac{\log x}{2\log 2}.$$

Solution. For any set of primes S define $f_S(x)$ to be the number of integers n such that $1 \leq n \leq x$ with $\gamma(n) \subset S$ where $\gamma(n)$ is the set of primes dividing n. Suppose that S is a finite set with t elements. Write such an n in the form $n = r^2 s$ with s squarefree. Since $1 \leq r^2 s \leq x$, we see that $r \leq \sqrt{x}$ and there are at most 2^t choices for s corresponding to the various subsets of S since s is squarefree. Thus $f_S(x) \leq 2^t\sqrt{x}$.

Put $\pi(x) = m$ so that $p_{m+1} > x$. If $S = \{p_1, \ldots, p_m\}$, then $f_S(x) = x$. Then

$$x \leq 2^m\sqrt{x} = 2^{\pi(x)}\sqrt{x}.$$

Thus $\sqrt{x} \leq 2^{\pi(x)}$ and hence $\frac{1}{2}\log x \leq \pi(x)\log 2$, or equivalently,

$$\pi(x) \geq \frac{\log x}{2\log 2}.$$

Exercise 1.1.25 Let $\psi(x) = \sum_{p^\alpha \leq x} \log p$ where the summation is over prime powers $p^\alpha \leq x$.

(i) For $0 \leq x \leq 1$, show that $x(1-x) \leq \frac{1}{4}$. Deduce that

$$\int_0^1 x^n(1-x)^n\, dx \leq \frac{1}{4^n}$$

for every natural number n.

(ii) Show that $e^{\psi(2n+1)}\int_0^1 x^n(1-x)^n\, dx$ is a positive integer. Deduce that $\psi(2n+1) \geq 2n\log 2$.

(iii) Prove that $\psi(x) \geq \frac{1}{2}x\log 2$ for $x \geq 6$. Deduce that

$$\pi(x) \geq \frac{x\log 2}{2\log x}$$

for $x \geq 6$.

Solution. Clearly $4x^2 - 4x + 1 = (2x-1)^2 \geq 0$ so (i) is now immediate. The integral $\int_0^1 x^n(1-x)^n\,dx$ consists of a sum of rational numbers whose denominators are less than $2n+1$. Since $\mathrm{lcm}(1,2,\ldots,2n+1) = e^{\psi(2n+1)}$, we find

$$e^{\psi(2n+1)} \int_0^1 x^n(1-x)^n\,dx$$

is a positive integer. Thus, $e^{\psi(2n+1)} \geq 2^{2n}$. This proves (ii).

For (iii), choose n so that $2n-1 \leq x < 2n+1$. Then, by (ii),

$$\psi(x) \geq \psi(2n-1) \geq (2n-2)\log 2 > (x-3)\log 2.$$

For $x \geq 6$, $x - 3 > x/2$ so that $\psi(x) > x\log 2/2$. Since $\psi(x) \leq \pi(x)\log x$, we deduce that

$$\pi(x) \geq \frac{x\log 2}{2\log x}$$

for $x \geq 6$.

Exercise 1.1.26 By observing that

$$\prod_{n<p\leq 2n} p \;\Big|\; \binom{2n}{n},$$

show that

$$\pi(x) \leq \frac{9x\log 2}{\log x}$$

for every integer $x \geq 2$.

Solution. Since

$$\prod_{n<p\leq 2n} p \;\Big|\; \binom{2n}{n},$$

we deduce that

$$\sum_{n<p\leq 2n} \log p \leq 2n\log 2$$

because

$$\binom{2n}{n} \leq 2^{2n}.$$

Therefore

$$\theta(2n) - \theta(n) \leq 2n\log 2,$$

where

$$\theta(n) = \sum_{p\leq n} \log p.$$

An easy induction shows that $\theta(2^r) \leq 2^{r+1}\log 2$ for every positive integer r. given an integer $x \geq 2$, determine r so that

$$2^r \leq x < 2^{r+1}.$$

Then
$$\theta(x) \leq \theta(2^{r+1}) \leq 2^{r+2} \log 2 \leq 4x \log 2.$$
We deduce, in particular,
$$\sum_{\sqrt{x}<p\leq x} \log p \leq 4x \log 2,$$
so that
$$\left(\tfrac{1}{2}\log x\right)\left(\pi(x) - \pi(\sqrt{x})\right) \leq 4x \log 2.$$
This means
$$\pi(x) - \pi(\sqrt{x}) \leq \frac{8x \log 2}{\log x}$$
and
$$\pi(x) \leq \frac{8x\log 2}{\log x} + \sqrt{x} \leq \frac{9x \log 2}{\log x}$$
because
$$\sqrt{x} \leq \frac{x \log 2}{\log x}$$
for $x \geq 10$, as is easily checked by examining the graph of
$$f(x) = \sqrt{x}\log 2 - \log x.$$
For $x \leq 10$, the inequality is verified directly.

1.2 Applications of Unique Factorization

Exercise 1.2.1 Suppose that $a, b, c \in \mathbb{Z}$. If $ab = c^2$ and $(a, b) = 1$, then show that $a = d^2$ and $b = e^2$ for some $d, e \in \mathbb{Z}$. More generally, if $ab = c^g$ then $a = d^g$ and $b = e^g$ for some $d, e \in \mathbb{Z}$.

Solution. Write $a = p_1^{\alpha_1} p_2^{\alpha_2} \cdots p_r^{\alpha_r}$ and $b = q_1^{\beta_1} q_2^{\beta_2} \cdots q_s^{\beta_s}$ where p_i and q_j are primes for $1 \leq i \leq r$ and $1 \leq j \leq s$ and $p_i \neq q_j$ for any i, j since $(a, b) = 1$.
$$\begin{aligned} ab &= (p_1^{\alpha_1} \cdots p_r^{\alpha_r})(q_1^{\beta_1} \cdots q_s^{\beta_s}) \\ &= c^2 \\ &= p_1^{2\gamma_1} \cdots p_r^{2\gamma_r} \cdot q_1^{2\theta_1} \cdots q_s^{2\theta_s}, \end{aligned}$$
where $c = p_1^{\gamma_1} \cdots p_r^{\gamma_r} q_1^{\theta_1} \cdots q_s^{\theta_s}$.

By unique factorization, $\alpha_i = 2\gamma_i$ and $\beta_j = 2\theta_j$ for $1 \leq i \leq r$ and $1 \leq j \leq s$. Hence we can write $a = p_1^{2\gamma_1} \cdots p_r^{2\gamma_r}$ and $b = q_1^{2\theta_1} \cdots q_s^{2\theta_s}$. Hence $\exists d, e \in \mathbb{Z}$ such that $a = d^2$ and $b = e^2$ where $d = p_1^{\gamma_1} \cdots p_r^{\gamma_r}$ and $e = q_1^{\theta_1} \cdots q_s^{\theta_s}$.

The argument for gth powers is identical.

Exercise 1.2.2 Solve the equation $x^2 + y^2 = z^2$ where x, y, and z are integers and $(x, y) = (y, z) = (x, z) = 1$.

Solution. Assume that x and y are odd. Then both $x^2 \equiv 1 \pmod 4$ and $y^2 \equiv 1 \pmod 4$. Hence $z^2 \equiv 2 \pmod 4$. But there is no $z \in \mathbb{Z}$ satisfying $z^2 \equiv 2 \pmod 4$, so one of x or y is even.

Without loss of generality, suppose x is even and y is odd. Then z is odd. We have $x^2 = z^2 - y^2$, so

$$\begin{aligned} \frac{x^2}{4} &= \frac{z^2 - y^2}{4}, \\ \Rightarrow \quad \left(\frac{x}{2}\right)^2 &= \frac{(z+y)}{2}\frac{(z-y)}{2}. \end{aligned}$$

Since $(x, y) = (y, z) = (x, z) = 1$, we see that $((z+y)/2, (z-y)/2) = 1$. By Exercise 1.2.1, there exist $a, b \in \mathbb{Z}$ such that $(z+y)/2 = a^2$ and $(z-y)/2 = b^2$. Hence we have the two equations $z + y = 2a^2$ and $z - y = 2b^2$.

Thus the solution is $x = 2ab$, $y = a^2 - b^2$, and $z = a^2 + b^2$ where $(a, b) = 1$ and a and b have opposite parity since y and z are odd. Conversely, any such triple (x, y, z) gives rise to a solution.

Exercise 1.2.3 Show that $x^4 + y^4 = z^2$ has no nontrivial solution. Hence deduce, with Fermat, that $x^4 + y^4 = z^4$ has no nontrivial solution.

Solution. Suppose that $x^4 + y^4 = z^2$ has a nontrivial solution. Take $|z|$ to be minimal. By Exercise 1.2.2, we can write

$$\begin{aligned} x^2 &= 2ab, && (1.1) \\ y^2 &= b^2 - a^2, && (1.2) \\ z &= b^2 + a^2, && (1.3) \end{aligned}$$

with $(x, y) = 1$ and a and b having opposite parity.

Suppose that b is even. Then we see that

$$y^2 = b^2 - a^2 \equiv -1 \equiv 3 \pmod 4.$$

This is impossible. Hence a is even. Then $\exists c \in \mathbb{Z}$ such that $a = 2c$ and $(c, b) = 1$. Then $x^2 = 2 \cdot 2bc = 4bc$. Since $(b, c) = 1$, b and c are perfect squares by Exercise 1.2.1. Hence $\exists m, n \in \mathbb{Z}$ such that $b = m^2, c = n^2$ where $(m, n) = 1$. By (1.2), we see that $y^2 = b^2 - a^2 = m^4 - 4n^4$. Hence $(2n^2)^2 + y^2 = (m^2)^2$ and $(2n^2, y) = (y, m^2) = (2n^2, m^2) = 1$.

By Exercise 1.2.2, $2n^2 = 2\alpha\beta$, $y = \beta^2 - \alpha^2$, and $m^2 = \alpha^2 + \beta^2$ where $(\alpha, \beta) = 1$ and α and β have opposite parity. Thus we can see that $n^2 = \alpha\beta$. Hence by Exercise 1.2.1, $\exists p, q \in \mathbb{Z}$ such that $\alpha = p^2$ and $\beta = q^2$. Hence we have $m^2 = p^4 + q^4$. This is a solution of the equation $x^4 + y^4 = z^2$. But $m < b < |z|$ since $m^2 = b < b^2 + a^2 = z$. This is a contradiction to the minimality of $|z|$. Therefore $x^4 + y^4 = z^2$ has no nontrivial solution.

Now suppose that $x^4 + y^4 = z^4$ has a nontrivial solution. This would imply that $x^4 + y^4 = t^2$ where $t = z^2$ has a nontrivial solution. But we proved above that this is impossible, so $x^4 + y^4 = z^4$ has no nontrivial solution.

Exercise 1.2.4 Show that $x^4 - y^4 = z^2$ has no nontrivial solution.

Solution. Suppose that $x^4 - y^4 = (x^2+y^2)(x^2-y^2) = z^2$ has a nontrivial solution, and choose the solution such that $|x|$ is minimal. If x is even, then both y and z must be odd (since x, y, z are coprime). But then we can rewrite the equation as $(x^2)^2 = z^2 + (y^2)^2$ and we know from Exercise 1.2.2 that this equation has no solutions for x even. So x is odd.

Suppose that y is odd. We again write the equation as $(x^2)^2 = z^2+(y^2)^2$, and we see that by Exercise 1.2.2 we can write

$$z = 2ab, \quad y^2 = a^2 - b^2, \quad x^2 = a^2 + b^2,$$

for relatively prime integers a, b. Now,

$$a^4 - b^4 = (a^2 + b^2)(a^2 - b^2) = x^2y^2 = (xy)^2,$$

and we have found another solution to the equation $x^4 - y^4 = z^2$. But $a < \sqrt{a^2 + b^2} = x$, contradicting our assumption about x. We conclude that there are no solutions for y odd.

Now suppose that y is even. Then we can use Exercise 1.2.2 and write

$$y^2 = 2cd, \quad z = c^2 - d^2, \quad x^2 = c^2 + d^2,$$

where c, d are coprime and of opposite parity. Without loss, we assume that c is even, d odd. But then we have $(2c, d) = 1$, and we can use Exercise 1.2.1 which says that we can find integers s, t such that $2c = s^2, d = t^2$. In fact, s is even so we can write $s = 2u$, and thus $c = 2u^2$. Therefore we can write $x^2 = c^2+d^2 = (2u^2)^2+(t^2)^2$. We now deduce that we can find integers v, w such that $2u^2 = 2vw$, $t^2 = v^2 - w^2$, $x = v^2 + w^2$. Since $u^2 = vw$, we can write $v = a^2, w = b^2$. But looking back, we see that $t^2 = v^2 - w^2 = a^4 - b^4$, and since $a = \sqrt{v} < v^2 + w^2 = x$, which is a contradiction. So $x^4 - y^4 = z^2$ has no nontrivial solutions.

Exercise 1.2.5 Prove that if $f(x) \in \mathbb{Z}[x]$, then $f(x) \equiv 0 \pmod p$ is solvable for infinitely many primes p.

Solution. We will call p a prime divisor of f if $p \mid f(n)$ for some n. Clearly f always has a prime divisor. Hence it suffices to show that f has infinitely many prime divisors. Suppose that f has only finitely many prime divisors.

Let $f(x) = a_nx^n + a_{n-1}x^{n-1} + \cdots + a_1x + a_0$ and let $p_1, \ldots, p_k$ be the prime divisors of f. For simplicity, we will write $m = p_1 \cdots p_k$. Then

$$\begin{aligned} f(a_0m) &= a_n(a_0m)^n + \cdots + a_1(a_0m) + a_0 \\ &= a_0(a_na_0{}^{n-1}m^n + a_{n-1}a_0{}^{n-2}m^{n-1} + \cdots + a_1m + 1). \end{aligned}$$

Let $g(x) = a_n {a_0}^{n-1} x^n + a_{n-1} {a_0}^{n-2} x^{n-1} + \cdots + a_1 x + 1$. Then we can see that $(p_i, g(m)) = 1$ for $1 \le i \le k$. Hence $g(m)$ has a prime divisor different from p_i for $1 \le i \le k$. The prime divisor of $g(m)$ is also that of $f(a_0 m)$. Hence we can see that there is a new prime divisor of f different from p_i for $1 \le i \le k$. This is a contradiction. Therefore f has infinitely many prime divisors.

Exercise 1.2.6 Let q be prime. Show that there are infinitely many primes p so that $p \equiv 1 \pmod{q}$.

Solution. Let us consider the polynomial

$$f(x) = \frac{x^q - 1}{x - 1} = 1 + x + \cdots + x^{q-1},$$

and suppose that p is a prime divisor of $f(x)$. Then $x^q \equiv 1 \pmod{p}$ for some x. Let x_0 be an integer such that $f(x_0) \equiv 0 \pmod{p}$. Then ${x_0}^q \equiv 1 \pmod{p}$. If x_0 is not congruent to 1 (mod p), then q is the order of $x_0 \pmod{p}$ since q is a prime. Consider the multiplicative group $G = \{\overline{1}, \ldots, \overline{p-1}\}$. We see that $\overline{x_0} \in G$. Since q is the order of $x_0 \pmod{p}$, we can see that $q \mid (p-1)$. Hence $p - 1 \equiv 0 \pmod{q}$ and hence $p \equiv 1 \pmod{q}$. If $x_0 \equiv 1 \pmod{p}$, then $1 + x_0 + \cdots + x_0^{q-1} \equiv 0 \pmod{p}$ means $p = q$. Therefore, any prime divisor of f is either q or $\equiv 1 \pmod{q}$. By Exercise 1.2.5, there are infinitely many primes p such that $f(x) \equiv 0 \pmod{p}$ is solvable since $f(x) \in \mathbb{Z}[x]$. We conclude that there are infinitely many primes p such that $p \equiv 1 \pmod{q}$.

Exercise 1.2.7 Show that F_n divides $F_m - 2$ if n is less than m, and from this deduce that F_n and F_m are relatively prime if $m \ne n$.

Solution. Write $m = n + k$ where k is a nonzero positive integer. Then

$$\begin{aligned}
\frac{F_m - 2}{F_n} &= \frac{F_{n+k} - 2}{F_n} \\
&= \frac{2^{2^{n+k}} - 1}{2^{2^n} + 1} \\
&= \frac{(2^{2^n})^{2^k} - 1}{2^{2^n} + 1} \\
&= \frac{t^{2^k} - 1}{t + 1} = t^{2^k - 1} - t^{2^k - 2} + \cdots - 1,
\end{aligned}$$

where $t = 2^{2^n}$. Hence F_n divides $F_m - 2$.

Let d be the greatest common divisor of F_n and F_m. Then $d \mid F_n$, so $d \mid F_m - 2$ and $d \mid F_m$. Hence $d \mid 2$ and hence $d = 1$ or 2. Since F_n and F_m are odd, $d = 1$. Thus $(F_n, F_m) = 1$. Therefore F_n and F_m are relatively prime if $m \ne n$.

Exercise 1.2.8 Consider the nth Fermat number $F_n = 2^{2^n} + 1$. Prove that every prime divisor of F_n is of the form $2^{n+1}k + 1$.

Solution. Let p be a prime divisor of $2^{2^n} + 1$. Then $2^{2^n} \equiv -1 \pmod{p}$, so $(2^{2^n})^2 \equiv 1 \pmod{p}$. Hence we have $2^{2^{n+1}} \equiv 1 \pmod{p}$. We will show that the order of 2 (mod p) is 2^{n+1}. Let $x = \text{ord}_p\, 2$. Since $x \mid 2^{n+1}$ we can write $x = 2^m$ where m is an integer and $1 \leq m \leq n+1$. Hence for all $n \geq m$, $2^{2^n} \equiv 1 \pmod{p}$. But by assumption $2^{2^n} \equiv -1 \pmod{p}$, which implies that $2^{2^m} \equiv 1 \pmod{p}$ holds only if $m \geq n+1$. We now consider the multiplicative group $G = \{\overline{1}, \ldots, \overline{p-1}\}$. We must have $2^{n+1} \mid p-1$ since $\text{ord}_p\, 2 = 2^{n+1}$ and the order of G is $p-1$. Therefore we can write $p - 1 = 2^{n+1}k$ where k is an integer, and we conclude that $p = 2^{n+1}k + 1$.

Exercise 1.2.9 Given a natural number n, let $n = p_1^{\alpha_1} \cdots p_k^{\alpha_k}$ be its unique factorization as a product of prime powers. We define the squarefree part of n, denoted $S(n)$, to be the product of the primes p_i for which $\alpha_i = 1$. Let $f(x) \in \mathbb{Z}[x]$ be nonconstant and monic. Show that $\liminf S(f(n))$ is unbounded as n ranges over the integers.

Solution. By Exercise 1.2.5, we know that f has infinitely many prime divisors. Let p be such a prime and suppose $f(x_0) \equiv 0 \pmod{p}$. Observe that $f(x_0 + p) \equiv f(x_0) + pf'(x_o) \pmod{p^2}$. We define the discriminant of a monic polynomial f to be

$$\prod_{i>j} (r_i - r_j)^2,$$

where $r_1, \ldots, r_n$ are the roots of f. If $p \mid f'(x_0)$, then p would divide the discriminant of f. (Why? see Exercise 4.3.3.) Choosing p sufficiently large, we may assume this does not happen. In either case, we deduce that the squarefree part of $f(x_0)$ is divisible by p or the squarefree part of $f(x_0 + p)$ is divisible by p. If $S(f(n))$ were bounded, we have derived a contradiction.

1.3 The *ABC* Conjecture

Exercise 1.3.1 Assuming the *ABC* Conjecture, show that if $xyz \neq 0$ and $x^n + y^n = z^n$ for three mutually coprime integers x, y, and z, then n is bounded.

Solution. First observe that $\max\big(|x|, |y|, |z|\big) > 1$ for otherwise we have $xyz = 0$. By the *ABC* Conjecture, we have

$$\max\big(|x|^n, |y|^n, |z|^n\big) \leq \kappa(\varepsilon)\big(\text{rad}(xyz)\big)^{1+\varepsilon}.$$

Without any loss of generality, suppose that $\max\big(|x|, |y|, |z|\big) = |z|$. We deduce that $|z|^n \leq \kappa(\varepsilon)|z|^{3+3\varepsilon}$. Since $|z| > 1$, we conclude that n is bounded.

Exercise 1.3.2 Let p be an odd prime. Suppose that $2^n \equiv 1 \pmod{p}$ and $2^n \not\equiv 1 \pmod{p^2}$. Show that $2^d \not\equiv 1 \pmod{p^2}$ where d is the order of 2 (mod p).

Solution. Since $2^n \equiv 1 \pmod{p}$, we must have $d \mid n$. Write $n = de$. If $2^d = 1 + kp$ and $p \mid k$, then

$$\begin{aligned} 2^n = 2^{de} &= (1+kp)^e \\ &\equiv 1 + kpe \pmod{p^2} \\ &\equiv 1 \pmod{p^2}, \end{aligned}$$

a contradiction. This proves the result.

Exercise 1.3.3 Assuming the *ABC* Conjecture, show that there are infinitely many primes p such that $2^{p-1} \not\equiv 1 \pmod{p^2}$.

Solution. Let us write $2^n - 1 = u_n v_n$ where u_n is the squarefree part of $2^n - 1$ and v_n is the squarefull (or powerfull) part of $2^n - 1$. (Recall that a number N is called squarefull (powerfull) if for every prime $q \mid N$ we have $q^2 \mid N$. Thus for any number N, $N/S(N)$ is squarefull (or powerfull) with $S(N)$ the squarefree part of N.) Therefore $(u_n, v_n) = 1$.

We begin by showing that if $p \mid u_n$, then $2^{p-1} \not\equiv 1 \pmod{p^2}$. Indeed, we know that $p \mid 2^n - 1$ and $p^2 \nmid 2^n - 1$. (As defined earlier in this chapter, $p^\alpha \| n$ means that $p^\alpha \mid n$ but $p^{\alpha+1} \nmid n$. In this case, we would write $p \| 2^n - 1$.) By Exercise 1.3.2, $p^2 \nmid 2^d - 1$ where d is the order of $2 \pmod{p}$. Now $d \mid (p-1)$ by the little theorem of Fermat and Lagrange's theorem. Write $df = p - 1$. Then $2^d = 1 + kp$ with $p \nmid k$ so that $2^{p-1} = (1+kp)^f \equiv 1 + kfp \pmod{p^2}$. Since $f \mid p - 1$ and $p \nmid k$, we find that $2^{p-1} \not\equiv 1 \pmod{p^2}$ for every prime p dividing u_n.

Now suppose there are only finitely many such primes p. Since u_n is squarefree, this implies that u_n is bounded. Now consider the *ABC* equation:

$$(2^n - 1) + 1 = 2^n.$$

The *ABC* Conjecture implies that

$$2^n \leq \kappa(\varepsilon)\bigl(2 \operatorname{rad}(2^n - 1)\bigr)^{1+\varepsilon}.$$

But $\operatorname{rad}(2^n - 1) \leq u_n v_n^{1/2}$ and so

$$u_n v_n = 2^n - 1 < 2^n \leq \kappa(\varepsilon)\bigl(2 u_n v_n^{1/2}\bigr)^{1+\varepsilon}.$$

Since u_n is bounded, this implies that v_n is bounded, and hence n is bounded, a contradiction.

Remark. This is due to J. Silverman [Sil] who also obtains, assuming the *ABC* Conjecture, that the number of primes $p \leq x$ for which $2^{p-1} \not\equiv 1 \pmod{p^2}$ is $\gg \log x$.

Exercise 1.3.4 Show that the number of primes $p \leq x$ for which

$$2^{p-1} \not\equiv 1 \pmod{p^2}$$

is $\gg \log x / \log\log x$, assuming the *ABC* Conjecture.

Solution. If for any n, we have $u_n = 1$, then the ABC Conjecture implies, as above, that n is bounded. Thus for n sufficiently large (say $n > N$), $u_n > 1$. For each prime q satisfying $N < q \leq (\log x)/\log 2$, we have $u_q > 1$. Moreover, for any two distinct primes q_1 and q_2, $\gcd(2^{q_1} - 1, 2^{q_2} - 1) = 1$ because $p \mid 2^{q_1} - 1$ and $p \mid 2^{q_2} - 1$ implies that the order of 2 (mod p) divides q_1 and q_2 and so it divides their gcd, which is 1. This implies that $p \mid 1$, which is a contradiction.

Thus, for every prime $p \mid u_q$, we find $2^{p-1} \not\equiv 1 \pmod{p^2}$. In addition, the $u_q's$ are mutually coprime. In this way we obtain

$$\begin{aligned}
\pi\left(\frac{\log x}{\log 2}\right) &\geq \frac{\frac{\log x}{\log 2}\log 2}{2\log\left(\frac{\log x}{\log 2}\right)} \quad \text{by Exercise 1.1.25} \\
&= \frac{\log x}{2(\log\log x - \log\log 2)} \\
&\gg \frac{\log x}{\log\log x},
\end{aligned}$$

primes $p < x$ such that $2^{p-1} \not\equiv 2 \pmod{p^2}$.

Exercise 1.3.5 Show that if the Erdös conjecture above is true, then there are infinitely many primes p such that $2^{p-1} \not\equiv 1 \pmod{p^2}$.

Solution. Suppose for $p > p_0$ that $2^{p-1} \equiv 1 \pmod{p^2}$. Let $t = \prod_{p \leq p_0} p$. Then

$$\phi(t) = \prod_{p \leq p_0} (p-1).$$

Now consider the sequence $c_n = 2^{nt\phi(t)} - 1$. We claim that c_n is powerfull. Indeed if $2 < p \leq p_0$, then by Euler's theorem $p^2 \mid c_n$. If $p > p_0$, and $p \mid c_n$, then $p^2 \mid v_{nt\phi(t)}$ by the argument in Exercise 1.3.3. Thus $p^2 \mid c_n$ and so c_n is squarefull. For n even, say $n = 2k$, we deduce that both $2^{kt\phi(t)} - 1$ and $2^{kt\phi(t)} + 1$ are powerfull. But then, so is $2^{kt\phi(t)}$, contrary to the Erdös conjecture.

Exercise 1.3.6 Assuming the ABC Conjecture, prove that there are only finitely many n such that $n-1, n, n+1$ are squarefull.

Solution. If $n-1, n, n+1$ are squarefull, consider the ABC equation

$$(n^2 - 1) + 1 = n^2.$$

Then,

$$\begin{aligned}
n^2 &\leq \kappa(\varepsilon)\big(\text{rad}(n^2(n^2-1))\big)^{1+\varepsilon} \\
&\leq \kappa(\varepsilon)\big(n^{1/2}\sqrt{n-1}\sqrt{n+1}\big)^{1+\varepsilon}
\end{aligned}$$

since n, $n-1$, and $n+1$ are all squarefull. The inequality implies that n is bounded. (This is due to A. Granville.)

Exercise 1.3.7 Suppose that a and b are odd positive integers satisfying

$$\operatorname{rad}(a^n - 2) = \operatorname{rad}(b^n - 2)$$

for every natural number n. Assuming ABC, prove that $a = b$. (This problem is due to H. Kisilevsky.)

Solution. Suppose without loss that $a < b$. Hence $\log b > \log a$ so we can choose $\varepsilon > 0$ so that $\log b > (1 + \varepsilon) \log a$. Now apply the ABC Conjecture to the equation $(b^n - 2) + 2 = b^n$. Then

$$\begin{aligned} b^n &\leq \kappa(\varepsilon)\big(2b \operatorname{rad}(b^n - 2)\big)^{1+\varepsilon} \\ &\leq \kappa(\varepsilon)\big(2b \operatorname{rad}(a^n - 2)\big)^{1+\varepsilon} \\ &\leq \kappa(\varepsilon)\big(2ba^n\big)^{1+\varepsilon}. \end{aligned}$$

Taking nth roots and letting $n \to \infty$ gives $\log b \leq (1 + \varepsilon) \log a$, which is a contradiction. This completes the proof.

Of course, we may consider the equation $\operatorname{rad}(a^n - c) = \operatorname{rad}(b^n - c)$ for a fixed integer c coprime to a and b. The above argument applies in this context as well. Recently, R. Schoof and C. Corrales-Rodrigáñez [Sc] established this result in the special case $c = 1$ without assuming ABC.

It is also worth observing that we do not need the equation

$$\operatorname{rad}(a^n - 2) = \operatorname{rad}(b^n - 2)$$

satisfied for all natural numbers n, but just an infinite subsequence.

1.4 Supplementary Problems

Exercise 1.4.1 Show that every proper ideal of $\mathbb{Z}$ is of the form $n\mathbb{Z}$ for some integer n.

Solution. Suppose there is an ideal I for which this is not true. Then show that there exist elements $a, b \in I$ such that $\gcd(a, b) = 1$.

Exercise 1.4.2 An ideal I is called *prime* if $ab \in I$ implies $a \in I$ or $b \in I$. Prove that every prime ideal of $\mathbb{Z}$ is of the form $p\mathbb{Z}$ for some prime integer p.

Solution. If I is an ideal, then it is of the form $n\mathbb{Z}$ for some integer n by the previous question. Then $ab \in I$ implies that $n \mid ab$. But then since I is prime, either $a \in I$ or $b \in I$, so $n \mid a$ or $n \mid b$, implying that n is prime.

Exercise 1.4.3 Prove that if the number of prime Fermat numbers is finite, then the number of primes of the form $2^n + 1$ is finite.

Solution. Consider primes of the form $2^n + 1$. If n has an odd factor, say $n = rs$ with r odd, then $2^{rs} + 1$ is divisible by $2^s + 1$, and is therefore not prime.

Exercise 1.4.4 If $n > 1$ and $a^n - 1$ is prime, prove that $a = 2$ and n is prime.

Solution. If $a > 2$, then $a^n - 1$ is divisible by $a - 1$. So assume $a = 2$. Then if n has a factor, say k, then $2^k - 1 \mid 2^n - 1$. Therefore if $a^n - 1$ is prime, $a = 2$ and n is prime. Numbers of this form are called *Mersenne numbers*.

Exercise 1.4.6 Prove that if p is an odd prime, any prime divisor of $2^p - 1$ is of the form $2kp + 1$, with k a positive integer.

Solution. Suppose q is a prime divisor of $2^p - 1$. Then q must be odd. We note that $2^p \equiv 1 \pmod q$ and also, by Fermat's little Theorem, $2^{q-1} \equiv 1 \pmod q$. Then $p \mid (q-1)$ since p is prime. Then $q \equiv 1 \pmod p$ so $q = mp + 1$ but since q is odd, $m = 2k$ for some k, and so $q = 2kp + 1$.

Exercise 1.4.7 Show that there are no integer solutions to the equation $x^4 - y^4 = 2z^2$.

Solution. We will consider only solutions with $\gcd(x, y) = 1$, since any common factor of x, y will also divide z. Therefore any solution to this equation with $\gcd(x, y) \neq 1$ will lead to a solution with $\gcd(x, y) = 1$.

We notice that since the right-hand side of the equation is even, x and y are either both even or both odd. Since x, y are coprime, they must be odd. Then $x^4 - y^4 \equiv 0 \pmod 4$ and so z is even. We can factor the equation as

$$(x^2 + y^2)(x^2 - y^2) = 2z^2.$$

We note that $x^2 + y^2 \equiv 2 \pmod 4$ and $x^2 - y^2 \equiv 0 \pmod 4$, so $(x^2 + y^2)/2$ is odd. Now we rewrite our equation as

$$\left(\frac{x^2 + y^2}{2}\right)(x^2 - y^2) = z^2.$$

If there is an integer δ such that $\delta \mid (x^2 + y^2)/2$ and $\delta \mid x^2 - y^2$, then $\delta \mid 2x^2$ and $\delta \mid 2y^2$. But x, y are relatively prime and so $\delta \mid 2$. We know that $2 \nmid (x^2 + y^2)/2$ so $\delta = 1$ and our two factors are coprime.

This implies

$$\begin{aligned} x^2 + y^2 &= 2a^2, \\ x^2 - y^2 &= 4b^2, \end{aligned}$$

since $x^2 - y^2 \equiv 0 \pmod 4$. We now factor this second equation above as

$$\left(\frac{x + y}{2}\right)\left(\frac{x - y}{2}\right) = b^2.$$

It is easy to see that the two factors are coprime, and so we can write

$$x + y = 2c^2,$$
$$x - y = 2d^2.$$

Now we notice that we have

$$(x+y)^2 + (x-y)^2 = 2x^2 + 2y^2 = 2(x^2 + y^2).$$

Given our expressions above, this translates into the equation

$$4c^4 + 4d^4 = 4a^2,$$

but we know that $x^4 + y^4 = z^2$ has no solutions in $\mathbb{Z}$. Thus, the given equation has no solution.

Exercise 1.4.8 Let p be an odd prime number. Show that the numerator of

$$1 + \frac{1}{2} + \frac{1}{3} + \cdots + \frac{1}{p-1}$$

is divisible by p.

Solution. Look at the sum modulo p.

Exercise 1.4.9 Let p be an odd prime number greater than 3. Show that the numerator of

$$1 + \frac{1}{2} + \frac{1}{3} + \cdots + \frac{1}{p-1}$$

is divisible by p^2.

Solution. Pair up $1/i$ and $1/(p-i)$ and consider the sum mod p.

Exercise 1.4.10 (Wilson's Theorem) Show that $n > 1$ is prime if and only if n divides $(n-1)! + 1$.

Solution. When n is prime, consider $(n-1)! \pmod{n}$ by pairing each residue class with its multiplicative inverse.

Exercise 1.4.11 For each $n > 1$, let Q be the product of all numbers $a < n$ which are coprime to n. Show that $Q \equiv \pm 1 \pmod{n}$.

Solution. Q is clearly congruent to the product of elements of order 2. Now pair up a and $(n-a)$.

Exercise 1.4.12 In the previous exercise, show that $Q \equiv 1 \pmod{n}$ whenever n is odd and has at least two prime factors.

Solution. Clearly $Q \equiv (-1)^s \pmod{n}$ where $2s$ is the number of elements a satisfying $a^2 \equiv 1 \pmod{n}$. Use the Chinese Remainder Theorem (see Exercise 5.3.13) to determine s.

Exercise 1.4.13 Use Exercises 1.2.7 and 1.2.8 to show that there are infinitely many primes $\equiv 1 \pmod{2^r}$ for any given r.

Solution. If $p \mid F_n$, then $p \equiv 1 \pmod{2^{n+1}}$. For each $n \geq r$, we have $p \equiv 1 \pmod{2^r}$. By Exercise 1.2.7 these primes are all distinct.

Exercise 1.4.14 Suppose p is an odd prime such that $2p + 1 = q$ is also prime. Show that the equation

$$x^p + 2y^p + 5z^p = 0$$

has no solutions in integers.

Solution. Consider the equation mod q. Then $x^p \equiv \pm 1$ or $0 \pmod{q}$.

Exercise 1.4.15 If x and y are coprime integers, show that if

$$(x+y) \text{ and } \frac{x^p + y^p}{x+y}$$

have a common prime factor, it must be p.

Solution. Suppose a prime q is a common factor. Then

$$0 \equiv \frac{x^p + y^p}{x+y} \equiv x^{p-1} - x^{p-2}y + \cdots + y^{p-1} \equiv px^{p-1} \pmod{q}.$$

Exercise 1.4.16 (Sophie Germain's trick) Let p be a prime such that $2p + 1 = q > 3$ is also prime. Show that

$$x^p + y^p + z^p = 0$$

has no integral solutions with $p \nmid xyz$.

Solution. By the previous exercise, $x + y = a^p$ and $(x^p + y^p)/(x + y) = c^p$ for some integers a and c. By symmetry, $y + z = b^p$, $x + z = d^p$. If $q \nmid xyz$, then $x^p + y^p + z^p \equiv 0 \pmod{q}$ is impossible since $\pm 1 \pm 1 \pm 1 \equiv 0 \pmod{q}$ is impossible. Now suppose $q \mid xyz$. If $q \mid x$, then $q \nmid yz$ so that $2x + y + z \equiv b^p \equiv a^p + d^p \pmod{q}$ which again is impossible if a, b, and d are coprime to q. Thus one of these must be divisible by q. It is easy to see that this must be b. Thus, $y + z \equiv 0 \pmod{q}$. Since

$$\frac{y^p + z^p}{y + z}$$

is also a pth power, t^p (say), we obtain the congruence

$$t^p \equiv py^{p-1} \pmod{q}.$$

Since q does not divide t, we deduce that

$$py^{p-1} \equiv \pm 1 \pmod{q}.$$

Also, $x + y = a^p$ implies $y \equiv a^p \pmod{q}$ so that y is a pth power mod q which is coprime to q. Thus, $p \equiv \pm 1 \pmod{q}$, a contradiction.

Exercise 1.4.17 Assuming ABC, show that there are only finitely many consecutive cubefull numbers.

Solution. If $n-1$ and n are cubefull, then apply ABC to $n-(n-1)=1$.

Exercise 1.4.18 Show that

$$\sum_p \frac{1}{p} = +\infty,$$

where the summation is over prime numbers.

Solution. Clearly,

$$\sum_{n\le x}\frac{1}{n} \le \prod_{p\le x}\left(1-\frac{1}{p}\right)^{-1} = \prod_{p\le x}\left(1+\frac{1}{p}+\frac{1}{p^2}+\cdots\right)$$

since every natural number $n \le x$ can be written as a product of primes $p \le x$. Now take logs. Then

$$\sum_{p\le x}\frac{1}{p} + O(1) \ge \log\left(\sum_{n\le x}\frac{1}{n}\right).$$

Since the harmonic series diverges, the result follows.

Exercise 1.4.19 (Bertrand's Postulate) (a) If $a_0 \ge a_1 \ge a_2 \ge \cdots$ is a decreasing sequence of real numbers tending to 0, show that

$$\sum_{n=0}^{\infty}(-1)^n a_n \le a_0 - a_1 + a_2.$$

(b) Let $T(x) = \sum_{n\le x}\psi(x/n)$, where $\psi(x)$ is defined as in Exercise 1.1.25. Show that

$$T(x) = x\log x - x + O(\log x).$$

(c) Show that

$$T(x) - 2T\left(\frac{x}{2}\right) = \sum_{n\le x}(-1)^{n-1}\psi\left(\frac{x}{n}\right) = (\log 2)x + O(\log x).$$

Deduce that

$$\psi(x) - \psi\left(\frac{x}{2}\right) \ge \tfrac{1}{3}(\log 2)x + O(\log x).$$

Solution. From

$$\sum_{n=0}^{\infty}(-1)^n a_n = a_0 - (a_1 - a_2) - (a_3 - a_4) - \cdots,$$

(a) is immediate.

To see (b), observe that

$$\sum_{n\leq x}\log n = \sum_{n\leq x}\left(\sum_{p^\alpha|n}\log p\right) = \sum_{m\leq x}\psi\left(\frac{x}{m}\right) = T(x).$$

By comparing areas,

$$\sum_{n\leq x}\log n = \int_1^x (\log t)\, dt + O(\log x)$$

implies (b).

The first part of (c) is now clear. Since $\psi(x/n)$ is a decreasing function of n, we apply (a) to get

$$\psi(x) - \psi\left(\frac{x}{2}\right) + \psi\left(\frac{x}{3}\right) \geq (\log 2)x + O(\log x).$$

By Exercise 1.1.14, $\psi(x) \leq 2x\log 2$. Therefore,

$$\psi(x) - \psi\left(\frac{x}{2}\right) \geq \tfrac{1}{3}(\log 2)x + O(\log x).$$

Hence, there is a prime between $x/2$ and x for x sufficiently large.

(This simple proof is due to S. Ramanujan. We can deduce $\psi(x) \leq 2x\log 2$ directly from (a) and (b) without using the solution to Exercise 1.1.26.)

Chapter 2

Euclidean Rings

2.1 Preliminaries

Exercise 2.1.2 Let D be squarefree. Consider $R = \mathbb{Z}[\sqrt{D}]$. Show that every element of R can be written as a product of irreducible elements.

Solution. We define a map $n : R \to \mathbb{N}$ such that for $a + b\sqrt{D} \in R$,

$$n(a + b\sqrt{D}) = |a^2 - Db^2|.$$

We must check that this map satisfies conditions (i) and (ii) from the previous example.

(i) For $a + b\sqrt{D}, c + d\sqrt{D} \in R$,

$$\begin{aligned} n[(a + b\sqrt{D})(c + d\sqrt{D})] &= n[(ac + bdD) + (ad + bc)\sqrt{D}] \\ &= |(ac + bdD)^2 - (ad + bc)^2 D| \\ &= |(a^2 - b^2 D)(c^2 - d^2 D)| \\ &= n(a + b\sqrt{D})n(c + d\sqrt{D}), \end{aligned}$$

so condition (i) is satisfied.

(ii) If $r = a + b\sqrt{D}$ is a unit in R, then $\exists s = c + d\sqrt{D} \in R$ such that $rs = 1$. But by condition (i), since $1 = n(1)$, $1 = n(r)n(s)$. Since our map n only takes on values in the positive integers, then $n(r) = n(s) = 1$ for all units of R. The converse is clear.

Since we have found a map n which satisfies the conditions of Example 2.1.1, we can deduce that every element of R can be written as a product of irreducible elements.

Exercise 2.1.3 Let $R = \mathbb{Z}[\sqrt{-5}]$. Show that $2, 3, 1 + \sqrt{-5}$, and $1 - \sqrt{-5}$ are irreducible in R, and that they are not associates.

Solution. We define a map $n : R \to \mathbb{N}$ such that $n(a + b\sqrt{-5}) = a^2 + 5b^2$.

If 2 is not irreducible, then there are elements $r, s \in R$ such that $rs = 2$, with r, s not units. But then $n(r)n(s) = n(2) = 4$, and since r, s are not units, it must be that $n(r) = n(s) = 2$. Then we must find integers a, b such that $a^2 + 5b^2 = 2$, which is clearly impossible, so 2 must be irreducible.

If 3 is not irreducible then we can find $r, s \in R$ with $rs = 3$, and r, s not units. Since $n(3) = 9$, we must have $n(r) = n(s) = 3$. But by the same argument as above, we see that this is impossible.

$n(1 + \sqrt{-5}) = 6$. The only proper divisors of 6 are 2 and 3, and so if $1 + \sqrt{-5}$ is not irreducible, then we can find $r \in R$, r not a unit and $r \mid (1 + \sqrt{-5})$ with either $n(r) = 2$ or $n(r) = 3$. But we showed above that this is not possible, so $1 + \sqrt{-5}$ is irreducible. Since $n(1 - \sqrt{-5}) = 6$, then $1 - \sqrt{-5}$ must also be irreducible.

If two elements of R are associates, then they must have the same norm, a fact which follows immediately from the condition that all units have norm 1. If $a + b\sqrt{-5}$ is a unit, then $a^2 + 5b^2 = 1$. This will only occur when $a = \pm 1$, and so the only units of $\mathbb{Z}[\sqrt{-5}]$ are 1 and -1. Of $2, 3, 1 \pm \sqrt{-5}$, we see that the only two which could possibly be associates are $1 + \sqrt{-5}$ and $1 - \sqrt{-5}$ because they have the same norm. However, if we multiply $1 + \sqrt{-5}$ by either of the units of $\mathbb{Z}[\sqrt{-5}]$, we will not get $1 - \sqrt{-5}$, and so they cannot be associates.

We conclude that $2, 3, 1 + \sqrt{-5}$ and $1 - \sqrt{-5}$ are all irreducible and are not associates.

Exercise 2.1.4 Let R be a domain satisfying (i) above. Show that (ii) is equivalent to (ii*): if π is irreducible and π divides ab, then $\pi \mid a$ or $\pi \mid b$.

Solution. Suppose R satisfies both (i) and (ii) above. Let $\pi \in R$ be an irreducible element, and suppose that $\pi \mid ab$, where

$$a = \tau_1 \tau_2 \cdots \tau_r,$$

$$b = \gamma_1 \gamma_2 \cdots \gamma_s,$$

and τ_i, γ_j are irreducible.

We know that $\pi \mid ab = \tau_1 \cdots \tau_r \gamma_1 \cdots \gamma_s$, so it follows that $ab = \pi \lambda_1 \cdots \lambda_n$ where each λ_i is irreducible. By condition (ii), $\pi \sim \tau_i$ for some i, or $\pi \sim \gamma_j$ for some j. Thus, if $\pi \mid ab$, then $\pi \mid a$ or $\pi \mid b$.

Now suppose that R is a domain satisfying conditions (i) and (ii*) above, and suppose that we have an element a which has two different factorizations into irreducibles: $a = \tau_1 \cdots \tau_r$ and $a = \pi_1 \cdots \pi_s$. Consider τ_1. We know that $\tau_1 \mid a$, and so $\tau_1 \mid \pi_1 \cdots \pi_s$. By (ii*) we know that $\tau_1 \mid \pi_i$ for some i, and since both are irreducible, they must be associates.

We can now remove both τ_1 and π_i from our factorization of a. We next consider τ_2. Following the same process, we can pair up τ_2 with its associate, and we can continue to do this until we have paired up each of the irreducible factors τ_i with its associate π_j. It is clear that if we continue in this fashion, we must have $r = s$.

Exercise 2.1.5 Show that if π is an irreducible element of a principal ideal domain, then (π) is a maximal ideal (where (x) denotes the ideal generated by the element x).

Solution. We define $\gcd(a, b)$, the greatest common divisor of $a, b \in R$, to be an element d such that the ideal (a, b) equals the ideal (d). It is unique up to units. For a unique factorization domain, this definition coincides with the usual one. We note that d must divide both a and b since they are in the ideal (d).

If π is irreducible, we consider the ideal (π, α) where α is any element not in (π). Since α is not a multiple of π, and π is irreducible, then the only common divisors of α and π will be units. Then $\gcd(\pi, \alpha) = 1$. In other words, the ideal generated by π is a maximal ideal.

Exercise 2.1.8 If F is a field, prove that $F[x]$, the ring of polynomials in x with coefficients in F, is Euclidean.

Solution. We define a map $\phi : F[x] \to \mathbb{N}$ such that for $f \in F[x]$, $\phi(f) = \deg(f)$. Now consider any two polynomials $f(x), g(x) \in F[x]$.

If $\deg(g) > \deg(f)$, then we can certainly write $f(x) = 0 \cdot g(x) + f(x)$, which satisfies the Euclidean condition. Then we can assume that $m = \deg(g) \leq \deg(f) = n$, and write $f(x) = a_0 + a_1 x + \cdots + a_n x^n$, and $g(x) = b_0 + b_1 x + b_2 x^2 + \cdots + b_m x^m$, where $a_n, b_m \neq 0$ and $m \leq n$. We proceed by induction on the degree of f. That is, we will prove by induction on the degree of f that we can write $f(x) = q(x)g(x) + r(x)$, where $r = 0$ or $\deg(r) < \deg(g)$.

Define a new polynomial

$$h(x) = a_n b_m^{-1} g(x) x^{n-m}.$$

Observe that the leading term of $h(x)$ is $a_n b_m^{-1} b_m x^m x^{n-m} = a_n x^n$ which is the leading term of $f(x)$, so that if $f_1(x) = f(x) - h(x)$, either $f_1(x) = 0$ or $\deg(f_1) < \deg(f)$. The theorem holds for $f_1(x)$, so we can write $f_1(x) = f(x) - h(x) = q(x)g(x) + r(x)$, where $r = 0$ or $\deg(r) < \deg(g) = m$. Now $f(x) = q(x)g(x) + h(x) + r(x)$ and since $h(x)$ is a multiple of $g(x)$, the result follows.

2.2 Gaussian Integers

Exercise 2.2.1 Show that $\mathbb{Z}[i]$ is Euclidean.

Solution. We define a map $\phi : \mathbb{Z}[i] \to \mathbb{N}$ such that $\phi(a + bi) = a^2 + b^2$. Now, given any two elements of $\mathbb{Z}[i]$, say $\alpha = a + bi$ and $\gamma = c + di$, can we find $q, r \in \mathbb{Z}$ such that $a + bi = q(c + di) + r$, where $r = 0$ or $\phi(r) < \phi(c + di)$?

Since we cannot divide α and γ in the ring $\mathbb{Z}[i]$, we move temporarily to the ring $\mathbb{Q}[i] = \{r + si \mid r, s \in \mathbb{Q}\}$. In this ring,

$$\begin{aligned}
\frac{\alpha}{\gamma} &= \frac{(a+bi)}{(c+di)} \\
&= \frac{(ac+bd)}{(c^2+d^2)} + \frac{(bc-ad)}{(c^2+d^2)}i \\
&= r + si,
\end{aligned}$$

with $r, s \in \mathbb{Q}$. We can now choose $m, n \in \mathbb{Z}$ such that $|r - m| \leq 1/2$, and $|s - n| \leq 1/2$. We set $q = m + ni$. Then $q \in \mathbb{Z}[i]$, and $\alpha = q\gamma + r$ for some suitable r, with

$$\begin{aligned}
\phi(r) &= \phi(\alpha - q\gamma) \\
&= \phi(\alpha/\gamma - q)\phi(\gamma) \\
&= [(r-m)^2 + (s-n)^2]\phi(\gamma) \\
&\leq \left(\tfrac{1}{4} + \tfrac{1}{4}\right)\phi(\gamma) \\
&= \tfrac{1}{2}\phi(\gamma) \\
&< \phi(\gamma).
\end{aligned}$$

We have shown that our map ϕ satisfies the properties specified above, and so $\mathbb{Z}[i]$ is Euclidean.

Exercise 2.2.2 Prove that if p is a positive prime, then there is an element $x \in \mathbb{F}_p := \mathbb{Z}/p\mathbb{Z}$ such that $x^2 \equiv -1 \pmod{p}$ if and only if either $p = 2$ or $p \equiv 1 \pmod{4}$. (Hint: Use Wilson's theorem, Exercise 1.4.10.)

Solution. If $p = 2$, then $1 \equiv -1 \pmod{2}$, so $1^2 = 1 \equiv -1 \pmod{2}$. Hence we can take $x = 1$. Conversely, if $1 \equiv -1 \pmod{p}$, we can see that $p = 2$ since $1 = ap - 1$ for some integer a which implies $ap = 2$.

We will show that in any field $\mathbb{F}_p$ where 1 is not congruent to $-1 \pmod{p}$, $x^2 \equiv -1 \pmod{p}$ for an element x if and only if x has order 4 in the group of units of the field. Suppose that $x^2 \equiv -1 \pmod{p}$. Then the first four powers of x are $x, -1, -x, 1$. Hence x has order 4.

Conversely, suppose that x has order 4. Then, $x^4 = (x^2)^2 \equiv 1 \pmod{p}$, so $(x^2)^2 - 1 \equiv 0 \pmod{p}$. Hence $(x^2+1)(x^2-1) \equiv 0 \pmod{p}$. Since x is an element of a field, $x^2 + 1 \equiv 0 \pmod{p}$ or $x^2 - 1 \equiv 0 \pmod{p}$. However, if $x^2 - 1 \equiv 0 \pmod{p}$, x has order 2. Hence $x^2 \equiv -1 \pmod{p}$.

If $p \neq 2$, then $\mathbb{F}_p$ is a field where 1 is not congruent to $-1 \pmod{p}$. Hence the existence of an element x such that $x^2 \equiv -1 \pmod{p}$ is equivalent to the existence of an element of order 4 in the group of units of $\mathbb{F}_p$. Let U_p be the group of units of $\mathbb{F}_p$. Then $|U_p| = p - 1$, and since the order of any element divides the order of the group, if U_p has an element of order 4, then we have $4 \mid p - 1$ or, equivalently, $p \equiv 1 \pmod{4}$. Conversely, if we suppose that $p \equiv 1 \pmod 4$, then we can write $p = 4k + 1$ where k is an integer and U_p is a cyclic group of order $p - 1 = 4k$. If g is a generator of U_p, then g has order $4k$. So g^k has order 4. Hence we can see that if 1 is not congruent to -1 mod p, $x^2 \equiv -1 \pmod p$ occurs if and only if x has order 4 in the group of units of the field, which occurs if and only if $p \equiv 1 \pmod 4$.

Alternate Solution: Wilson's theorem gives

$$(p-1)! \equiv -1 \pmod p.$$

We can pair up k and $(p - k)$ in the product above so that

$$k(p-k) \equiv -k^2 \pmod p$$

implies

$$(-1)^{(p-1)/2} \left(\frac{p-1}{2}\right)!^{\,2} \equiv -1 \pmod p.$$

Thus, if $p \equiv 1 \pmod 4$, there is an $x \in \mathbb{F}_p$ so that $x^2 \equiv -1 \pmod p$. The converse follows from Fermat's little Theorem:

$$1 \equiv (x^2)^{(p-1)/2} \equiv (-1)^{(p-1)/2} \pmod p$$

so that $(p-1)/2$ is even. That is, $p \equiv 1 \pmod 4$.

We will provide another alternative proof of this fact in Chapter 7, using quadratic residues.

Exercise 2.2.3 Find all integer solutions to $y^2 + 1 = x^3$ with $x, y \neq 0$.

Solution. If x is even, then $x^3 \equiv 0 \pmod 8$, which implies in turn that $y^2 \equiv 7 \pmod 8$. However, if $y \equiv 1, 3, 5, 7 \pmod 8$, then $y^2 \equiv 1 \pmod 8$. So x must be odd, and y even.

We can factor this equation in the ring $\mathbb{Z}[i]$ to obtain $(y+i)(y-i) = x^3$. If $\exists \delta$ such that $\delta \mid (y+i)$ and $\delta \mid (y-i)$, then $\delta \mid 2i$, which implies that $\delta \mid 2$. But this would mean that x is divisible by 2, which we know is not true. Therefore, we know that $(y+i)$ and $(y-i)$ are relatively prime in $\mathbb{Z}[i]$, and that they must both be cubes.

We know that we can write $y + i = e_1(a + bi)^3$ and $y - i = e_2(c + di)^3$ where $a, b, c, d \in \mathbb{Z}$ and e_1, e_2 are units in $\mathbb{Z}[i]$. However, the only units of $\mathbb{Z}[i]$ are ± 1 and $\pm i$, and these are all cubes, so without loss, assume that $e_1 = e_2 = 1$.

Next, we expand our factorization for $y + i$ to get

$$y + i = a^3 + 3a^2bi + 3ab^2 - b^3 i.$$

Comparing the imaginary parts, we get $1 = 3a^2b - b^3 = b(3a^2 - b^2)$, with $a, b \in \mathbb{Z}$. The only integers which multiply together to give 1 are ± 1, so we know that $b = \pm 1$. If $b = 1$, then we have $1 = 3a^2 - 1$, implying $3a^2 = 2$, which has no integer solutions, so $b \neq 1$. If $b = -1$, then we get $1 = -3a^2 + 1$, and so $3a^2 = 0$, and a must be 0. However, if we substitute $a = 0$ back into our original equation for $y + i$, we find that $y = 0$, which we did not allow, and so $b \neq -1$.

We conclude that the equation $y^2 + 1 = x^3$ has no integer solutions with $x, y \neq 0$.

Exercise 2.2.4 If π is an element of R such that when $\pi \mid ab$ with $a, b \in R$, then $\pi \mid a$ or $\pi \mid b$, then we say that π is prime. What are the primes of $\mathbb{Z}[i]$?

Solution. Given $\pi = (a+bi) \in \mathbb{Z}[i]$, we define the complex conjugate of π to be the element $\overline{\pi} = (a-bi)$. We note that $n(\pi) = a^2+b^2 = \pi\overline{\pi}$, and so given any prime π in $\mathbb{Z}[i]$, we know that π divides $n(\pi)$. Using this information, we observe that we can find all the Gaussian primes by examining the irreducible factors of the primes of $\mathbb{Z}$. For, let $n(\pi) = p_1 p_2 \cdots p_k$ be the prime decomposition of $n(\pi)$. We know that $\pi \mid n(\pi)$, so $\pi \mid p_i$ for some i. If $\pi \mid p_i$ and $\pi \mid p_j$, with $p_i \neq p_j$, then $\pi \mid 1$, since $\gcd(p_i, p_j) = 1$. But then π would be a unit, and thus not irreducible. So, by examining all of the divisors of the primes in $\mathbb{Z}$, we will discover all of the primes of $\mathbb{Z}[i]$, once and only once each.

We let π be a prime in $\mathbb{Z}[i]$, and p the prime in $\mathbb{Z}$ such that $\pi \mid p$. By the properties of the map n, $n(\pi) \mid n(p) = p^2$, so $n(\pi) = p$ or $n(\pi) = p^2$. If we let $\pi = a + bi$, then $a^2 + b^2 = p$, or $a^2 + b^2 = p^2$.

All the primes of $\mathbb{Z}$ are congruent to 1, 2 or 3 (mod p), and we will examine these cases separately.

Case 1. $p \equiv 3 \pmod 4$.

We just proved that if $\pi = a + bi$ is prime, then either $a^2 + b^2 = p$, or $a^2 + b^2 = p^2$ for some integer prime p. Let us assume that the first of these possibilities is true. We know that p is odd, so one of a, b is even. Let us say that a is even, and b odd, so that $a = 2x$ and $b = 2y + 1$ for some $x, y \in \mathbb{Z}$. Then

$$\begin{aligned} a^2 + b^2 &= 4x^2 + 4y^2 + 4y + 1 \\ &= 4(x^2 + y^2 + y) + 1 \\ &\equiv 1 \pmod 4. \end{aligned}$$

Since we had assumed that $p \equiv 3 \pmod 4$, we have a contradiction. So $a^2 + b^2 = p^2$, which means that $n(\pi) = n(p)$, and so p and π must be associates.

Therefore, primes in $\mathbb{Z}$ that are congruent to 3 (mod 4) and their associates are prime in the ring $\mathbb{Z}[i]$.

Case 2. $p \equiv 2 \pmod 4$.

There is, of course, only one such integer prime: 2. Assume we have a prime π which divides 2. Since $2 = (1 + i)(1 - i)$, then $\pi \mid (1 + i)$ or

$\pi \mid (1-i)$. But $n(1+i) = n(1-i) = 2$, and it is easy to show that $(1+i)$ and $(1-i)$ are irreducible in $\mathbb{Z}[i]$ and so they are prime. So, if $\pi \mid 2$, then $\pi \sim (1+i)$ or $\pi \sim (1-i)$.

Case 3. $p \equiv 1 \pmod 4$.

We recall Wilson's Theorem, Exercise 1.4.10, which states that if p is a prime, then $(p-1)! \equiv -1 \pmod p$. We will in fact be using a corollary of this theorem, which states that if p is a prime number of the form $1+4m$, then $p \mid (n^2+1)$, where $(2m)! = n$. (We can also apply Exercise 2.2.2.)

If $p \mid (n^2+1) = (n+i)(n-i)$ and $\pi \mid p$, then $\pi \mid (n+i)$ or $\pi \mid (n-i)$. If p were to divide $(n \pm i)$, then $p \mid n$ and $p \mid 1$, which is clearly not the case since p is not a unit. We conclude that p and π are not associates, so $n(\pi) \neq n(p)$, which implies that $n(\pi) = a^2 + b^2 = p$. Thus, if $p \equiv 1 \pmod 4$, then p does not remain prime in $\mathbb{Z}[i]$. We can deduce that if $\pi = a \pm bi$ and $a^2 + b^2 = p$, then π is prime in $\mathbb{Z}[i]$.

Exercise 2.2.5 A positive integer a is the sum of two squares if and only if $a = b^2c$ where c is not divisible by any positive prime $p \equiv 3 \pmod 4$.

Solution. Suppose that a is the sum of two squares. Let $a = s^2 + t^2$ and let $(s,t) = b$. Then $a = (bx)^2 + (by)^2 = b^2(x^2+y^2)$ where $(x,y) = 1$. Let $c = x^2 + y^2$. Then we have $a = b^2c$ where c is the sum of two relatively prime squares.

By Exercise 2.2.2, c is not divisible by any prime $p \equiv 3 \pmod 4$. In fact, suppose that $p \mid x^2 + y^2$. Then, $x^2 + y^2 \equiv 0 \pmod p$, $x^2 \equiv -y^2 \pmod p$ and so $((y^{-1})^2 \cdot x^2) = (y^{-1} \cdot x)^2 \equiv -1 \pmod p$. By Exercise 2.2.2, either $p = 2$ or $p \equiv 1 \pmod 4$. Hence c is not divisible by any prime $p \equiv 3 \pmod 4$.

Now suppose that we have an integer a which we can write as $a = b^2c$, and suppose that c is not divisible by any positive prime $p \equiv 3 \pmod 4$. Then c is a product of primes each of which, by Exercise 2.2.4, is a sum of two squares. Then $b^2 = n(b)$ and $c = n(t+ri)$ where t and r are integers.

Then

$$\begin{aligned} b^2 \cdot c &= n(b(t+ri)) \\ &= n(bt + bri) \\ &= b^2 \cdot t^2 + b^2 \cdot r^2 \\ &= (bt)^2 + (br)^2. \end{aligned}$$

Hence $a = b^2c$ is written as the sum of two squares.

2.3 Eisenstein Integers

Exercise 2.3.1 Show that $\mathbb{Z}[\rho]$ is a ring.

Solution. First observe that $\mathbb{Z}[\rho]$ is a subset of the complex numbers, so associativity, distributivity, and commutativity are immediate for addition and multiplication. Also, $0, 1 \in \mathbb{Z}[\rho]$, so we have additive and multiplicative identities. If $a + b\rho \in \mathbb{Z}[\rho]$, then $-a - b\rho \in \mathbb{Z}[\rho]$, so we have additive inverses. It remains to verify closure under addition and multiplication; if $a, b, c, d \in \mathbb{Z}$, then $(a+b\rho)+(c+d\rho) = (a+c)+(b+d)\rho \in \mathbb{Z}[\rho]$, so we have closure under addition. Also $(a+b\rho)(c+d\rho) = ac+(ad+bc)\rho+bd\rho^2$. We will therefore have closure under multiplication if $\rho^2 \in \mathbb{Z}[\rho]$. But $\rho^2 = -1 - \rho$, so $\mathbb{Z}[\rho]$ is a commutative ring with unit.

Exercise 2.3.2 (a) Show that $\mathbb{Z}[\rho]$ is Euclidean.
(b) Show that the only units in $\mathbb{Z}[\rho]$ are ± 1, $\pm\rho$, and $\pm\rho^2$.

Solution. (a) Define $\phi : \mathbb{Z}[\rho] \to \mathbb{N}$ so that $\phi(a + b\rho) = a^2 - ab + b^2 = \alpha\overline{\alpha}$ for $\alpha \in \mathbb{Z}[\rho]$. We consider $\alpha, \beta \in \mathbb{Z}[\rho], \beta \neq 0$. We have

$$\frac{\alpha}{\beta} = \frac{\alpha\overline{\beta}}{\beta\overline{\beta}}.$$

Now $\beta\overline{\beta} \in \mathbb{Z}$ and $\alpha\overline{\beta} \in \mathbb{Z}[\rho]$, so

$$\frac{\alpha\overline{\beta}}{\beta\overline{\beta}} = s + t\rho$$

for some $s, t \in \mathbb{Q}$. We set m and n to be the integers closest to s and t, respectively, i.e., choose m and n so $|m - s| \leq 1/2$ and $|n - t| \leq 1/2$. We set $q = m + n\rho$. Now,

$$\begin{aligned} \phi\left(\frac{\alpha}{\beta} - q\right) &= (s-m)^2 - (s-m)(t-n) + (t-n)^2 \\ &\leq \tfrac{1}{4} + \tfrac{1}{4} + \tfrac{1}{4} \\ &< 1. \end{aligned}$$

So, writing $r = \alpha - q\beta$, then if $r \neq 0$, $\phi(r) = \phi(\beta)\phi(\alpha/\beta - q) < \phi(\beta)$, and with the map ϕ, for any $\alpha, \beta \in \mathbb{Z}[\rho]$, we can write $\alpha = q\beta + r$ where $r = 0$ or $\phi(r) < \phi(\beta)$. Thus, $\mathbb{Z}[\rho]$ is Euclidean.

(b) Observe that ϕ is a multiplicative map into the natural numbers, so that if η is a unit of $\mathbb{Z}[\rho]$, then $\phi(\eta) = 1$. We thus see immediately that $\pm 1, \pm\rho, \pm\rho^2$ are all units (it is easy to see that they are distinct). Suppose $\eta = a+b\rho$ were a unit of $\mathbb{Z}[\rho]$. Then $a^2 - ab + b^2 = 1$ and $(2a-b)^2 + 3b^2 = 4$. From this equation it is clear that $b = 0$ or ± 1 are the only possible integer values b could take, since $3b^2$ must be less than 4. To each solution for b there are two corresponding solutions for a, and thus at most six distinct pairs (a, b) in total. By the pigeonhole principle the list given above includes all the units of $\mathbb{Z}[\rho]$.

Exercise 2.3.3 Let $\lambda = 1 - \rho$. Show that λ is irreducible, so we have a factorization of 3 (unique up to unit).

Solution. We have that $\phi(\lambda) = 3$. If $d \mid \lambda$, then $\phi(d) \mid 3$, i.e., $\phi(d) = 1$ or 3, so d is either a unit or an associate of λ, and λ is irreducible.

Exercise 2.3.4 Show that $\mathbb{Z}[\rho]/(\lambda)$ has order 3.

Solution. Suppose $\alpha \in \mathbb{Z}[\rho]$, so that $\alpha = a + b\rho$ for integers a, b. Then $\alpha = a + b - b(1-\rho) = a + b - b\lambda \equiv a + b \pmod{\lambda}$. Considered mod 3, $a + b$ could have residues 0, 1, or 2. Since $\lambda \mid 3$ (see Exercise 2.3.3, above), then α will have one of these residues mod λ. Since $\phi(\lambda)$ does not divide $\phi(1) = 1$, or $\phi(2) = 4$, none of these classes are equivalent mod λ, and so we have three distinct residue classes, which we may denote by 0 and ± 1.

2.4 Some Further Examples

Exercise 2.4.2 Show that $\mathbb{Z}[\sqrt{-2}]$ is Euclidean.

Solution. We define a norm $\phi : \mathbb{Z}[\sqrt{-2}] \to \mathbb{N}$ by $\phi(a + b\sqrt{-2}) = a^2 + 2b^2$. For α, $\beta \in \mathbb{Z}[\sqrt{-2}]$, we consider $\alpha/\beta = \alpha\overline{\beta}/\beta\overline{\beta}$. Notice that $\beta\overline{\beta} = \phi(\beta)$ so $\beta\overline{\beta} \in \mathbb{Z}$. Also, $\overline{\beta} \in \mathbb{Z}[\sqrt{-2}]$, so $\alpha\overline{\beta} \in \mathbb{Z}[\sqrt{-2}]$, so $\alpha/\beta = \alpha\overline{\beta}/\beta\overline{\beta} = c + d\sqrt{-2}$ for some c, $d \in \mathbb{Q}$. We choose m and n as the closest integers to c and d, i.e., so that $|m - c| \leq 1/2$ and $|n - d| \leq 1/2$. We write $q = m + n\sqrt{-2}$. We have that $\phi(\alpha/\beta - q) = (c - m)^2 + 2(d - n)^2 \leq 1/4 + 1/2 < 1$. So we write $\alpha = q\beta + r$ and $r = \alpha - q\beta$. If $r \neq 0$, then $\phi(r) = \phi(\beta)\phi(\alpha/\beta - q) < \phi(\beta)$. We conclude that $\mathbb{Z}[\sqrt{-2}]$ is Euclidean.

Exercise 2.4.3 Solve $y^2 + 2 = x^3$ for $x, y \in \mathbb{Z}$.

Solution. Write $(y + \sqrt{-2})(y - \sqrt{-2}) = x^3$. If y were even, then x would be also, but if x is even, then $x^3 \equiv 0 \pmod 8$ whereas 8 does not divide $y^2 + 2$. So y and x are both odd. Observe that $(y + \sqrt{-2})$ and $(y - \sqrt{-2})$ are relatively prime, since if d divided both, then d would divide $2\sqrt{-2}$ and would thus have even norm, which is not possible since y is odd. Thus $(y + \sqrt{-2})$ is a cube multiplied by a unit. The only units of $\mathbb{Z}[\sqrt{-2}]$ are 1 and -1, which are both cubes. Without loss, assume that the unit in question is 1. We write

$$\begin{aligned}(y + \sqrt{-2}) &= (a + b\sqrt{-2})^3 \\ &= a^3 - 6ab^2 + (3a^2b - 2b^3)\sqrt{-2}.\end{aligned}$$

Comparing real and imaginary parts, we find that

$$y = a^3 - 6ab^2$$

and

$$\begin{aligned} 1 &= (3a^2b - 2b^3) \\ &= b(3a^2 - 2b^2). \end{aligned}$$

Thus, $b \mid 1$ so $b = \pm 1$. It follows that $a = \pm 1$. Substituting into the equation for y, we find that $y = \pm 5$. Thus, the only solution to the given equation is $x = 3, y = \pm 5$.

Exercise 2.4.5 Show that $\mathbb{Z}[\sqrt{2}]$ is Euclidean.

Solution. We define a norm $\phi : \mathbb{Z}[\sqrt{2}] \to \mathbb{N}$ by $\phi(a + b\sqrt{2}) = |a^2 - 2b^2|$. Let $\alpha, \beta \in \mathbb{Z}[\sqrt{2}]$. We write $\beta = c + d\sqrt{2}$ and consider

$$\frac{\alpha}{\beta} = \frac{\alpha(c - d\sqrt{2})}{\beta(c - d\sqrt{2})}.$$

Notice that $|\beta(c-d\sqrt{2})| = \phi(\beta)$ so $\beta(c-d\sqrt{2}) \in \mathbb{Z}$. Also, $(c-d\sqrt{2}) \in \mathbb{Z}[\sqrt{2}]$, so $\alpha(c - d\sqrt{2}) \in \mathbb{Z}[\sqrt{2}]$, so

$$\frac{\alpha}{\beta} = \frac{\alpha(c - d\sqrt{2})}{\beta(c - d\sqrt{2})} = t + u\sqrt{2}$$

for some t, $u \in \mathbb{Q}$. We choose m and n as the closest integers to t and u, i.e. so that $|m - t| \leq 1/2$ and $|n - u| \leq 1/2$. We write $q = m + n\sqrt{2}$. We have that

$$\begin{aligned} \phi(\alpha/\beta - q) &= \mid (t - m)^2 - 2(u - n)^2 \mid \\ &\leq \mid (t - m)^2 \mid + \mid 2(u - n)^2 \mid \\ &\leq \tfrac{1}{4} + \tfrac{1}{2} \\ &< 1. \end{aligned}$$

We write $\alpha = q\beta + r$, so $r = \alpha - q\beta$. If $r \neq 0$, then $\phi(r) = \phi(\beta)\phi(\alpha/\beta - q) < \phi(\beta)$. We conclude that $\mathbb{Z}[\sqrt{2}]$ is Euclidean.

Exercise 2.4.6 Let $\varepsilon = 1+\sqrt{2}$. Write $\varepsilon^n = u_n + v_n\sqrt{2}$. Show that $u_n^2 - 2v_n^2 = \pm 1$.

Solution. Since ϕ is multiplicative and we have $\phi(\varepsilon) = | - 1|$, then

$$\phi(\varepsilon^n) = |(-1)^n| = |u_n^2 - 2v_n^2| = 1.$$

This gives infinitely many solutions to $x^2 - 2y^2 = \pm 1$. It is easy to see that all of these solutions are distinct: $\varepsilon \in \mathbb{Z}[\sqrt{2}]$ and $\varepsilon > 1$ so $\varepsilon^{n+1} > \varepsilon^n$ for all positive n.

Exercise 2.4.7 Show that there is no unit η in $\mathbb{Z}[\sqrt{2}]$ such that $1 < \eta < 1 + \sqrt{2}$. Deduce that every unit (greater than zero) of $\mathbb{Z}[\sqrt{2}]$ is a power of $\varepsilon = 1 + \sqrt{2}$.

Solution. Since -1 is a unit, for any unit ξ, $-\xi$ is also a unit, and negative and positive units are in one-to-one correspondence; we shall only consider the positive units of $\mathbb{Z}[\sqrt{2}]$. We write η as $a + b\sqrt{2}$. Since η is a unit, $\phi(\eta) = (a + b\sqrt{2})(a - b\sqrt{2}) = \pm 1$. By assumption $(a + b\sqrt{2}) > 1$ and $|(a + b\sqrt{2})(a - b\sqrt{2})| = 1$, so it follows that

$$-1 < (a - b\sqrt{2}) < 1.$$

Also, by assumption, $1 < a + b\sqrt{2} < 1 + \sqrt{2}$. So, adding these two inequalities gives

$$0 < 2a < 2 + \sqrt{2}.$$

Since $a \in \mathbb{Z}$ this implies that $a = 1$. Notice now that there is no integer b such that

$$1 < 1 + b\sqrt{2} < 1 + \sqrt{2}.$$

If any unit, ψ, did exist which was not some power of ε, then by our Euclidean algorithm we would be able to divide by $(1 + \sqrt{2})^k$, where k is chosen so that $(1 + \sqrt{2})^k < \psi < (1 + \sqrt{2})^{k+1}$ and this would produce a new unit ψ' where $1 < \psi' < 1 + \sqrt{2}$. So the only positive units of $\mathbb{Z}[\sqrt{2}]$ are those of the form $(1 + \sqrt{2})^n$; there are infinitely many.

2.5 Supplementary Problems

Exercise 2.5.1 Show that $R = \mathbb{Z}[(1 + \sqrt{-7})/2]$ is Euclidean.

Solution. Given $\alpha, \beta \in R$, we want to find $\gamma, \delta \in R$ such that $\alpha = \beta\gamma + \delta$, with $N(\delta) < N(\beta)$. This is equivalent to showing that we can find a γ with $N(\alpha/\beta - \gamma) < 1$.

Now, $\alpha/\beta = x + y\sqrt{-7}$ with $x, y \in \mathbb{Q}$. Let $\gamma = (u + v\sqrt{-7})/2$ with $u, v \in \mathbb{Z}$ and $u \equiv v \pmod 2$. We want

$$N\left(x + y\sqrt{-7} - \left(\frac{u + v\sqrt{-7}}{2}\right)\right) = \left(x - \frac{u}{2}\right)^2 + 7\left(y - \frac{v}{2}\right)^2 < 1$$

or, equivalently,

$$(2x - u)^2 + 7(2y - v)^2 < 4.$$

First consider $2y$. Choose for v either $[2y]$ or $[2y] + 1$, so that $2y - v \leq 1/2$. Now choose for u either $[2x]$ or $[2x] + 1$, whichever has the same parity as v. Then $2x - u \leq 1$. Then

$$(2x - u)^2 + 7(2y - v)^2 \leq 1 + \tfrac{7}{4} = \tfrac{11}{4} < 4.$$

We have found a γ which works, and proved that $\mathbb{Z}[(1 + \sqrt{-7})/2]$ is Euclidean.

Exercise 2.5.2 Show that $\mathbb{Z}[(1 + \sqrt{-11})/2]$ is Euclidean.

Solution. Proceed as in Exercise 2.5.1. Given α, β, we wish to find γ such that $N(\alpha/\beta - \gamma) < 1$. Let $\alpha/\beta = x + y\sqrt{-11}, x, y \in \mathbb{Q}$, and $\gamma = (u + v\sqrt{-11})/2$ with $u, v \in \mathbb{Z}$ and $u \equiv v \pmod 2$. We want

$$N\left(x + y\sqrt{-11} - \left(\frac{u + v\sqrt{-11}}{2}\right)\right) < 1,$$

or

$$(2x - u)^2 + 11(2y - v)^2 < 4.$$

As in the previous exercise, choose v first to be the integer which is closest to $2y$, and then choose u to be the integer closest to $2x$ which also has the same parity as v. Then $(2x - u) \leq 1$ and $(2y - v) \leq 1/2$, so

$$(2x - u)^2 + 11(2y - v)^2 \leq 1 + \tfrac{11}{4} = \tfrac{15}{4} < 4.$$

Therefore $\mathbb{Z}[(1 + \sqrt{-11})/2]$ is Euclidean.

Exercise 2.5.3 Find all integer solutions to the equation $x^2 + 11 = y^3$.

Solution. In the ring $\mathbb{Z}[(1 + \sqrt{-11})/2]$, we can factor the equation as

$$(x - \sqrt{-11})(x + \sqrt{-11}) = y^3.$$

Now, suppose that $\delta \mid (x - \sqrt{-11})$ and $\delta \mid (x + \sqrt{-11})$ (which implies that $\delta \mid y$). Then $\delta \mid 2x$ and $\delta \mid 2\sqrt{-11}$ which means that $\delta \mid 2$ because otherwise, $\delta \mid \sqrt{-11}$, meaning that $11 \mid x$ and $11 \mid y$, which we can see is not true by considering congruences mod 11^2. Then $\delta = 1$ or 2, since 2 has no factorization in this ring. We will consider these cases separately.

Case 1. $\delta = 1$.

Then the two factors of y^3 are coprime and we can write

$$(x + \sqrt{-11}) = \varepsilon\left(\frac{a + b\sqrt{-11}}{2}\right)^3,$$

where $a, b \in \mathbb{Z}$ and $a \equiv b \pmod 2$. Since the units of $\mathbb{Z}[(1 + \sqrt{-11})/2]$ are ± 1, which are cubes, then we can bring the unit inside the brackets and rewrite the above without ε. We have

$$8(x + \sqrt{-11}) = (a + b\sqrt{-11})^3 = a^3 + 3ab^2\sqrt{-11} - 33ab^2 - 11b^3\sqrt{-11}$$

and so, comparing real and imaginary parts, we get

$$\begin{aligned} 8x &= a^3 - 33ab^2 = a(a^2 - 33b^2), \\ 8 &= 3a^2b - 11b^3 = b(3a^2 - 11b^2). \end{aligned}$$

This implies that $b \mid 8$ and so we have 8 possibilities: $b = \pm 1, \pm 2, \pm 4, \pm 8$. Substituting these back into the equations to find a, x, and y, and remembering that $a \equiv b \pmod 2$ and that $a, x, y \in \mathbb{Z}$ will give all solutions to the equation.

Case 2. $\delta = 2$.

If $\delta = 2$, then y is even and x is odd. We can write $y = 2y_1$, which gives the equation

$$\left(\frac{x+\sqrt{-11}}{2}\right)\left(\frac{x-\sqrt{-11}}{2}\right) = 2y_1^3.$$

Since 2 divides the right-hand side of this equation, it must divide the left-hand side, so

$$2 \left| \left(\frac{x+\sqrt{-11}}{2}\right)\right.$$

or

$$2 \left| \left(\frac{x-\sqrt{-11}}{2}\right)\right. .$$

However, since x is odd, 2 divides neither of the factors above. We conclude that $\delta \neq 2$, and thus we found all the solutions to the equation in our discussion of Case 1.

Exercise 2.5.4 Prove that $\mathbb{Z}[\sqrt{3}]$ is Euclidean.

Solution. Given $\alpha, \beta \in \mathbb{Z}[\sqrt{3}]$ we want to find $\gamma, \delta \in \mathbb{Z}[\sqrt{3}]$ such that $\alpha = \beta\gamma + \delta$, with $N(\delta) < N(\beta)$. Put another way, we want to show that $N(\alpha/\beta - \gamma) < 1$. Let $\alpha/\beta = x + y\sqrt{3}, x, y \in \mathbb{Q}$. Let $\gamma = u + v\sqrt{3}$, with $u, v \in \mathbb{Z}$.

Now, $N(\alpha/\beta - \gamma) = |(x-u)^2 - 3(y-v)^2|$. This will be maximized when $(x-u)$ is small and $(y-v)$ is large. Choose for u and v the closest integers to x and y, respectively. Then the minimum value for $(x-u)$ is 0, while the maximum value for $(y-v)$ is 1/2. Then $N(\alpha/\beta - \gamma) \leq |-3/4| < 1$. The conclusion follows.

Exercise 2.5.5 Prove that $\mathbb{Z}[\sqrt{6}]$ is Euclidean.

Solution. Assume that $\mathbb{Z}[\sqrt{6}]$ is not Euclidean. This means that there is at least one $x + y\sqrt{6} \in \mathbb{Q}(\sqrt{6})$ such that there is no $\gamma = u + v\sqrt{6} \in \mathbb{Z}[\sqrt{6}]$ such that $|(x-u)^2 - 6(y-v)^2| < 1$. Without loss, we can suppose that $0 \leq x \leq 1/2$, and $0 \leq y \leq 1/2$. We assert that there exist such a pair (x, y) such that

$$(x-u)^2 \geq 1 + 6(y-v)^2,$$

or

$$6(y-v)^2 \geq 1 + (x-u)^2,$$

for every $u, v \in \mathbb{Z}$. In particular, we will use the following inequalities:

$$\begin{aligned}
&\text{either (a)} & x^2 &\geq 1 + 6y^2 & &\text{or (b)} & 6y^2 &\geq 1 + x^2, & &(2.1)\\
&\text{either (a)} & (1-x)^2 &\geq 1 + 6y^2 & &\text{or (b)} & 6y^2 &\geq 1 + (1-x)^2, & &(2.2)\\
&\text{either (a)} & (1+x)^2 &\geq 1 + 6y^2 & &\text{or (b)} & 6y^2 &\geq 1 + (1+x)^2. & &(2.3)
\end{aligned}$$

If $x = y = 0$, then both first inequalities fail, so we can rule out this case. Next, we look at the first two inequalities on the left. Since $x^2, (1-x)^2 \leq 1$ and $1+6y^2 \geq 1$ and x, y are not both 0, these two inequalities fail so (2.1 (b)) and (2.2 (b)) must be true. Now consider (2.3 (a)). If $(1+x)^2 \geq 1+6y^2$ and $6y^2 \geq 1+(1-x)^2$ as we just showed, then

$$(1+x)^2 \geq 1+6y^2 \geq 2+(1-x)^2$$

which implies that $4x \geq 2$ and since $x \leq 1/2$, we conclude that $x = 1/2$. Substituting this into the previous inequalities, we get that

$$\tfrac{9}{4} \geq 1+6y^2 \geq \tfrac{9}{4},$$

so $6y^2 = \frac{5}{4}$. Let $y = r/s$ with $\gcd(r,s) = 1$. We now have that $24r^2 = 5s^2$. Since $r \nmid s$, then $r^2 \mid 5$, so $r = 1$. But then $24 = 5s^2$, a contradiction. Therefore, (2.3 (b)) is true, which implies that

$$6y^2 \geq 1+(1+x)^2 \geq 2.$$

However, since $y \leq 1/2$, $6y^2 \geq 2$ implies that $6 \geq 8$, a contradiction. Then neither (2.3 (a)) nor (2.3 (b)) are true, so $\mathbb{Z}[\sqrt{6}]$ must be Euclidean.

Exercise 2.5.6 Show that $\mathbb{Z}[(1+\sqrt{-19})/2]$ is not Euclidean for the norm map.

Solution. If a ring R is Euclidean, then given any $\alpha, \beta \in R$ we can find δ, γ such that $\alpha = \beta\gamma + \delta$ with $\delta = 0$ or $N(\delta) < N(\beta)$. Another way of describing this condition is to say that given any $\beta \in R$, we can find a representative for each nonzero residue class of $R/(\beta)$ such that the representative has norm less than the norm of β. We will try to find an element of $R = \mathbb{Z}[(1+\sqrt{-19})/2]$ for which this is not true.

Consider $\beta = 2$. $N(2) = 4$. We want to find all other elements of R with norm strictly less than 4.

$$N\left(\frac{a+b\sqrt{-19}}{2}\right) = \frac{a^2+19b^2}{4} < 4,$$

$$\Rightarrow \qquad a^2+19b^2 < 16.$$

First note that if $b > 0$, there are no solutions to this inequality. For $b = 0$, we can have $a = 0, \pm 2$, since $a \equiv b \pmod 2$. Thus, there are just three elements with norm less than 4. However, there are more than three residue classes of $R/(2)$ (check this!). Therefore, the ring $R = \mathbb{Z}[(1+\sqrt{-19})/2]$ is non-Euclidean with respect to the norm map.

Exercise 2.5.7 Prove that $\mathbb{Z}[\sqrt{-10}]$ is not a unique factorization domain.

Solution. Consider the elements $2+\sqrt{-10}, 2-\sqrt{-10}, 2, 7$. Show that they are all irreducible and are not associates. Then note that

$$(2+\sqrt{-10})(2-\sqrt{-10}) = 14,$$

$$2 \cdot 7 = 14.$$

Exercise 2.5.8 Show that there are only finitely many rings $\mathbb{Z}[\sqrt{d}]$ with $d \equiv 2$ or 3 (mod 4) which are norm Euclidean.

Solution. If $\mathbb{Z}[\sqrt{d}]$ is Euclidean for the norm map, then for any $\delta \in \mathbb{Q}(\sqrt{d})$, we can find $\alpha \in \mathbb{Z}[\sqrt{d}]$ such that

$$|N(\delta - \alpha)| < 1.$$

Write $\delta = r + s\sqrt{d}$, $\alpha = a + b\sqrt{d}$, $a, b \in \mathbb{Z}$, $r, s \in \mathbb{Q}$. Then

$$|(r-a)^2 - d(s-b)^2| < 1.$$

In particular, take $r = 0$, $s = t/d$ where t is an integer to be chosen later. Then

$$\left| a^2 - d\left(b - \frac{t}{d}\right)^2 \right| < 1$$

so that $|(bd-t)^2 - da^2| < d$. Since $(bd-t)^2 - da^2 \equiv t^2 \pmod{d}$, there are integers x and z such that

$$z^2 - dx^2 \equiv t^2 \pmod{d},$$

with $|z^2 - dx^2| < d$.

In case $d \equiv 3 \pmod 4$, we choose an odd integer t such that

$$5d < t^2 < 6d,$$

which we can do if d is sufficiently large. Then $z^2 - dx^2 = t^2 - 5d$ or $t^2 - 6d$. Then one of the equations

$$z^2 - t^2 = d(x^2 - 5)$$

or

$$z^2 - t^2 = d(x^2 - 6)$$

is true. We consider this modulo 8. Then $t^2 \equiv 1 \pmod 8$ since t is odd. Also, $x^2, z^2 \equiv 0, 1$, or $4 \pmod 8$ and $d \equiv 3$ or $7 \pmod 8$. We are easily led to $t^2 - z^2 \equiv 0, 1$, or $5 \pmod 8$. This means

$$d(x^2 - 5) \equiv 5, 4, \text{ or } 1 \pmod 8$$

or

$$d(x^2 - 6) \equiv 6, 5, 2, \text{ or } 1 \pmod 8.$$

All of these congruences are impossible. In case $d \equiv 2 \pmod 4$, we choose t odd satisfying $2d < t^2 < 3d$ and proceed as above.

(The case $d \equiv 1 \pmod 4$ is more difficult and has been handled by Heilbronn who was the first to show that there are only finitely many real quadratic fields which are norm-Euclidean.)

A more general and analogous result for imaginary quadratic fields will be proved in Exercise 4.5.21 in Chapter 4.

Exercise 2.5.9 Find all integer solutions of $y^2 = x^3 + 1$.

Solution. We will determine all integer solutions of $y^2 - 1 = x^3$. From $(y-1)(y+1) = x^3$, we see that if $(y-1, y+1) = 1$, then $y - 1 = u^3$, $y + 1 = v^3$ (say). Thus,

$$2 = v^3 - u^3 = (v-u)(v^2 + vu + u^2)$$

from which we deduce that

$$v - u = \pm 1, \quad v^2 + vu + u^2 = \pm 2$$

or

$$v - u = \pm 2, \quad v^2 + vu + u^2 = \pm 1.$$

This gives rise to four cases. The only case that leads to a solution is $v - u = 2$ and $v^2 + vu + u^2 = 1$. This yields the solution $(x, y) = (-1, 0)$.

Now suppose $(y-1, y+1) = 2$. This gives rise to two cases

$$y - 1 = 2u^3, \quad y + 1 = 4v^3 \qquad \text{and} \qquad y - 1 = 4u^3, \quad y + 1 = 2v^3.$$

In the first case, we are led to $u^3 + 1 = 2v^3$ and in the second case, we get $2u^3 + 1 = v^3$. As -1 is a cube, both equations are covered if we can determine all integer solutions of

$$x^3 + y^3 = 2z^3.$$

We will use a "descent" argument to determine all coprime solutions.

To this end, we consider the ring of Eisenstein integers $\mathbb{Z}[\rho]$ where $\rho^2 + \rho + 1 = 0$. We recall a few facts about this ring. It is well-known that this is a Euclidean domain for the norm map given by

$$N(a + b\rho) = a^2 + ab + b^2.$$

Its unit group is $\{\pm 1, \pm\rho, \pm\rho^2\}$. It is also easily checked that $1, \rho, \rho^2$ represent all the distinct coprime residue classes modulo $2\mathbb{Z}[\rho]$. We see that the cube of every coprime residue class is 1 (modulo 2). If u is a unit $\equiv 1 \pmod 2$, then $u = \pm 1$. Now we claim that any coprime solution (x, y, z) of

$$x^3 + y^3 = 2uz^3$$

in $\mathbb{Z}[\rho]$ satisfies $N(xyz) = 1$. Suppose not. Let (x, y, z) be such that $N(xyz)$ is minimal and ≥ 2. We may let

$$A = x + y, \quad B = \rho x + \rho^2 y, \quad C = \rho^2 x + \rho y$$

so that

$$ABC = 2uz^3, \quad A + B + C = 0.$$

Let $d = (A, B, C)$ so that the above equation becomes

$$\frac{A}{d}\frac{B}{d}\frac{C}{d} = 2u\left(\frac{z}{d}\right)^3.$$

Now 2 is an irreducible element in $\mathbb{Z}[\rho]$ and $A/d, B/d, C/d$ are mutually coprime (as their sum is zero) so it can divide only one of them, say C/d without any loss of generality. Thus, we may write

$$A/d = u_1\alpha^3, \quad B/d = u_2\beta^3, \quad C/d = -2u_3\gamma^3,$$

with u_1, u_2, u_3 units. Also, $\alpha\beta\gamma \neq 0$ for otherwise, $z = 0$ and $x = \pm y$, which are not coprime solutions. Hence,

$$u_1\alpha^3 + u_2\beta^3 = 2u_3\gamma^3,$$

and dividing by the unit u_1 gives the equation

$$\alpha^3 + u'\beta^3 = 2u_4\gamma^3$$

for some units u' and u_4. Observe that $(\beta, 2) = 1$ for otherwise, $2|\alpha$ and $2|\gamma$ which implies that α, β, γ are not coprime, a contradiction. Reducing the above equation mod 2 shows that u' is a cube mod 2, and by our remark above u' must be ± 1. Thus u' is a cube and we have

$$\alpha^3 + \beta^3 = 2u\gamma^3.$$

Notice that by our choice of (x, y, z)

$$N(xyz)^3 \leq N(\alpha\beta\gamma)^3 = N(ABC/d^3) = N(z)^3/N(d)^3$$

which means that $N(xyd)^3 \leq 1$. Thus, x, y, d are units. Hence, $x^3 = \pm 1$ and $y^3 = \pm 1$ and z is also a unit. Thus, $N(xyz) = 1$ contrary to our choice. This proves our claim.

Therefore, the only solution for $u^3 + 1 = 2v^3$ is $u^3 = \pm 1$. This leads to the solutions $(x, y) = (2, 3), (1, 0)$ for the equation $y^2 - 1 = x^3$. In the other case of $2u^3 + 1 = v^3$, we have $v = \pm 1$ which leads to $(0, 1), (2, -3)$. We get a final set of five integer solutions for $y^2 - 1 = x^3$.

Exercise 2.5.10 Let $x_1, ..., x_n$ be indeterminates. Evaluate the determinant of the $n \times n$ matrix whose (i, j)-th entry is x_i^{j-1}. (This is called the *Vandermonde determinant.*)

Solution. Let $V(x_1, ..., x_n)$ denote the value of the determinant. If we fix $x_2, ..., x_n$, we may view the determinant as a polynomial in x_1 of degree $n - 1$. Since the determinant is zero if $x_1 = x_i$ for $i \geq 2$, the roots of the polynomial are $x_2, ..., x_n$. It is easy to see that the leading coefficient is

$$(-1)^{n-1}V(x_2, ..., x_n)$$

so that the determinant is

$$(-1)^{n-1}V(x_2, ..., x_n)\prod_{j=2}^{n}(x_1 - x_j).$$

By induction, we see that

$$V(x_1, ..., x_n) = (-1)^{\binom{n}{2}}\prod_{j>i}(x_i - x_j).$$

Chapter 3

Algebraic Numbers and Integers

3.1 Basic Concepts

Exercise 3.1.2 Show that if $r \in \mathbb{Q}$ is an algebraic integer, then $r \in \mathbb{Z}$.

Solution. Let $r = c/d$, $(c, d) = 1$, be an algebraic integer. Then r is the root of a monic polynomial in $\mathbb{Z}[x]$, say $f(x) = x^n + b_{n-1}x^{n-1} + \cdots + b_0$. So

$$\begin{aligned} f(r) = \left(\frac{c}{d}\right)^n + b_{n-1}\left(\frac{c}{d}\right)^{n-1} + \cdots + b_0 &= 0 \\ \Leftrightarrow \qquad c^n + b_{n-1}c^{n-1}d + \cdots + b_0 d^n &= 0. \end{aligned}$$

This implies that $d \mid c^n$, which is true only when $d = \pm 1$. So $r = \pm c \in \mathbb{Z}$.

Exercise 3.1.3 Show that if $4 \mid (d+1)$, then

$$\frac{-1 \pm \sqrt{-d}}{2}$$

is an algebraic integer.

Solution. Consider the monic polynomial

$$x^2 + x + \frac{d+1}{4} \in \mathbb{Z}[x]$$

when $4 \mid d + 1$. The roots of this polynomial, which by definition are algebraic integers, are

$$x = \frac{-1 \pm \sqrt{1 - 4\frac{d+1}{4}}}{2} = \frac{-1 \pm \sqrt{-d}}{2}.$$

Exercise 3.1.6 Find the minimal polynomial of $\sqrt{n}$ where n is a squarefree integer.

Solution. If $n = 1$, the minimal polynomial is $x - 1$. If $n \neq 1$, then $x^2 - n$ is irreducible and has $\sqrt{n}$ as a root. Thus, the minimal polynomial is either linear or quadratic. If it is linear, we obtain that $\sqrt{n}$ is rational, a contradiction. Thus, $x^2 - n$ is the minimal polynomial of $\sqrt{n}$ when $n \neq 1$.

Exercise 3.1.7 Find the minimal polynomial of $\sqrt{2}/3$.

Solution. It is $x^2 - 2/9$ since $\sqrt{2}/3$ is a root, and $\sqrt{2}/3$ is not rational.

3.2 Liouville's Theorem and Generalizations

Exercise 3.2.4 Show that $\sum_{n=1}^{\infty} 2^{-3^n}$ is transcendental.

Solution. Suppose that

$$\alpha = \sum_{n=1}^{\infty} \frac{1}{2^{3^n}}$$

is algebraic. We proceed as in Example 3.2.2 and consider the partial sum:

$$\sum_{n=1}^{k} \frac{1}{2^{3^n}} = \frac{p_k}{q_k},$$

with $q_k = 2^{3^k}$. As before,

$$\left|\alpha - \frac{p_k}{q_k}\right| = \left|\sum_{n=k+1}^{\infty} \frac{1}{2^{3^n}}\right| \leq \frac{S}{2^{3^{k+1}}}.$$

But since α is algebraic, by Roth's theorem we have the inequality

$$\frac{S}{q_k^3} \geq \frac{c(\alpha)}{q_k^{2+\varepsilon}}.$$

But again we can choose k to be as large as we want, and so for ε sufficiently small, this inequality does not hold. Thus, α is transcendental.

Exercise 3.2.5 Show that, in fact, $\sum_{n=1}^{\infty} 2^{-f(n)}$ is transcendental when

$$\lim_{n\to\infty} \frac{f(n+1)}{f(n)} > 2.$$

Solution. Suppose

$$\alpha = \sum_{n=1}^{\infty} \frac{1}{2^{f(n)}}$$

is algebraic. Following the same argument as above, we get the inequalities

$$\frac{S}{q_{k+1}} \geq \left|\alpha - \frac{p_k}{q_k}\right| \geq \frac{c(\alpha)}{q_k^{2+\varepsilon}},$$

where $q_k = 2^{f(k)}$. Now, for k sufficiently large,

$$\frac{f(k+1)}{f(k)} > 2+\delta \quad \Rightarrow \quad f(k+1) > (2+\delta)f(k).$$

So, for large k,

$$\frac{q_{k+1}}{q_k} = \frac{2^{f(k+1)}}{2^{f(k)}} > 2^{(1+\delta)f(k)} = q_k^{(1+\delta)}$$

which implies that $q_{k+1} > q_k^{2+\delta}$. By Roth's theorem, we can deduce that

$$\begin{aligned} \frac{c(\alpha)}{q_k^{2+\varepsilon}} &\leq \frac{S}{q_{k+1}}, \\ \Rightarrow \quad \frac{c(\alpha)}{q_k^{2+\varepsilon}} &\leq \frac{S}{q_k^{2+\delta}}, \\ \Rightarrow \quad q_k^{\delta} &\leq \frac{S}{c(\alpha)} q_k^{\varepsilon}. \end{aligned}$$

As $q_k \to \infty$, this implies $\delta \leq \varepsilon$, a contradiction for $\varepsilon < \delta/2$ (say).

3.3 Algebraic Number Fields

Exercise 3.3.3 Let α be an algebraic number and let $p(x)$ be its minimal polynomial. Show that $p(x)$ has no repeated roots.

Solution. Suppose α is a repeated root of $p(x)$. Then we can write

$$p(x) = (x-\alpha)^2 g(x),$$

for some polynomial $g(x) \in \mathbb{C}[x]$, and

$$p'(x) = 2(x-\alpha)g(x) + (x-\alpha)^2 g'(x).$$

So $p'(\alpha) = 0$ and from Theorem 3.1.4, $p(x) \mid p'(x)$. But $\deg(p') < \deg(p)$, and we have a contradiction. If β is a repeated root of $p(x)$, then by the following exercise, β has the same minimal polynomial and repeating the above argument with β leads to a contradiction. Thus p has no repeated roots.

Exercise 3.3.4 Let α, β be algebraic numbers such that β is conjugate to α. Show that β and α have the same minimal polynomial.

Solution. Let $p(x)$ be the minimal polynomial of α, and let $q(x)$ be the minimal polynomial of β. By the definition of conjugate roots, β is a common root of $p(x)$ and $q(x)$.

Using the division algorithm, we can write $p(x) = a(x)q(x) + r(x)$ for some $a(x), r(x) \in \mathbb{Q}[x]$ and either $r = 0$ or $\deg(r) < \deg(q)$. But

$$p(\beta) = a(\beta)q(\beta) + r(\beta) = 0$$

and $q(\beta) = 0$ so $r(\beta)$ must also be 0. Since q is the minimal polynomial for β, $r = 0$. Thus $p(x) = a(x)q(x)$, but, by Theorem 3.1.4, p is irreducible, and both $p(x)$ and $q(x)$ are monic, so $p(x) = q(x)$.

Exercise 3.3.6 Let $K = \mathbb{Q}(\theta)$ be of degree n over $\mathbb{Q}$. Let $\omega_1, \ldots, \omega_n$ be a basis of K as a vector space over $\mathbb{Q}$. Show that the matrix $\Omega = (\omega_i^{(j)})$ is invertible.

Solution.

$$\Omega = \begin{pmatrix} \omega_1 & \omega_1^{(2)} & \cdots & \omega_1^{(n)} \\ \omega_2 & \omega_2^{(2)} & \cdots & \omega_2^{(n)} \\ \vdots & \vdots & & \vdots \\ \omega_n & \omega_n^{(2)} & \cdots & \omega_n^{(n)} \end{pmatrix}.$$

Since θ is an algebraic number of degree n, $\alpha_1 = 1, \alpha_2 = \theta, \ldots, \alpha_n = \theta^{n-1}$ also forms a basis for K over $\mathbb{Q}$. Let $A = (\alpha_i^{(j)})$. Then,

$$\det A = \begin{vmatrix} 1 & 1 & \cdots & 1 \\ \theta & \theta^{(2)} & \cdots & \theta^{(n)} \\ \vdots & \vdots & & \vdots \\ \theta^{n-1} & \theta^{(2)n-1} & \cdots & \theta^{(n)n-1} \end{vmatrix}$$

which is the Vandermonde determinant. So A is invertible. Further,

$$\begin{aligned} \omega_i^{(j)} &= \sum_{k=1}^{n} (b_{ik}\alpha_k)^{(j)} \\ &= \sum_{k=1}^{n} b_{ik}\alpha_k^{(j)}, \end{aligned}$$

where $1 \leq i, j \leq n$, and $b_{ik} \in \mathbb{Q}$.

Since the set $\{\omega_1, \ldots, \omega_n\}$, as well as the set $\{\alpha_1, \ldots, \alpha_n\}$, are linearly independent sets, it follows that both the rows and columns of the matrix $B = (b_{ik})$ are linearly independent. Hence B is invertible and $\Omega = BA$ and from elementary linear algebra, $\det \Omega = \det B \det A \neq 0$. Thus Ω is invertible.

Exercise 3.3.7 Let α be an algebraic number. Show that there exists $m \in \mathbb{Z}$ such that $m\alpha$ is an algebraic integer.

Solution. Let $p(x) \in \mathbb{Q}[x]$ be the minimal polynomial of α. So,

$$p(\alpha) = \alpha^n + a_{n-1}\alpha^{n-1} + \cdots + a_1\alpha + a_0 = 0.$$

Choose $m \in \mathbb{Z}$ so that $ma_0, ma_1, \ldots, ma_{n-1}$ are all integers. Now,

$$\begin{aligned} m^n\alpha^n + m^n a_{n-1}\alpha^{n-1} + \cdots + m^n a_1\alpha + m^n a_0 &= 0, \\ \Leftrightarrow \quad (m\alpha)^n + ma_{n-1}(m\alpha)^{n-1} + \cdots + m^{n-1}a_1(m\alpha) + m^n a_0 &= 0. \end{aligned}$$

Let $g(x) = x^n + ma_{n-1}x^{n-1} + \cdots + m^{n-1}a_1x + m^n a_0$, then $g(x)$ is a monic polynomial in $\mathbb{Z}[x]$ and $g(m\alpha) = 0$. Thus $m\alpha$ is an algebraic integer.

Exercise 3.3.8 Show that $\mathbb{Z}[x]$ is not (a) Euclidean or (b) a PID.

Solution. (a) Consider the elements $2, x \in \mathbb{Z}[x]$. Clearly there is no way to write $x = a(x)2 + r(x)$ where both the conditions (i) $a(x), r(x) \in \mathbb{Z}[x]$ and (ii) $\deg(r) < \deg(2) = 0$ or $r = 0$ are satisfied.

(b) Again consider the two polynomials $2, x \in \mathbb{Z}[x]$. Clearly $(x) \not\subseteq (2)$ and $(2) \not\subseteq (x)$. Now, if $(x, 2) = (\alpha)$ for some $\alpha \in \mathbb{Z}[x]$, then $\alpha \mid 2$ and $\alpha \mid x$. But if $\alpha \mid 2$, then $\alpha \in \mathbb{Z}$, so $\alpha = \pm 1$ or $\alpha = \pm 2$. However, $\pm 1 \notin (x, 2)$ and $\pm 2 \nmid x$. So the ideal generated by x and 2 is not generated by a single element in $\mathbb{Z}[x]$. $\mathbb{Z}[x]$ is not a PID.

Exercise 3.3.11 Let $f(x) = x^n + a_{n-1}x^{n-1} + \cdots + a_1x + a_0$, and assume that for p prime $p \mid a_i$ for $0 \le i < k$ and $p^2 \nmid a_0$. Show that $f(x)$ has an irreducible factor of degree at least k. (The case $k = n$ is referred to as Eisenstein's criterion for irreducibility.)

Solution. We will prove this by induction on n, the degree of $f(x)$. The case when $n = 1$ is trivial, so let us assume that the above statement is true for any polynomial of degree less than n.

If $f(x)$ is irreducible, there is nothing to prove, so assume that $f(x)$ is not irreducible. Then we can write

$$\begin{aligned} f(x) &= g(x)h(x) \\ &= (b_0 + b_1x + \cdots + b_rx^r)(c_0 + c_1x + \cdots + c_tx^t). \end{aligned}$$

Since $p \mid a_0$ and $p^2 \nmid a_0$, and $a_0 = b_0c_0$, we deduce $p \mid b_0$ or $p \mid c_0$ but not both.

Suppose, without loss of generality, that $p \mid b_0$. We next consider $a_1 = b_0c_1 + b_1c_0$. Since $p \mid a_1$ and $p \mid b_0$, but $p \nmid c_0$, then $p \mid b_1$. Continuing in this fashion, we get that $p \mid b_i$ for $0 \le i < k$. If $r < k$, then we can factor out a p from each of the coefficients in $g(x)$, but this is absurd since then p would divide every coefficient in $f(x)$, and f is monic. Therefore $k \le r$, $p \mid b_i$, $0 \le i < k$, $p^2 \nmid b_0$. Also $b_rc_t = 1$ implies that $g(x)$ or $-g(x)$ is monic. In any event, we have another polynomial which satisfies the conditions set out above but which has degree less than n. Thus, by induction, $g(x)$ has an irreducible factor of degree greater than or equal to k, and this factor is also an irreducible factor of $f(x)$.

Exercise 3.3.12 Show that $f(x) = x^5 + x^4 + 3x^3 + 9x^2 + 3$ is irreducible over $\mathbb{Q}$.

Solution. By applying Exercise 3.3.11 to $f(x)$ with $p = 3$, we deduce that if $f(x)$ is not irreducible, then we can factor it into the product of a polynomial of degree 4 and a polynomial of degree 1, and so $f(x)$ has a rational root. However, we showed in Exercise 3.1.2 that if $r \in \mathbb{Q}$ is an algebraic integer, then $r \in \mathbb{Z}$. Thus, $f(x)$ must have an integral root, and this root must divide the constant term which is 3. The only choices are then $\pm 1, \pm 3$, and it is easy to check that these are not roots of the polynomial in question.

We conclude that $f(x)$ is irreducible since it has no rational root.

3.4 Supplementary Problems

Exercise 3.4.1 Show that

$$\sum_{n=0}^{\infty} \frac{1}{a^{n!}}$$

is transcendental for $a \in \mathbb{Z}$, $a \geq 2$.

Solution. Suppose it is algebraic, and call the sum α. Look at the partial sum

$$\alpha_k = \sum_{n=0}^{k} \frac{1}{a^{n!}} = \frac{p_k}{q_k},$$

with $q_k = a^{k!}$. Then

$$\begin{aligned} |\alpha - \alpha_k| &= \left| \sum_{n=k+1}^{\infty} \frac{1}{a^{n!}} \right| \\ &\leq \frac{1}{a^{(k+1)!}} M, \end{aligned}$$

where

$$M = 1 + \frac{1}{a^{k+2}} + \left(\frac{1}{a^{k+2}} \right)^2 + \cdots,$$

an infinite geometric series with a finite sum. Thus,

$$\left| \sum_{n=k+1}^{\infty} \frac{1}{a^{n!}} \right| \leq \frac{M}{q_k^{k+1}}.$$

If α is algebraic of degree n, then Liouville's theorem tells us that we can find a constant $c(\alpha)$ such that

$$\frac{M}{q_k^{k+1}} \geq \left| \alpha - \frac{p_k}{q_k} \right| \geq \frac{c(\alpha)}{q_k^n}.$$

However, we can choose k as large as we wish to obtain a contradiction.

Exercise 3.4.2 Show that

$$\sum_{n=1}^{\infty} \frac{1}{a^{3^n}}$$

is transcendental for $a \in \mathbb{Z}, a \geq 2$.

Solution. Suppose that

$$\alpha = \sum_{n=1}^{\infty} \frac{1}{a^{3^n}}$$

is algebraic. Consider the partial sum

$$\alpha_k = \sum_{n=1}^{k} \frac{1}{a^{3^n}} = \frac{p_k}{q_k},$$

with $q_k = a^{3^k}$. We have

$$\left|\alpha - \frac{p_k}{q_k}\right| = \left|\sum_{n=k+1}^{\infty} \frac{1}{a^{3^n}}\right| \leq \frac{S}{a^{3^{k+1}}},$$

where $S = 1 + 1/a + 1/a^2 + \cdots$. Then by Roth's theorem,

$$\frac{S}{q_k^3} \geq \frac{c(\alpha, \varepsilon)}{q_k^{2+\varepsilon}}.$$

But, we can choose k to be as large as we want to produce a contradiction.

Exercise 3.4.3 Show that

$$\sum_{n=1}^{\infty} \frac{1}{a^{f(n)}}$$

is transcendental when

$$\lim_{n\to\infty} \frac{f(n+1)}{f(n)} > 2.$$

Solution. Suppose that

$$\alpha = \sum_{n=1}^{\infty} \frac{1}{a^{f(n)}}$$

is algebraic. Following the same argument as in the previous exercise, we get

$$\frac{S}{q_{k+1}} \geq \left|\alpha - \frac{p_k}{q_k}\right| \geq \frac{c(\alpha)}{q_k^{2+\varepsilon}},$$

where $q_k = a^{f(k)}$. For k sufficiently large,

$$\frac{f(k+1)}{f(k)} > 2 + \delta,$$

and so

$$\frac{q_{k+1}}{q_k} = \frac{a^{f(k+1)}}{a^{f(k)}} > a^{(1+\delta)f(k)} = q_k^{(1+\delta)}.$$

This implies that $q_{k+1} > q_k^{2+\delta}$. By Roth's theorem, we can deduce that

$$\frac{c(\alpha)}{q_k^{2+\varepsilon}} \leq \frac{S}{q_{k+1}} \quad \Rightarrow \quad \frac{c(\alpha)}{q_k^{2+\varepsilon}} \leq \frac{S}{q_k^{2+\delta}},$$
$$\Rightarrow \quad q_k^{\delta} \leq \frac{S}{c(\alpha)} q_k^{\varepsilon}.$$

As $q_k \to \infty$, we find $\delta \leq \varepsilon$, a contradiction for sufficiently small ε.

Exercise 3.4.4 Prove that $f(x) = x^6 + 7x^5 - 12x^3 + 6x + 2$ is irreducible over $\mathbb{Q}$.

Solution. By Exercise 3.3.11, since $2 \mid a_i$ for $0 \leq i < 5$ then $f(x)$ has a factor of degree at least 5. This means that the polynomial is either irreducible or it has a rational root. We showed earlier that if a polynomial in $\mathbb{Z}[x]$ has a rational root, then the root is actually an integer. We also know that any roots of a polynomial will divide its constant term, which in this case is 2. It suffices to check that $\pm 1, \pm 2$, are not roots to deduce that $f(x)$ is irreducible.

Exercise 3.4.5 Using Thue's theorem, show that $f(x, y) = x^6 + 7x^5y - 12x^3y^3 + 6xy^5 + 8y^6 = m$ has only a finite number of solutions for $m \in \mathbb{Z}^*$.

Solution. Use the previous exercise to prove that the polynomial is irreducible. The result follows from Thue's theorem and Example 3.2.3.

Exercise 3.4.6 Let ζ_m be a primitive mth root of unity. Show that

$$\prod_{\substack{0 \leq i,j \leq m-1 \\ i \neq j}} (\zeta_m^i - \zeta_m^j) = (-1)^{m-1} m^m.$$

Solution. Since

$$x^m - 1 = \prod_{i=0}^{m-1} (x - \zeta_m^i),$$

we see that the constant term is

$$(-1)^m \prod_{i=0}^{m-1} \zeta_m^i = -1.$$

Differentiating $x^m - 1$ above via the product rule, and setting $x = \zeta_m^j$, we see that

$$m\zeta_m^{j(m-1)} = \prod_{\substack{i=0 \\ i \neq j}}^{m-1} (\zeta_m^j - \zeta_m^i).$$

Taking the product over j gives the result.

Exercise 3.4.7 Let

$$\phi_m(x) = \prod_{\substack{1 \le i \le m \\ (i,m)=1}} (x - \zeta_m^i)$$

denote the mth cyclotomic polynomial. Prove that

$$x^m - 1 = \prod_{d|m} \phi_d(x).$$

Solution. Every mth root of unity is a primitive dth root of unity for some $d \mid m$. Conversely, every dth root of unity is also an mth root of unity for $d \mid m$. The result is now immediate.

Exercise 3.4.8 Show that $\phi_m(x) \in \mathbb{Z}[x]$.

Solution. We induct on m. For $m = 1$, this is clear. Suppose we have proved it true for $\phi_r(x)$ with $r < m$. Then setting

$$v(x) = \prod_{\substack{d|m \\ d<m}} \phi_d(x),$$

we have by induction $v(x) \in \mathbb{Z}[x]$. Since $v(x)$ is monic, and $v(x) \mid (x^m - 1)$, we find by long division that $(x^m - 1)/v(x) = \phi_m(x) \in \mathbb{Z}[x]$.

Exercise 3.4.9 Show that $\phi_m(x)$ is irreducible in $\mathbb{Q}[x]$ for every $m \ge 1$.

Solution. Let $f(x)$ be the minimal polynomial of ζ_m and suppose $\phi_m(x) = f(x)g(x)$ with $f(x), g(x) \in \mathbb{Q}[x]$. By Gauss' lemma (see Theorem 2.1.9) we may suppose that $f(x), g(x) \in \mathbb{Z}[x]$. Let p be coprime to m. Then ζ_m^p is again a primitive mth root of unity. Thus

$$f(\zeta_m^p)g(\zeta_m^p) = 0.$$

Suppose $f(\zeta_m^p) \neq 0$. Then $g(\zeta_m^p) = 0$. Since $g(x^p) \equiv g(x)^p \pmod p$ we deduce that $g(x)$ and $f(x)$ have a common root in $\overline{\mathbb{F}_p}$, a contradiction since $x^m - 1$ has no multiple roots in $\overline{\mathbb{F}_p}$. Thus, $f(\zeta_m^p) = 0$ for any $(p, m) = 1$. It follows that $f(\zeta_m^i) = 0$ for any $(i, m) = 1$. Therefore $\deg(f) = \varphi(m) = \deg(\phi_m)$.

Exercise 3.4.10 Let I be a subset of the positive integers $\le m$ which are coprime to m. Set

$$f(x) = \prod_{i \in I} (x - \zeta_m^i).$$

Suppose that $f(\zeta_m) = 0$ and $f(\zeta_m^p) \neq 0$ for some prime p. Show that $p \mid m$. (This observation gives an alternative proof for the irreducibility of $\phi_m(x)$.)

Solution. Let $K = \mathbb{Q}(\zeta_m)$. Then $f(\zeta_m^p)$ divides

$$\prod_{0 \leq i \leq m-1} (\zeta_m^p - \zeta_m^i)$$

in the ring $\mathcal{O}_K$. Hence $N_{K/\mathbb{Q}}(f(\zeta_m^p))$ divides m by Exercise 3.4.6 above. Since

$$f(x)^p - f(x^p) \in p\mathbb{Z}[x],$$

we see upon setting $x = \zeta_m$ that $p \mid f(\zeta_m^p)$ and hence $p \mid m$, as desired.

Exercise 3.4.11 Consider the equation $x^3 + 3x^2y + xy^2 + y^3 = m$. Using Thue's theorem, deduce that there are only finitely many integral solutions to this equation.

Solution. To use Thue's theorem, we must show that $f(x, y) = x^3 + 3x^2 + xy^2 + y^3$ is irreducible. If this polynomial is reducible, then so is the polynomial $f(x, 1) = x^3 + 3x^2 + x + 1$. However, since $f(x, 1)$ has degree 3, then if it reduces it will have a factor of degree 1, a rational root. We have already shown that all rational roots of a monic polynomial in $\mathbb{Z}[x]$ are actually integers, and all roots must divide the constant term of a polynomial. The only possibilities for such a root are $x = \pm 1$. A quick calculation shows that neither of these two are in fact a root of $f(x, 1)$, and so $f(x, 1)$ is irreducible, implying that $f(x, y)$ is irreducible. We can now apply the results of Example 3.2.3.

Exercise 3.4.12 Assume that n is an odd integer, $n \geq 3$. Show that $x^n + y^n = m$ has only finitely many integral solutions.

Solution. If (x_0, y_0) is a solution to $x^n + y^n = m$, then $(x_0 + y_0) \mid m$, since

$$x^n + y^n = (x + y)(x^{n-1} - x^{n-2y} + \cdots + y^{n-1}).$$

Suppose $|x| \geq m$. Then the distance between x and the nearest nth power will be greater than m, and x cannot satisfy the above equation. We then have a bound on the size of x along with the constraint that $x + y \mid m$. There can only be a finite number of pairs which satisfy these two constraints.

Exercise 3.4.13 Let ζ_m denote a primitive mth root of unity. Show that $\mathbb{Q}(\zeta_m)$ is normal over $\mathbb{Q}$.

Solution. ζ_m is a root of the mth cyclotomic polynomial, which we have shown to be irreducible. Thus, the conjugate fields are $\mathbb{Q}(\zeta_m^j)$ where $(j, m) = 1$ and these are identical with $\mathbb{Q}(\zeta_m)$.

Exercise 3.4.14 Let a be squarefree and greater than 1, and let p be prime. Show that the normal closure of $\mathbb{Q}(a^{1/p})$ is $\mathbb{Q}(a^{1/p}, \zeta_p)$.

Solution. The polynomial $x^p - a$ is irreducible (by Eisenstein's criterion). The conjugates of $a^{1/p}$ are $\zeta_p^j a^{1/p}$. If K is the normal closure of $\mathbb{Q}(a^{1/p})$, it must contain all the pth roots of unity. The result is now immediate.

Chapter 4

Integral Bases

4.1 The Norm and the Trace

Exercise 4.1.2 Let $K = \mathbb{Q}(i)$. Show that $i \in \mathcal{O}_K$ and verify that $\operatorname{Tr}_K(i)$ and $\mathrm{N}_K(i)$ are integers.

Solution. We know that i is a root of the irreducible polynomial $x^2 + 1$, and so its conjugates are $i, -i$.

Thus, $\operatorname{Tr}_K(i) = i - i = 0 \in \mathbb{Z}$ and $\mathrm{N}_K(i) = i(-i) = 1 \in \mathbb{Z}$.

Exercise 4.1.3 Determine the algebraic integers of $K = \mathbb{Q}(\sqrt{-5})$.

Solution. We first note that $1, \sqrt{-5}$ form a $\mathbb{Q}$-basis for K. Thus any $\alpha \in K$ looks like $\alpha = r_1 + r_2\sqrt{-5}$ with $r_1, r_2 \in \mathbb{Q}$. Since $[K : \mathbb{Q}] = 2$, we can deduce that the conjugates of α are $r_1 + r_2\sqrt{-5}$ and $r_1 - r_2\sqrt{-5}$. Then $\operatorname{Tr}_K(\alpha) = 2r_1$ and

$$\begin{aligned} \mathrm{N}_K(\alpha) &= (r_1 + r_2\sqrt{-5})(r_1 - r_2\sqrt{-5}) \\ &= r_1^2 + 5r_2^2. \end{aligned}$$

By Lemma 4.1.1, if $\alpha \in \mathcal{O}_K$, then the trace and norm are integers. Also, α is a root of the monic polynomial $x^2 - 2r_1x + r_1^2 + 5r_2^2$ which is in $\mathbb{Z}[x]$ when the trace and norm are integers. We conclude that for $\alpha = r_1 + r_2\sqrt{-5}$ to be in $\mathcal{O}_K$, it is necessary and sufficient that $2r_1$ and $r_1^2 + 5r_2^2$ be integers. This implies that r_1 has a denominator at most 2, which forces the same for r_2. Then by setting $r_1 = g_1/2$ and $r_2 = g_2/2$ we must have $(g_1^2 + 5g_2^2)/4 \in \mathbb{Z}$ or, equivalently, $g_1^2 + 5g_2^2 \equiv 0 \pmod 4$. Thus, as all squares are congruent to 0 or 1 (mod 4), we conclude that g_1 and g_2 are themselves even, and thus $r_1, r_2 \in \mathbb{Z}$. We conclude then that $\mathcal{O}_K = \mathbb{Z} + \mathbb{Z}\sqrt{-5}$.

Exercise 4.1.5 Show that the definition of nondegeneracy above is independent of the choice of basis.

Solution. If $f_1, \dots, f_n$ is another basis and $A = (B(f_i, f_j))$, then

$$A = P^T B P,$$

where P is the change of basis matrix from $e_1, \dots, e_n$ to $f_1, \dots, f_n$. Since P is nonsingular, $\det A \neq 0$ if and only if $\det B \neq 0$.

4.2 Existence of an Integral Basis

Exercise 4.2.1 Show that $\exists \omega_1^*, \omega_2^*, \dots, \omega_n^* \in K$ such that

$$\mathcal{O}_K \subseteq \mathbb{Z}\omega_1^* + \mathbb{Z}\omega_2^* + \cdots + \mathbb{Z}\omega_n^*.$$

Solution. Let $\omega_1, \omega_2, \dots, \omega_n$ be a $\mathbb{Q}$-basis for K, and recall from Exercise 3.3.7 that for any $\alpha \in K$ there is a nonzero integer m such that $m\alpha \in \mathcal{O}_K$. Thus we can assume that $\omega_1, \omega_2, \dots, \omega_n$ are in $\mathcal{O}_K$. Now, as the bilinear pairing $B(x, y)$ defined previously was nondegenerate, we can find a dual basis $\omega_1^*, \omega_2^*, \dots, \omega_n^*$ satisfying $B(\omega_i, \omega_j^*) = \delta_{ij}$. If we write $\omega_j^* = \sum c_{kj}\omega_k$ we have

$$\begin{aligned} \delta_{ij} &= \operatorname{Tr}_K(\omega_i \omega_j{}^*) \\ &= \operatorname{Tr}_K(\omega_i \sum c_{kj}\omega_k) \\ &= \sum c_{kj} \operatorname{Tr}_K(\omega_i \omega_k). \end{aligned}$$

If we introduce now the matrices, $C = (c_{ij}), \Omega = (\omega_i^{(j)})$, then the above becomes

$$I_n = \Omega\Omega^T C \quad \Rightarrow \quad C^{-1} = \Omega\Omega^T.$$

We conclude that C is nonsingular and that $\omega_1{}^*, \omega_2{}^*, \dots, \omega_n{}^*$ forms a $\mathbb{Q}$-basis for K.

Let α be an arbitrary element of $\mathcal{O}_K$. We write

$$\alpha = \sum_{j=1}^{n} a_j \omega_j^* \quad \text{with} \quad a_j \in \mathbb{Q}$$

so

$$\alpha\omega_i = \sum_{j=1}^{n} a_j \omega_i \omega_j^* \quad \forall i,$$

and

$$\operatorname{Tr}_K(\alpha\omega_i) = \sum a_j \operatorname{Tr}_K(\omega_i \omega_j^*) = a_i \quad \forall i.$$

But $\alpha\omega_i \in \mathcal{O}_K$ implies the left-hand side above is in $\mathbb{Z}$, and thus $a_i \in \mathbb{Z}$ for all i. It follows then that $\mathcal{O}_K \subseteq \mathbb{Z}\omega_1^* + \mathbb{Z}\omega_2^* + \cdots + \mathbb{Z}\omega_n^*$.

Exercise 4.2.3 Show that $\mathcal{O}_K$ has an integral basis.

Solution. We apply the results of Theorem 4.2.2 with $M = \mathbb{Z}\omega_1^* + \mathbb{Z}\omega_2^* + \cdots + \mathbb{Z}\omega_n^*$ and $N = \mathcal{O}_K$. It follows directly from the theorem that there exist $\omega_1, \omega_2, \ldots, \omega_n \in \mathcal{O}_K$ such that $\mathcal{O}_K = \mathbb{Z}\omega_1 + \mathbb{Z}\omega_2 + \cdots + \mathbb{Z}\omega_n$.

Exercise 4.2.4 Show that $\det(\mathrm{Tr}(\omega_i\omega_j))$ is independent of the choice of integral basis.

Solution. Let $\omega_1, \omega_2, \ldots, \omega_n$ and $\theta_1, \theta_2, \ldots, \theta_n$ be two distinct integral bases for an algebraic number field K. We can write

$$\begin{aligned} \omega_i &= \sum_{j=1}^{n} c_{ij}\theta_j, \\ \theta_i &= \sum_{j=1}^{n} d_{ij}\omega_j, \end{aligned}$$

for all i, where c_{ij} and d_{ij} are all integers. Then (c_{ij}) and $(c_{ij})^{-1}$ both have entries in $\mathbb{Z}$. So $\det(c_{ij})$, $\det(c_{ij})^{-1} \in \mathbb{Z}$, meaning that $\det(c_{ij}) = \pm 1$.

Then

$$\begin{aligned} \mathrm{Tr}(\omega_i\omega_j) &= \mathrm{Tr}\left(\left(\sum_l c_{il}\theta_l\right)\left(\sum_m c_{jm}\theta_m\right)\right) \\ &= \sum_{l,m} c_{il}c_{jm}\,\mathrm{Tr}\left(\theta_l\theta_m\right). \end{aligned}$$

Now if we define $\Omega = (\omega_i^{(j)}), C = (c_{ij}), \Theta = (\theta_i^{(j)})$, then we can write the above as the matrix equation $\Omega^T\Omega = C(\Theta^T\Theta)C^T$ from which it follows that the determinants of Ω and Θ are equal, up to sign. Hence, $\det(\Theta^T\Theta) = \det(\Omega^T\Omega)$.

Exercise 4.2.5 Show that the discriminant is well-defined. In other words, show that given $\omega_1, \omega_2, \ldots, \omega_n$ and $\theta_1, \theta_2, \ldots, \theta_n$, two integral bases for K, we get the same discriminant for K.

Solution. Just as above, we have $\Omega^T\Omega = C(\Theta^T\Theta)C^T$ for some matrix C with determinant ± 1. Then $d_K = (\det\Omega)^2 = (\det\Theta)^2(\det C)^2 = (\det\Theta)^2$. This proves that the discriminant does not depend upon the choice of integral basis.

Exercise 4.2.6 Show that

$$d_{K/\mathbb{Q}}(1, a, \ldots, a^{n-1}) = \prod_{i>j}\left(\sigma_i(a) - \sigma_j(a)\right)^2.$$

We denote $d_{K/\mathbb{Q}}(1, a, \ldots, a^{n-1})$ by $d_{K/\mathbb{Q}}(a)$.

Solution. First we note that $\sigma_i(a)$ takes a to its ith conjugate, $a^{(i)}$. Define the matrix $\Omega = \left(\sigma_i(a^j)\right)$. Then it is easy to see that

$$\Omega = \begin{pmatrix} 1 & a & \cdots & a^{n-1} \\ \vdots & \vdots & & \vdots \\ 1 & a^{(n)} & \cdots & a^{(n)n-1} \end{pmatrix},$$

which is a Vandermonde matrix, and so

$$\det \Omega = \prod_{i>j} \left(a^{(i)} - a^{(j)}\right) = \prod_{i>j} \left(\sigma_i(a) - \sigma_j(a)\right).$$

It follows that

$$\begin{aligned} d_{K/\mathbb{Q}}(a) &= \left[\det(\sigma_i(a^j))\right]^2 \\ &= \prod_{i<j} \left(\sigma_i(a) - \sigma_j(a)\right)^2. \end{aligned}$$

Exercise 4.2.7 Suppose that $u_i = \sum_{j=1}^n a_{ij} v_j$ with $a_{ij} \in \mathbb{Q}, v_j \in K$. Show that $d_{K/\mathbb{Q}}(u_1, u_2, \ldots, u_n) = \left(\det(a_{ij})\right)^2 d_{K/\mathbb{Q}}(v_1, v_2, \ldots, v_n)$.

Solution. By definition, $d_{K/\mathbb{Q}}(u_1, u_2, \ldots, u_n) = \left[\det(\sigma_i(u_j))\right]^2$.

$$\sigma_i(u_j) = \sigma_i\left(\sum_{k=1}^n a_{jk} v_k\right) = \sum_{k=1}^n a_{jk} \sigma_i(v_k).$$

If we define the matrices $U = (\sigma_i(u_j)), A = (a_{ij}), V = (\sigma_i(v_j))$, then it is clear that $U = VA^T$ and so $(\det U)^2 = (\det VA^T)^2$, and we get the desired result:

$$d_{K/\mathbb{Q}}(u_1, u_2, \ldots, u_n) = (\det(a_{ij}))^2 d_{K/\mathbb{Q}}(v_1, v_2, \ldots, v_n).$$

Exercise 4.2.8 Let $a_1, a_2, \ldots, a_n \in \mathcal{O}_K$ be linearly independent over $\mathbb{Q}$. Let $N = \mathbb{Z}a_1 + \mathbb{Z}a_2 + \cdots + \mathbb{Z}a_n$ and $m = [\mathcal{O}_K : N]$. Prove that

$$d_{K/\mathbb{Q}}(a_1, a_2, \ldots, a_n) = m^2 d_K.$$

Solution. Let $\alpha_1, \alpha_2, \ldots, \alpha_n$ be an integral basis of $\mathcal{O}_K$. Theorem 4.2.2 says that N has a basis $\beta_1, \beta_2, \ldots, \beta_n$ such that $\beta_i = \sum_{j \geq i} p_{ij} \alpha_j$. Then from Exercise 4.2.7,

$$\begin{aligned} d_{K/\mathbb{Q}}(\beta_1, \ldots, \beta_n) &= \left(\det(p_{ij})\right)^2 d_{K/\mathbb{Q}}(\alpha_1, \ldots, \alpha_n) \\ &= m^2 d_K. \end{aligned}$$

Reasoning as in Exercise 4.2.5, we deduce

$$d_{K/\mathbb{Q}}(\beta_1, \ldots, \beta_n) = d_{K/\mathbb{Q}}(a_1, \ldots, a_n),$$

which proves the result.

4.3 Examples

Exercise 4.3.2 Let $m \in \mathbb{Z}$, $\alpha \in \mathcal{O}_K$. Prove that $d_{K/\mathbb{Q}}(\alpha + m) = d_{K/\mathbb{Q}}(\alpha)$.

Solution. By definition, $d_{K/\mathbb{Q}}(\alpha) = \prod_{i<j}(\alpha^{(i)} - \alpha^{(j)})^2$. We note that the ith conjugate of $\alpha + m$ is simply $\alpha^{(i)} + m$, and so

$$\begin{aligned} d_{K/\mathbb{Q}}(\alpha + m) &= \prod_{i<j}\left(\alpha^{(i)} + m - (\alpha^{(j)} + m)\right)^2 \\ &= \prod_{i<j}\left(\alpha^{(i)} - \alpha^{(j)}\right)^2 \\ &= d_{K/\mathbb{Q}}(\alpha), \end{aligned}$$

as desired.

Exercise 4.3.3 Let α be an algebraic integer, and let $f(x)$ be the minimal polynomial of α. If f has degree n, show that $d_{K/\mathbb{Q}}(\alpha) = (-1)^{\binom{n}{2}} \prod_{i=1}^{n} f'(\alpha^{(i)})$.

Solution. Let $f(x)$ be the minimal polynomial of α. Then if $\alpha^{(1)}, \ldots, \alpha^{(n)}$ are the conjugates of α, $f(x) = \prod_{k=1}^{n}(x - \alpha^{(k)})$. Then

$$f'(x) = \sum_{k=1}^{n} \frac{f(x)}{(x - \alpha^{(k)})}$$

and

$$f'(\alpha^{(i)}) = \prod_{k \neq i}(\alpha^{(i)} - \alpha^{(k)}).$$

Therefore

$$\begin{aligned} \prod_{i=1}^{n} f'(\alpha^{(i)}) &= \prod_{i=1}^{n}\prod_{k \neq i}\left(\alpha^{(i)} - \alpha^{(k)}\right) \\ &= \prod_{i<k}\left[-\left(\alpha^{(i)} - \alpha^{(k)}\right)^2\right] \\ &= (-1)^{\binom{n}{2}} d_{K/\mathbb{Q}}(\alpha). \end{aligned}$$

Exercise 4.3.5 If $D \equiv 1 \pmod 4$, show that every integer of $\mathbb{Q}(\sqrt{D})$ can be written as $(a + b\sqrt{D})/2$ where $a \equiv b \pmod 2$.

Solution. By Example 4.3.4, an integral basis is given by $1, (1 + \sqrt{D})/2$. Thus every integer is of the form

$$c + d\left(\frac{1 + \sqrt{D}}{2}\right) = \frac{(2c + d) + d\sqrt{D}}{2} = \frac{a + b\sqrt{D}}{2}.$$

Then we see that $a = 2c + d, b = d$ satisfies $a \equiv b \pmod 2$.

Conversely, if $a \equiv b \pmod 2$, writing $d = b$ and $a = 2c + d$ for some c, we find

$$\frac{a + b\sqrt{D}}{2} = c + d\left(\frac{1 + \sqrt{D}}{2}\right)$$

is an integer of $\mathbb{Q}(\sqrt{D})$.

Exercise 4.3.7 Let ζ be any primitive pth root of unity, and $K = \mathbb{Q}(\zeta)$. Show that $1, \zeta, \ldots, \zeta^{p-2}$ form an integral basis of K.

Solution. The minimal polynomial of ζ is the pth cyclotomic polynomial,

$$\Phi(x) = \frac{x^p - 1}{x - 1} = 1 + x^2 + \cdots + x^{p-1}.$$

We want to show that this is irreducible. Consider instead the polynomial $F(x) = \Phi(x+1)$. Clearly F will be irreducible over $\mathbb{Q}$ if and only if Φ is.

$$F(x) = \frac{(x+1)^p - 1}{x} = x^{p-1} + px^{p-2} + \binom{p}{2}x^{p-3} + \cdots + \binom{p}{p-2}x + p.$$

This is Eisensteinian with respect to p and so $F(x)$ (and thus $\Phi(x)$) is irreducible. The conjugates of ζ are $\zeta, \zeta^2, \ldots, \zeta^{p-1}$. We can deduce that $[K : \mathbb{Q}] = p - 1$.

Now,

$$\begin{aligned} \Phi'(x) &= \frac{px^{p-1}(x-1) - (x^p - 1)}{(x-1)^2} \\ &= \frac{px^{p-1} - (1 + x + \cdots + x^{p-1})}{x - 1}, \end{aligned}$$

and so

$$\Phi'(\zeta^k) = \frac{p\zeta^{-k}}{\zeta^k - 1}.$$

Using Exercise 4.3.3, we can compute

$$\begin{aligned} d_{K/\mathbb{Q}}(\zeta) &= \pm \prod_{k=1}^{p-1} \frac{p\zeta^{-k}}{\zeta^k - 1} \\ &= \pm p^{p-1} \frac{1}{\prod_{k=1}^{p-1}(\zeta^k - 1)} \\ &= \pm p^{p-1} \frac{1}{\prod_{k=1}^{p-1}(1 - \zeta^k)} \\ &= \pm p^{p-2}, \end{aligned}$$

since $\prod_{k=1}^{p-1}(1 - \zeta^k) = \Phi(1) = p$. We know that $d_{K/\mathbb{Q}}(\zeta) = p^{p-2} = m^2 d_K$ and also that $p \nmid m$ because F, the minimal polynomial for $\zeta - 1$, is p-Eisensteinian, and $d_{K/\mathbb{Q}}(\zeta - 1) = d_{K/\mathbb{Q}}(\zeta)$. Then m must be 1, meaning that $\mathcal{O}_K = \mathbb{Z}[1, \zeta, \ldots, \zeta^{p-2}]$.

4.4 Ideals in $\mathcal{O}_K$

Exercise 4.4.1 Let $\mathfrak{a}$ be a nonzero ideal of $\mathcal{O}_K$. Show that $\mathfrak{a} \cap \mathbb{Z} \neq \{0\}$.

Solution. Let α be a nonzero algebraic integer in $\mathfrak{a}$ satisfying the minimal polynomial $x^r + a_{r-1}x^{r-1} + \cdots + a_0 = 0$ with $a_i \in \mathbb{Z}\ \forall i$ and a_0 not zero. Then $a_0 = -(\alpha^r + \cdots + a_1\alpha)$. The left-hand side of this equation is in $\mathbb{Z}$, while the right-hand side is in $\mathfrak{a}$.

Exercise 4.4.2 Show that $\mathfrak{a}$ has an integral basis.

Solution. Let $\mathfrak{a}$ be an ideal of $\mathcal{O}_K$, and let $\omega_1, \omega_2, \ldots, \omega_n$ be an integral basis for $\mathcal{O}_K$. Note that for any ω_i in $\mathcal{O}_K$, $a_0\omega_i = -(\alpha^r + \cdots + a_1\alpha)\omega_i \in \mathfrak{a}$. Therefore $\mathfrak{a}$ has finite index in $\mathcal{O}_K$ and $\mathfrak{a} \subseteq \mathcal{O}_K = \mathbb{Z}\omega_1 + \mathbb{Z}\omega_2 + \cdots + \mathbb{Z}\omega_n$ has maximal rank. Then since $\mathfrak{a}$ is a submodule of $\mathcal{O}_K$, by Theorem 4.2.2 there exists an integral basis for $\mathfrak{a}$.

Exercise 4.4.3 Show that if $\mathfrak{a}$ is a nonzero ideal in $\mathcal{O}_K$, then $\mathfrak{a}$ has finite index in $\mathcal{O}_K$.

Solution. Surely, if $\mathcal{O}_K = \mathbb{Z}\omega_1 + \mathbb{Z}\omega_2 + \cdots + \mathbb{Z}\omega_n$, then by the preceding two exercises we can pick a rational integer a such that

$$a\mathcal{O}_K = a\mathbb{Z}\omega_1 + a\mathbb{Z}\omega_2 + \cdots + a\mathbb{Z}\omega_n \subset \mathfrak{a} \subset \mathcal{O}_K.$$

But $a\mathcal{O}_K$ obviously has index a^n in $\mathcal{O}_K$. Thus, the index of $\mathfrak{a}$ in $\mathcal{O}_K$ must be finite.

Exercise 4.4.4 Show that every nonzero prime ideal in $\mathcal{O}_K$ contains exactly one integer prime.

Solution. If $\wp$ is a prime ideal of $\mathcal{O}_K$, then certainly it contains an integer, from Exercise 4.4.1. By the definition of a prime ideal, if $ab \in \wp$, either $a \in \wp$ or $b \in \wp$. So $\wp$ must contain some rational prime. Now, if $\wp$ contained two distinct rational primes p, q, say, then it would necessarily contain their greatest common denominator which is 1. But this contradicts the assumption of nontriviality. So every prime ideal of $\mathcal{O}_K$ contains exactly one integer prime.

Exercise 4.4.5 Let $\mathfrak{a}$ be an integral ideal with basis $\alpha_1, \ldots, \alpha_n$. Show that

$$[\det(\alpha_i^{(j)})]^2 = (N\mathfrak{a})^2 d_K.$$

Solution. Since $\mathfrak{a}$ is a submodule of index $N\mathfrak{a}$ in $\mathcal{O}_K$, this is immediate from Exercise 4.2.8.

4.5 Supplementary Problems

Exercise 4.5.1 Let K be an algebraic number field. Show that $d_K \in \mathbb{Z}$.

Solution. By definition

$$d_K = \det(\omega_i^{(j)})^2 = \det\big(\mathrm{Tr}(\omega_i\omega_j)\big),$$

where $\omega_1, \ldots, \omega_n$ is an integral basis of $\mathcal{O}_K$. Since $\mathrm{Tr}(\omega_i\omega_j) \in \mathbb{Z}$, the determinant is an integer.

Exercise 4.5.2 Let $K/\mathbb{Q}$ be an algebraic number field of degree n. Show that $d_K \equiv 0$ or $1 \pmod 4$. This is known as Stickelberger's criterion.

Solution. Let $\omega_1, \ldots, \omega_n$ be an integral basis of $\mathcal{O}_K$. By definition,

$$d_K = \det\big(\sigma_i(\omega_j)\big)^2,$$

where $\sigma_1, \ldots, \sigma_n$ are the distinct embeddings of K into $\overline{\mathbb{Q}}$. Now write

$$\det(\sigma_i(\omega_j)) = P - N,$$

where P is the contribution arising from the even permutations and N the odd permutations in the definition of the determinant. Then

$$d_K = (P-N)^2 = (P+N)^2 - 4PN.$$

Since $\sigma_i(P+N) = P+N$, and $\sigma_i(PN) = PN$ we see that $P+N$ and PN are integers. Reducing mod 4 gives the result.

Exercise 4.5.3 Let $f(x) = x^n + a_{n-1}x^{n-1} + \cdots + a_1x + a_0$ with $a_i \in \mathbb{Z}$ be the minimal polynomial of θ. Let $K = \mathbb{Q}(\theta)$. If for each prime p such that $p^2 \mid d_{K/\mathbb{Q}}(\theta)$ we have $f(x)$ Eisensteinian with respect to p, show that $\mathcal{O}_K = \mathbb{Z}[\theta]$.

Solution. By Example 4.3.1, the index of θ is not divisible by p for any prime p satisfying $p^2 \mid d_{K/\mathbb{Q}}(\theta)$. By Exercise 4.2.8,

$$d_{K/\mathbb{Q}}(\theta) = m^2 d_K,$$

where $m = \big[\mathcal{O}_K : \mathbb{Z}[\theta]\big]$. Hence $m = 1$.

Exercise 4.5.4 If the minimal polynomial of α is $f(x) = x^n + ax + b$, show that for $K = \mathbb{Q}(\alpha)$,

$$d_{K/\mathbb{Q}}(\alpha) = (-1)^{\binom{n}{2}}\left(n^n b^{n-1} + a^n(1-n)^{n-1}\right).$$

Solution. By Exercise 4.3.3,

$$d_{K/\mathbb{Q}}(\alpha) = (-1)^{\binom{n}{2}} \prod_{i=1}^{n} f'(\alpha^{(i)}),$$

where $\alpha^{(1)}, \ldots, \alpha^{(n)}$ are the conjugates of α. Now

$$\begin{aligned} f'(x) &= nx^{n-1} + a \\ &= \frac{1}{x}(nx^n + ax) \end{aligned}$$

so that

$$f'(\alpha^{(i)}) = \frac{(-n(a\alpha^{(i)} + b) + a\alpha^{(i)})}{\alpha^{(i)}}.$$

Hence

$$\begin{aligned} \prod_{i=1}^{n} f'(\alpha^{(i)}) &= (-1)^n b^{-1} \prod_{i=1}^{n} (a(1-n)\alpha^{(i)} - nb) \\ &= b^{-1} a^n (1-n)^n f\left(\frac{nb}{a(1-n)}\right). \end{aligned}$$

Exercise 4.5.5 Determine an integral basis for $K = \mathbb{Q}(\theta)$ where $\theta^3 + 2\theta + 1 = 0$.

Solution. By applying the previous exercise, the discriminant of θ is -59, which is squarefree. Therefore $\mathcal{O}_K = \mathbb{Z}[\theta]$.

Exercise 4.5.6 (Dedekind) Let $K = \mathbb{Q}(\theta)$ where $\theta^3 - \theta^2 - 2\theta - 8 = 0$.

(a) Show that $f(x) = x^3 - x^2 - 2x - 8$ is irreducible over $\mathbb{Q}$.

(b) Consider $\beta = (\theta^2 + \theta)/2$. Show that $\beta^3 - 3\beta^2 - 10\beta - 8 = 0$. Hence β is integral.

(c) Show that $d_{K/\mathbb{Q}}(\theta) = -4(503)$, and $d_{K/\mathbb{Q}}(1, \theta, \beta) = -503$. Deduce that $1, \theta, \beta$ is a $\mathbb{Z}$-basis of $\mathcal{O}_K$.

(d) Show that every integer x of K has an even discriminant.

(e) Deduce that $\mathcal{O}_K$ has no integral basis of the form $\mathbb{Z}[\alpha]$.

Solution. Note that if (a) is not true, then f has a linear factor and by the rational root theorem, this factor must be of the form $x - a$ where $a \mid 8$. A systematic check rules out this possibility. (b) can be checked directly.

(c) This is easy to deduce from the formula

$$d_{K/\mathbb{Q}}(1, \theta, \beta) = \tfrac{1}{4} d_{K/\mathbb{Q}}(\theta)$$

as a simple computation shows.

For (d), write $x = A + B\theta + C\beta, A, B, C \in \mathbb{Z}$. Since

$$\begin{aligned} \beta^2 &= 6 + 2\theta + 3\beta, \\ \theta^2 &= 2\beta - \theta, \\ \theta\beta &= 2\beta + 4, \end{aligned}$$

we find

$$x^2 = (a^2 + 6C^2 + 8BC) + \theta(2C^2 - B^2 + 2AB) + \beta(2B^2 + 3C^2 + 2AC + 4BC)$$

so that

$$d_{K/\mathbb{Q}}(1, x, x^2) \equiv -503(BC)^2(3C + B)^2 \pmod{2},$$

which is an even number in all cases.

By (d), $d_{K/\mathbb{Q}}(\alpha)$ is even and hence is not equal to -503, which proves (e).

Exercise 4.5.7 Let $m = p^a$, with p prime and $K = \mathbb{Q}(\zeta_m)$. Show that

$$(1 - \zeta_m)^{\varphi(m)} = p\mathcal{O}_K.$$

Solution. First note that

$$\frac{x^m - 1}{x^{m/p} - 1} = \prod_{\substack{1 \le b < m \\ (b,m)=1}} (x - \zeta_m^b)$$

so that taking the limit as x goes to 1 of both sides gives

$$\begin{aligned} p &= \prod_{\substack{1 \le b < m \\ (b,m)=1}} (1 - \zeta_m^b) \\ &= (1 - \zeta_m)^{\varphi(m)} \prod_{\substack{1 \le b < m \\ (b,m)=1}} \frac{1 - \zeta_m^b}{1 - \zeta_m}. \end{aligned}$$

This latter quantity is a unit since

$$\frac{1 - \zeta_m}{1 - \zeta_m^b} = \frac{1 - \zeta_m^{ab}}{1 - \zeta_m^b} = 1 + \zeta_m^b + \cdots + \zeta_m^{b(a-1)}$$

for any a satisfying $ab \equiv 1 \pmod{m}$.

Exercise 4.5.8 Let $m = p^a$, with p prime, and $K = \mathbb{Q}(\zeta_m)$. Show that

$$d_{K/\mathbb{Q}}(\zeta_m) = \frac{(-1)^{\varphi(m)/2} m^{\varphi(m)}}{p^{m/p}}.$$

Solution. We need to compute the Vandermonde determinant given by

$$(-1)^{\varphi(m)/2} \prod_{\substack{1\leq a,b<m \\ (a,m)=1 \\ (b,m)=1 \\ a\neq b}} (\zeta_m^a - \zeta_m^b).$$

Let

$$\theta = \prod_{\substack{1\leq b<m \\ (b,m)=1}} (\zeta_m - \zeta_m^b).$$

Clearly $\theta = \phi_m'(\zeta_m)$ and $\mathrm{N}_{K/\mathbb{Q}}(\theta)$ is the discriminant we seek. Since

$$\phi_m(x) = \frac{x^m - 1}{x^{m/p} - 1},$$

we find

$$\phi_m'(\zeta_m) = \frac{m\zeta_m^{m-1}}{\zeta_m^{m/p} - 1},$$

and the norm of this element is

$$\frac{m^m}{\mathrm{N}_{K/\mathbb{Q}}(\zeta_m^{m/p} - 1)}.$$

Because $\eta = \zeta_m^{m/p}$ is a primitive pth root of unity,

$$\mathrm{N}_{K/\mathbb{Q}}(\zeta_m^{m/p} - 1) = \mathrm{N}_{\mathbb{Q}(\eta)/\mathbb{Q}}(\eta - 1)^{m/p}.$$

In Exercise 4.3.7, we saw that $\mathrm{N}_{\mathbb{Q}(\eta)/\mathbb{Q}}(\eta - 1) = p$.

Exercise 4.5.9 Let $m = p^a$, with p prime. Show that $\{1, \zeta_m, \ldots, \zeta_m^{\varphi(m)-1}\}$ is an integral basis for the ring of integers of $K = \mathbb{Q}(\zeta_m)$.

Solution. Clearly $\mathbb{Z}[\zeta_m] \subseteq \mathcal{O}_K$. We want to prove the reverse inclusion. Let $\lambda = 1 - \zeta_m$. Since $\lambda^j = (1-\zeta_m)^j \in \mathbb{Z}[\zeta_m]$ and $\zeta_m^j = (1-\lambda)^j \in \mathbb{Z}[\zeta_m]$, we see that $\mathbb{Z}[\zeta_m] = \mathbb{Z}[\lambda]$. Thus, it suffices to show $\mathbb{Z}[\lambda] \supseteq \mathcal{O}_K$. Let $\alpha \in \mathcal{O}_K$ and write

$$\alpha = \sum_{j=0}^{\varphi(m)-1} a_j\lambda^j = \sum_{j=0}^{\varphi(m)-1} b_j\zeta_m^j, \qquad a_j, b_j \in \mathbb{Q}.$$

It suffices to show $a_j \in \mathbb{Z}$ for $j = 0, 1, \ldots, \varphi(m) - 1$. Let $\sigma_c(\zeta_m) = \zeta_m^c$. Then

$$\sigma_c(\alpha) = \sum_{j=0}^{\varphi(m)-1} b_j\sigma_c(\zeta_m)^j = \sum_{j=0}^{\varphi(m)-1} b_j\zeta_m^{cj}$$

for each $(c, m) = 1$. We solve for b_j using Cramer's rule. Moreover, by the previous question, we see that

$$b_j = \frac{c_j}{p^{d_j}} \quad c_j \in \mathbb{Z}.$$

Thus, a_j has at most a power of p in its denominator. Let n be the least nonnegative integer such that $p^n a_j \in \mathbb{Z}$ for $j = 0, 1, \ldots, \varphi(m) - 1$. Suppose $n > 0$. Let k be the smallest nonnegative integer such that p does not divide $p^n a_k$. Then

$$a_r \lambda^r \in p\mathcal{O}_K = \lambda^{\varphi(m)}\mathcal{O}_K$$

(by the penultimate question) for $r = 0, 1, \ldots, k-1$. Since α is an integer, $p^n \alpha \in p\mathcal{O}_K = \lambda^{\varphi(m)}\mathcal{O}_K$ so that

$$\sum_{j=k}^{\varphi(m)-1} p^n a_j \lambda^{j-k} \in \lambda\mathcal{O}_K$$

so that $p^n a_k \in \lambda\mathcal{O}_K \cap \mathbb{Z} = p\mathbb{Z}$, a contradiction.

Exercise 4.5.10 Let $K = \mathbb{Q}(\zeta_m)$ where $m = p^a$. Show that

$$d_K = \frac{(-1)^{\varphi(m)/2} m^{\varphi(m)}}{p^{m/p}}.$$

Solution. This is immediate from the previous two questions.

Exercise 4.5.11 Show that $\mathbb{Z}[\zeta_n + \zeta_n^{-1}]$ is the ring of integers of $\mathbb{Q}(\zeta_n + \zeta_n^{-1})$, where ζ_n denotes a primitive nth root of unity, and $n = p^\alpha$.

Solution. Suppose $\alpha = a_0 + a_1(\zeta_n + \zeta_n^{-1}) + \cdots + a_N(\zeta_n + \zeta_n^{-1})^N$ is an algebraic integer with $N \leq \frac{1}{2}\phi(n) - 1$ and the $a_i \in \mathbb{Q}$. By subtracting those terms with $a_i \in \mathbb{Z}$, we may suppose $a_N \notin \mathbb{Z}$. Multiplying by ζ_n^N and expanding the result as a polynomial in ζ_n, we find that

$$\zeta_n^N \alpha = a_N + \cdots + a_N \zeta_n^{2N}$$

is an algebraic integer in $\mathbb{Q}(\zeta_n)$. Therefore, it lies in $\mathbb{Z}[\zeta_n]$. Since

$$2N \leq \phi(n) - 2 \leq \phi(n) - 1,$$

we conclude $a_N \in \mathbb{Z}$, contrary to our assumption above. (This is also true for arbitrary n by applying Exercise 4.5.25.)

Exercise 4.5.12 Let K and L be algebraic number fields of degree m and n, respectively, over $\mathbb{Q}$. Let $d = \gcd(d_K, d_L)$. Show that if $[KL : \mathbb{Q}] = mn$, then $\mathcal{O}_{KL} \subseteq 1/d\mathcal{O}_K\mathcal{O}_L$.

Solution. Let $\{\alpha_1, \ldots, \alpha_m\}$ be a $\mathbb{Z}$-basis for $\mathcal{O}_K$ and let $\{\beta_1, \ldots, \beta_n\}$ be a $\mathbb{Z}$-basis for $\mathcal{O}_L$. Then $\alpha_i\beta_j$, $1 \leq i \leq m$, $1 \leq j \leq n$, is a $\mathbb{Q}$-basis for KL over $\mathbb{Q}$ since $[KL : \mathbb{Q}] = mn$. Any $\omega \in \mathcal{O}_{KL}$ can therefore be written as

$$\omega = \sum_{i,j} \frac{m_{ij}}{r} \alpha_i \beta_j,$$

where r, $m_{ij} \in \mathbb{Z}$ and $\gcd(r, \gcd(m_{ij})) = 1$. It suffices to show that $r \mid d_K$ and by symmetry $r \mid d_L$ so that $r \mid d$. Since $[KL : \mathbb{Q}] = mn$, every embedding σ of K into $\mathbb{C}$ can be extended to KL acting trivially on L. Hence

$$\sigma(\omega) = \sum_{i,j} \frac{m_{ij}}{r} \sigma(\alpha_i)\beta_j.$$

Set $x_i = \sum_j m_{ij}\beta_j/r$. We then obtain m equations

$$\sum_{i=1}^{m} \sigma(\alpha_i)x_i = \sigma(\omega),$$

one for each $\sigma : K \hookrightarrow \mathbb{C}$. We solve for x_i by Cramer's rule: $x_i = \gamma_i/\delta$ where $\delta = \det(\sigma(\alpha_i))$. Since $\delta^2 = d_K$ we find

$$\delta\gamma_i = \sum_{j=1}^{n} \frac{\delta^2 m_{ij}}{r} \beta_j \in \mathcal{O}_K$$

since δ and each of γ_i are algebraic integers. Hence $d_K m_{ij}/r$ are all integers. It follows that r divides all $d_K m_{ij}$. Since $\gcd(r, \gcd(m_{ij})) = 1$, we deduce $r \mid d_K$.

Exercise 4.5.13 Let K and L be algebraic number fields of degree m and n, respectively, with $\gcd(d_K, d_L) = 1$. If $\{\alpha_1, \ldots, \alpha_m\}$ is an integral basis of $\mathcal{O}_K$ and $\{\beta_1, \ldots, \beta_n\}$ is an integral basis of $\mathcal{O}_L$, show that $\mathcal{O}_{KL}$ has an integral basis $\{\alpha_i\beta_j\}$ given that $[KL : \mathbb{Q}] = mn$. (In a later chapter, we will see that $\gcd(d_K, d_L) = 1$ implies that $[KL : \mathbb{Q}] = mn$.)

Solution. This is immediate from the previous question.

Exercise 4.5.14 Find an integral basis for $\mathbb{Q}(\sqrt{2}, \sqrt{-3})$.

Solution. If $K = \mathbb{Q}(\sqrt{2}), L = \mathbb{Q}(\sqrt{-3})$, then $d_K = 8$, $d_L = -3$ which are coprime. By the previous question, a $\mathbb{Z}$-basis for the ring of integers of $\mathbb{Q}(\sqrt{2}, \sqrt{-3})$ is given by

$$\left\{1, \sqrt{2}, \frac{1+\sqrt{-3}}{2}, \sqrt{2}\left(\frac{1+\sqrt{-3}}{2}\right)\right\}.$$

Exercise 4.5.15 Let p and q be distinct primes $\equiv 1 \pmod 4$. Let $K = \mathbb{Q}(\sqrt{p})$, $L = \mathbb{Q}(\sqrt{q})$. Find a $\mathbb{Z}$-basis for $\mathbb{Q}(\sqrt{p}, \sqrt{q})$.

Solution. We have $d_K = p$, $d_L = q$ which are coprime. Now invoke the penultimate question to deduce that

$$\left\{1, \frac{1+\sqrt{p}}{2}, \frac{1+\sqrt{q}}{2}, \left(\frac{1+\sqrt{p}}{2}\right)\left(\frac{1+\sqrt{q}}{2}\right)\right\}$$

is a $\mathbb{Z}$-basis for the ring of integers of $\mathbb{Q}(\sqrt{p}, \sqrt{q})$.

Exercise 4.5.16 Let K be an algebraic number field of degree n over $\mathbb{Q}$. Let $a_1, \ldots, a_n \in \mathcal{O}_K$ be linearly independent over $\mathbb{Q}$. Set

$$\Delta = d_{K/\mathbb{Q}}(a_1, \ldots, a_n).$$

Show that if $\alpha \in \mathcal{O}_K$, then $\Delta\alpha \in \mathbb{Z}[a_1, \ldots, a_n]$.

Solution. Write $\alpha = c_1a_1 + \cdots + c_na_n$ for some rational numbers c_i. By taking conjugates, we get a system of n equations and we can solve for c_i using Cramer's rule. Thus $c_j = A_jD/\Delta$ where $D^2 = \Delta$ and it is easy to see that A_jD is an algebraic integer. Therefore Δc_j is an algebraic integer lying in $\mathbb{Q}$ so $\Delta c_j \in \mathbb{Z}$, as required.

Exercise 4.5.17 (Explicit Construction of Integral Bases) Suppose K is an algebraic number field of degree n over $\mathbb{Q}$. Let $a_1, \ldots, a_n \in \mathcal{O}_K$ be linearly independent over $\mathbb{Q}$ and set

$$\Delta = d_{K/\mathbb{Q}}(a_1, \ldots, a_n).$$

For each i, choose the least natural number d_{ii} so that for some $d_{ij} \in \mathbb{Z}$, the number

$$w_i = \Delta^{-1}\sum_{j=1}^{i} d_{ij}a_j \in \mathcal{O}_K.$$

Show that $w_1, \ldots, w_n$ is an integral basis of $\mathcal{O}_K$.

Solution. First observe that there are integers c_{ij} so that

$$\Delta^{-1}\sum_{j=1}^{i} c_{ij}a_j \in \mathcal{O}_K$$

(e.g., $c_{ij} = \Delta$). Clearly $w_1, \ldots, w_n$ are linearly independent over $\mathbb{Q}$ because

$$d_{K/\mathbb{Q}}(w_1, \ldots, w_n) = \Delta^{-n}(d_{11}\cdots d_{nn})^2 d_{K/\mathbb{Q}}(a_1, \ldots, a_n)$$

by Exercise 4.2.7, and the right-hand side is nonzero. Observe now that if $\alpha \in \mathcal{O}_K$ can be written as

$$\alpha = \Delta^{-1}(c_1a_1 + \cdots + c_ja_j)$$

for some j, then $d_{jj} \mid c_j$. Indeed, write $c_j = sd_{jj} + r$, $0 \le r < d_{jj}$, so that

$$\alpha - sw_j = \Delta^{-1}\big((c_1 - d_{j1})a_1 + \cdots + ra_j\big) \in \mathcal{O}_K,$$

contrary to our choice of w_j if $r \neq 0$.

We now show by induction on j that every number of $\mathcal{O}_K$ of the form

$$\Delta^{-1}(x_1a_1 + \cdots + x_ja_j)$$

with $x_i \in \mathbb{Z}$ lies in $\mathbb{Z}[w_1, \ldots, w_n]$. For $j = 1$, there is nothing to prove because then $d_{11} \mid x_1$ and we are done. Assume that we have proved it for $j < k$. Then suppose

$$y = \Delta^{-1}(x_1a_1 + \cdots + x_ka_k) \in \mathcal{O}_K$$

with $x_i \in \mathbb{Z}$. Then $d_{kk} \mid x_k$ so that for some integer t,

$$y - tw_k = \Delta^{-1}(x_1'a_1 + \cdots + x_{k-1}'a_{k-1}) \in \mathcal{O}_K.$$

By induction, the right-hand side lies in $\mathbb{Z}[w_1, \ldots, w_n]$ and so does y. For $j = n$, this means that every number of $\mathcal{O}_K$ of the form $\Delta^{-1}(x_1a_1 + \cdots + x_na_n)$ with $x_i \in \mathbb{Z}$ lies in $\mathbb{Z}[w_1, \ldots, w_n]$. But by the previous exercise, every $\alpha \in \mathcal{O}_K$ can be so expressed.

Exercise 4.5.18 If K is an algebraic number field of degree n over $\mathbb{Q}$ and $a_1, \ldots, a_n \in \mathcal{O}_K$ are linearly independent over $\mathbb{Q}$, then there is an integral basis $w_1, \ldots, w_n$ of $\mathcal{O}_K$ such that

$$a_j = c_{j1}w_1 + \cdots + c_{jj}w_j,$$

$c_{ij} \in \mathbb{Z}, j = 1, \ldots, n$.

Solution. We take $w_1, \ldots, w_n$ as constructed in the previous exercise. This is an integral basis. Solving for a_j and noting that the matrix (d_{ij}) is lower triangular, we see that each a_j can be written as above. Moreover, the $c_{ij} \in \mathbb{Z}$ since $w_1, \ldots, w_n$ is an integral basis by construction.

Exercise 4.5.19 If $\mathbb{Q} \subseteq K \subseteq L$ and K, L are algebraic number fields, show that $d_K \mid d_L$.

Solution. Let $[K : \mathbb{Q}] = m$, $[L : K] = n$. Let $a_1, \ldots, a_m$ be an integral basis of K. Extend this to a basis of L (viewed as a vector space over $\mathbb{Q}$) so that $a_1, \ldots, a_{mn}$ is linearly independent over $\mathbb{Q}$. By Exercise 3.3.7, we may suppose each a_i is an algebraic integer. By the previous exercise, there is an integral basis $w_1, \ldots, w_{mn}$ of $\mathcal{O}_L$ such that

$$a_j = c_{j1}w_1 + \cdots + c_{jj}w_j, \qquad c_{ij} \in \mathbb{Z}.$$

Since the matrix (c_{ij}) is triangular, it is easy to see that $w_1, \ldots, w_m$ lie in K. Because the w_i are algebraic integers, and $a_1, \ldots, a_m$ is an integral basis of K, it follows that $w_1, \ldots, w_m$ is an integral basis of K. Now write down the definition of the discriminant of L. Let $\sigma_1, \ldots, \sigma_{mn}$ be the embeddings of L into $\mathbb{C}$ such that $\sigma_i(w_j) = w_j^{(i)}$ for $1 \leq j \leq m$.

We order the $\sigma_1, \ldots, \sigma_{mn}$ so that $\sigma_i(x) = \sigma_{i'}(x)$ for $i \equiv i' \pmod{m}$ and $x \in K$. Then

$$
\begin{aligned}
d_L &= \det\big(\sigma_i(w_j)\big)^2 \\
&= \begin{vmatrix}
w_1^{(1)} & \cdots & w_1^{(m)} & w_1^{(1)} & \cdots & w_1^{(m)} & \cdots & w_1^{(m)} \\
\vdots & & & & & & & \\
w_m^{(1)} & \cdots & w_m^{(m)} & w_m^{(1)} & \cdots & w_m^{(m)} & \cdots & w_m^{(m)} \\
w_{m+1}^{(1)} & \cdots & w_{m+1}^{(m)} & w_{m+1}^{(m+1)} & \cdots & w_{m+1}^{(2m)} & \cdots & w_{m+1}^{(mn)} \\
\vdots & & & & & & & \\
w_{mn}^{(1)} & \cdots & w_{mn}^{(m)} & w_{mn}^{(m+1)} & \cdots & & & w_{mn}^{(mn)}
\end{vmatrix}^2 \\
&= \begin{vmatrix}
w_1^{(1)} & \cdots & w_1^{(m)} & 0 & \cdots & 0 \\
\vdots & & & & & \\
w_m^{(1)} & \cdots & w_m^{(m)} & 0 & \cdots & 0 \\
w_{m+1}^{(1)} & \cdots & w_{m+1}^{(m)} & w_{m+1}^{(m+1)} - w_{m+1}^{(1)} & \cdots & w_{m+1}^{(mn)} - w_{m+1}^{(m)} \\
\vdots & & & & & \\
w_{mn}^{(1)} & \cdots & w_{mn}^{(m)} & w_{mn}^{(m+1)} - w_{mn}^{(1)} & \cdots & w_{mn}^{(mn)} - w_{mn}^{(m)}
\end{vmatrix}^2 \\
&= d_K \cdot a,
\end{aligned}
$$

where a is an algebraic integer. Since d_L/d_K is a rational number, we deduce that a is a rational integer.

Exercise 4.5.20 (The Sign of the Discriminant) Suppose K is a number field with r_1 real embeddings and $2r_2$ complex embeddings so that

$$r_1 + 2r_2 = [K : \mathbb{Q}] = n$$

(say). Show that d_K has sign $(-1)^{r_2}$.

Solution. Let $\omega_1, \ldots, \omega_n$ be an integral basis of K. Then

$$d_K = \det\big(\sigma_i(\omega_j)\big)^2,$$

where $\sigma_1, \ldots, \sigma_n$ are the embeddings of K. Clearly

$$\overline{\det\big(\sigma_i(\omega_j)\big)} = (-1)^{r_2} \det\big(\sigma_i(\omega_j)\big)$$

since complex conjugation interchanges r_2 rows.

If r_2 is even, then $\det\big(\sigma_i(\omega_j)\big)$ is real so that $d_K > 0$. If r_2 is odd, $\det\big(\sigma_i(\omega_j)\big)$ is purely imaginary so that $d_K < 0$.

Exercise 4.5.21 Show that only finitely many imaginary quadratic fields K are Euclidean.

Solution. If ψ is a Euclidean algorithm for $\mathcal{O}_K$, then let $\alpha_0 \in \mathcal{O}_K$ be such that $\psi(\alpha_0)$ is the minimum nonzero value of ψ. Then, every residue class mod α_0 is represented by 0 or an element $\alpha \in \mathcal{O}_K$ such that $\psi(\alpha) = 0$. Thus α_0 is a unit. In an imaginary quadratic field, there are only finitely many units. If $d_K \neq -3, -4$, the units are ± 1. Thus, if $d_K \neq -3, -4$, we find $\mathrm{N}_{K/\mathbb{Q}}(\alpha_0) \leq 3$, which implies that $\alpha_0 = \pm 1$. In particular, if α_1 is such that $\psi(\alpha_1) = \min \psi(\alpha)$ where the minimum ranges over $\psi(\alpha) \neq \psi(\alpha_0)$, then α_1 is not a unit, $\psi(\alpha_1) > \psi(\alpha_0)$ so that not every residue class mod α_1 can be represented by a class containing an element whose ψ-value is smaller than $\psi(\alpha_1)$.

Exercise 4.5.22 Show that $\mathbb{Z}[(1+\sqrt{-19})/2]$ is not Euclidean. (Recall that in Exercise 2.5.6 we showed this ring is not Euclidean for the norm map.)

Solution. The argument of the previous exercise shows that not all residue classes mod 2 and mod 3 are represented by elements of smaller ψ-value.

Exercise 4.5.23 (a) Let $A = (a_{ij})$ be an $m \times m$ matrix, $B = (b_{ij})$ an $n \times n$ matrix. We define the (Kronecker) tensor product $A \otimes B$ to be the $mn \times mn$ matrix obtained as

$$\begin{pmatrix} Ab_{11} & Ab_{12} & \cdots & Ab_{1n} \\ Ab_{21} & Ab_{22} & \cdots & Ab_{2n} \\ \vdots & \vdots & & \vdots \\ Ab_{n1} & Ab_{n2} & \cdots & Ab_{nn} \end{pmatrix},$$

where each block Ab_{ij} has the form

$$\begin{pmatrix} a_{11}b_{ij} & a_{12}b_{ij} & \cdots & a_{1m}b_{ij} \\ a_{21}b_{ij} & a_{22}b_{ij} & \cdots & a_{2m}b_{ij} \\ \vdots & \vdots & & \vdots \\ a_{m1}b_{ij} & a_{m2}b_{ij} & \cdots & a_{mm}b_{ij} \end{pmatrix}.$$

If C and D are $m \times m$ and $n \times n$ matrices, respectively, show that

$$(A \otimes B)(C \otimes D) = (AC) \otimes (BD).$$

(b) Prove that $\det(A \otimes B) = (\det A)^n (\det B)^m$.

Solution. Part (a) is a straightforward matrix multiplication computation. For part (b), we use linear algebra to find a matrix U such that $U^{-1}BU$ is upper triangular:

$$U^{-1}BU = \begin{pmatrix} c_{11} & c_{12} & \cdots & c_{1n} \\ 0 & c_{22} & \cdots & c_{2n} \\ \vdots & \vdots & & \vdots \\ 0 & 0 & \cdots & c_{nn} \end{pmatrix}.$$

Then $\det B = c_{11}c_{22}\cdots c_{nn}$. Also, by (a),

$$(I \otimes U)^{-1}(A \otimes B)(I \otimes U) = A \otimes (U^{-1}BU)$$

which is

$$\begin{pmatrix} Ac_{11} & Ac_{12} & \cdots & Ac_{1n} \\ 0 & Ac_{22} & \cdots & Ac_{2n} \\ \vdots & \vdots & & \vdots \\ 0 & 0 & \cdots & Ac_{nn} \end{pmatrix}.$$

Again, by linear algebra, we see that

$$\begin{aligned} \det(A \otimes B) &= \prod_{i=1}^{n} \det(Ac_{ii}) \\ &= \prod_{i=1}^{n} (c_{ii}^{m} \det A) \\ &= (\det B)^m (\det A)^n, \end{aligned}$$

as desired.

Exercise 4.5.24 Let K and L be algebraic number fields of degree m and n, respectively, with $\gcd(d_K, d_L) = 1$. Show that

$$d_{KL} = d_K^n \cdot d_L^m.$$

If we set

$$\delta(M) = \frac{\log |d_M|}{[M : \mathbb{Q}]}$$

deduce that $\delta(KL) = \delta(K) + \delta(L)$ whenever $\gcd(d_K, d_L) = 1$.

Solution. By a previous exercise, $\mathcal{O}_{KL}$ has integral basis $\{\alpha_i \beta_j\}$ where $\alpha_1, \ldots, \alpha_m$ is an integral basis of $\mathcal{O}_K$ and $\beta_1, \ldots, \beta_n$ is an integral basis of $\mathcal{O}_L$. Let

$$\begin{aligned} A &= \big(\mathrm{Tr}_{K/\mathbb{Q}}(\alpha_i \alpha_j)\big), \\ B &= \big(\mathrm{Tr}_{L/\mathbb{Q}}(\beta_i \beta_j)\big). \end{aligned}$$

Then it is easily verified that the discriminant of $KL/\mathbb{Q}$ is $\det(A \otimes B)$. By the previous exercise, this is $d_K^n d_L^m$. The second part of the question is immediate upon taking logarithms.

Exercise 4.5.25 Let ζ_m denote a primitive mth root of unity and let $K = \mathbb{Q}(\zeta_m)$. Show that $\mathcal{O}_K = \mathbb{Z}[\zeta_m]$ and

$$d_K = \frac{(-1)^{\phi(m)/2} m^{\varphi(m)}}{\prod_{p|m} p^{\phi(m)/(p-1)}}.$$

Solution. Factor

$$m = \prod_{p^\alpha \| m} p^\alpha.$$

Since the discriminants of $\mathbb{Q}(\zeta_{p^\alpha})$ for $p^\alpha \| m$ are coprime, we have by the previous exercise (and Supplementary Exercise 4.5.8)

$$\frac{\log |d_K|}{\phi(m)} = \sum_{p^\alpha \| m} \left(\alpha - \frac{1}{p-1} \right) \log p.$$

The sign of the determinant is $(-1)^{r_2} = (-1)^{\varphi(m)/2}$. The fact that $\mathcal{O}_K = \mathbb{Z}[\zeta_m]$ follows by an induction argument and Exercise 4.5.13.

Exercise 4.5.26 Let K be an algebraic number field. Suppose that $\theta \in \mathcal{O}_K$ is such that $d_{K/\mathbb{Q}}(\theta)$ is squarefree. Show that $\mathcal{O}_K = \mathbb{Z}[\theta]$.

Solution. Let $m = \left[\mathcal{O}_K : \mathbb{Z}[\theta]\right]$. Then, by Exercise 4.2.8,

$$d_{K/\mathbb{Q}}(\theta) = m^2 d_K.$$

If $d_{K/\mathbb{Q}}(\theta)$ is squarefree, $m = 1$.

Chapter 5

Dedekind Domains

5.1 Integral Closure

Exercise 5.1.1 Show that a nonzero commutative ring R with identity is a field if and only if it has no nontrivial ideals.

Solution. If $x \in R$, $x \neq 0$, is a nonunit, then $1 \notin (x)$, so (x) is a nontrivial ideal.

Suppose that R has a nontrivial ideal $\mathfrak{a}$. Let $x \in \mathfrak{a}, x \neq 0$. Then $(x) \subseteq \mathfrak{a}$. If x is a unit, then $1 \in (x) \subseteq \mathfrak{a}$, so $\mathfrak{a} = R$, a contradiction. Thus, x is not a unit, so R is not a field.

Exercise 5.1.3 Show that a finite integral domain is a field.

Solution. Let R be a finite integral domain. Let $x_1, x_2, \ldots, x_n$ be the elements of R. Suppose that $x_i x_j = x_i x_k$, for some $x_i \neq 0$.

Then $x_i(x_j - x_k) = 0$. Since R is an integral domain, $x_j = x_k$, so $j = k$. Thus, for any $x_i \neq 0$,

$$\{x_i x_1, x_i x_2, \ldots, x_i x_n\} = \{x_1, x_2, \ldots, x_n\}.$$

Since $1 \in R$, there exists x_j such that $x_i x_j = 1$. Therefore, x_i is invertible. Thus all nonzero elements are invertible, so R is a field.

Exercise 5.1.4 Show that every nonzero prime ideal $\wp$ of $\mathcal{O}_K$ is maximal.

Solution. $\mathcal{O}_K/\wp$ is finite from Exercise 4.4.3 and it is an integral domain from Theorem 5.1.2 (b). Thus, Exercise 5.1.3 shows that $\mathcal{O}_K/\wp$ is a field, which in turn implies that $\wp$ is a maximal ideal of $\mathcal{O}_K$.

Exercise 5.1.5 Show that every unique factorization domain is integrally closed.

Solution. Let R be a unique factorization domain, $\alpha \in Q(R)$.

Then $\alpha = a/b$, for some $a, b \in R$ with $(a, b) = 1$. If α is integral over R, then we have a polynomial equation

$$\alpha^n + c_{n-1}\alpha^{n-1} + \cdots + c_0 = 0, \qquad c_i \in R.$$

Thus, multiplying through by b^n and isolating a^n, we have

$$a^n = -b(c_{n-1}a^{n-1} + \cdots + c_1 a b^{n-2} + c_0 b^{n-1}).$$

Thus, $b | a^n$. But, $(a, b) = 1$. By unique factorization, $(a^n, b) = 1$. Therefore, b is a unit in R, so $a/b \in R$.

Thus, R is integrally closed.

5.2 Characterizing Dedekind Domains

Exercise 5.2.1 If $\mathfrak{a} \subsetneq \mathfrak{b}$ are ideals of $\mathcal{O}_K$, show that $N(\mathfrak{a}) > N(\mathfrak{b})$.

Solution. Define a map $f : \mathcal{O}_K/\mathfrak{a} \longrightarrow \mathcal{O}_K/\mathfrak{b}$ by $f(x + \mathfrak{a}) = x + \mathfrak{b}$. If $x + \mathfrak{a} = y + \mathfrak{a}$, then $x - y \in \mathfrak{a} \subseteq \mathfrak{b}$, so $x + \mathfrak{b} = y + \mathfrak{b}$. Thus, f is well-defined.

The function f is not one-to-one since for any $y \in \mathfrak{b} \backslash \mathfrak{a}$, $f(y + \mathfrak{a}) = 0$, but $y + \mathfrak{a} \neq 0$. It is onto since $x + \mathfrak{b} = f(x + \mathfrak{a})$. So we have a map from the finite set $\mathcal{O}_K/\mathfrak{a}$ to the finite set $\mathcal{O}_K/\mathfrak{b}$ which is onto but not one-to-one. Thus $|\mathcal{O}_K/\mathfrak{a}| > |\mathcal{O}_K/\mathfrak{b}|$, i.e., $N(\mathfrak{a}) > N(\mathfrak{b})$.

Exercise 5.2.2 Show that $\mathcal{O}_K$ is Noetherian.

Solution. Suppose that $\mathfrak{a}_1 \subsetneq \mathfrak{a}_2 \subsetneq \mathfrak{a}_3 \subsetneq \cdots$ is an ascending chain of ideals which does not terminate. Then $N(\mathfrak{a}_1) > N(\mathfrak{a}_2) > N(\mathfrak{a}_3) > \cdots$ but $N(\mathfrak{a}_i)$ is finite and positive for all i, so such a strictly decreasing sequence of positive integers must stop. Thus, the ascending chain of ideals must terminate.

Exercise 5.2.4 Show that any principal ideal domain is a Dedekind domain.

Solution. Let R be a principal ideal domain. R is Noetherian since every ideal is finitely generated. R is integrally closed since any principal ideal domain is a unique factorization domain, and so is integrally closed by Exercise 5.1.5.

Let $(p) \neq 0$ be a prime ideal and $(x) \supsetneq (p)$. Then, $p \in (x)$, so $p = xy$, for some $y \in R$. Thus, $xy \in (p)$, so $x \in (p)$ or $y \in (p)$. If $x \in (p)$, then $(x) = (p)$, and if $y \in (p)$, then $y = pq$, for some $q \in R$. This would imply that $p = xy = xqp$ and so $xq = 1$, since R is an integral domain. Thus, $(x) = R$. Therefore, (p) is maximal.

Thus, R is a Dedekind domain.

Exercise 5.2.5 Show that $\mathbb{Z}[\sqrt{-5}]$ is a Dedekind domain, but not a principal ideal domain.

Solution. $\mathbb{Z}[\sqrt{-5}]$ is not a unique factorization domain as was seen in Chapter 2 by taking $6 = 2 \times 3 = (1+\sqrt{-5})(1-\sqrt{-5})$, and so cannot be a principal ideal domain.

To see that it is a Dedekind domain, it is enough to show that it is the set of algebraic integers of the algebraic number field $K = \mathbb{Q}(\sqrt{-5})$. However, we have already proved this, in Exercise 4.1.3. So $\mathbb{Z}[\sqrt{-5}]$ is a Dedekind domain.

5.3 Fractional Ideals and Unique Factorization

Exercise 5.3.1 Show that any fractional ideal is finitely generated as an $\mathcal{O}_K$-module.

Solution. Let $\mathcal{A}$ be a fractional ideal of $\mathcal{O}_K$. Choose $m \in \mathbb{Z}$ such that $m\mathcal{A} \subseteq \mathcal{O}_K$. Since $\mathcal{A}$ is an $\mathcal{O}_K$-module, $m\mathcal{A}$ is an $\mathcal{O}_K$-module contained in $\mathcal{O}_K$ and so is an ideal of $\mathcal{O}_K$. Since $\mathcal{O}_K$ is Noetherian, $m\mathcal{A}$ is finitely generated as an ideal. If $m\mathcal{A}$ is generated as an ideal by $\{a_1, \ldots, a_n\}$, then $\mathcal{A}$ is generated by $\{m^{-1}a_1, \ldots, m^{-1}a_n\}$ as an $\mathcal{O}_K$-module. Thus, $\mathcal{A}$ is finitely generated as an $\mathcal{O}_K$-module.

Exercise 5.3.2 Show that the sum and product of two fractional ideals are again fractional ideals.

Solution. Let $\mathcal{A}$ and $\mathcal{B}$ be fractional ideals. Since $\mathcal{A}$ and $\mathcal{B}$ are both $\mathcal{O}_K$-modules, so are their sum and product.

Let $m\mathcal{A} \subseteq \mathcal{O}_K$, $n\mathcal{B} \subseteq \mathcal{O}_K$ with m, $n \in \mathbb{Z}$. Then

$$\begin{aligned} mn(\mathcal{AB}) &= (m\mathcal{A})(n\mathcal{B}) \\ &\subseteq \mathcal{O}_K, \end{aligned}$$

so $\mathcal{AB}$ is a fractional ideal. Also,

$$\begin{aligned} mn(\mathcal{A}+\mathcal{B}) &= n(m\mathcal{A}) + m(n\mathcal{B}) \\ &\subseteq n\mathcal{O}_K + m\mathcal{O}_K \\ &\subseteq \mathcal{O}_K, \end{aligned}$$

so $\mathcal{A}+\mathcal{B}$ is a fractional ideal.

Exercise 5.3.7 Show that any fractional ideal $\mathcal{A}$ can be written uniquely in the form

$$\frac{\wp_1 \cdots \wp_r}{\wp_1' \cdots \wp_s'},$$

where the $\wp_i$ and $\wp_j'$ may be repeated, but no $\wp_i = \wp_j'$.

Solution. Choose a nonzero element $c \in \mathbb{Z}$ such that $\mathfrak{b} := c\mathcal{A} \subseteq \mathcal{O}_K$. Let $(c) = \mathfrak{m}_1 \cdots \mathfrak{m}_s$, $\mathfrak{b} = \mathfrak{n}_1 \cdots \mathfrak{n}_t$, the $\mathfrak{m}_i$, $\mathfrak{n}_j$ prime.

Then $(c)\mathcal{A} = \mathfrak{b}$, so

$$\mathcal{A} = \frac{\mathfrak{n}_1 \cdots \mathfrak{n}_t}{\mathfrak{m}_1 \cdots \mathfrak{m}_s},$$

and cancelling the primes on the numerator that equal some prime on the denominator, we have that $\mathfrak{m}_i \neq \mathfrak{n}_j \; \forall i, j$. Also, if

$$\mathcal{A} = \frac{\mathfrak{a}_1 \cdots \mathfrak{a}_v}{\mathfrak{b}_1 \cdots \mathfrak{b}_w},$$

with no $\mathfrak{a}_i = \mathfrak{b}_j$, then

$$\mathfrak{a}_1 \cdots \mathfrak{a}_v \mathfrak{m}_1 \cdots \mathfrak{m}_s = \mathfrak{b}_1 \cdots \mathfrak{b}_w \mathfrak{n}_1 \cdots \mathfrak{n}_t.$$

By unique factorization and the fact that no $\mathfrak{b}_i$ is an $\mathfrak{a}_j$ and no $\mathfrak{m}_i$ is an $\mathfrak{n}_j$, the $\mathfrak{b}_i$'s must coincide up to reordering with the $\mathfrak{m}_j$'s and the $\mathfrak{a}_i$'s with the $\mathfrak{n}_j$'s, and so the factorization is unique.

Exercise 5.3.8 Show that, given any fractional ideal $\mathcal{A} \neq 0$ in K, there exists a fractional ideal $\mathcal{A}^{-1}$ such that $\mathcal{A}\mathcal{A}^{-1} = \mathcal{O}_K$.

Solution. Let

$$\mathcal{A} = \frac{\wp_1 \cdots \wp_r}{\wp_1' \cdots \wp_s'}.$$

Then

$$\mathcal{A}^{-1} = \frac{\wp_1' \cdots \wp_s'}{\wp_1 \cdots \wp_r}$$

is a fractional ideal with $\mathcal{A}\mathcal{A}^{-1} = \mathcal{O}_K$.

Exercise 5.3.9 Show that if $\mathfrak{a}$ and $\mathfrak{b}$ are ideals of $\mathcal{O}_K$, then $\mathfrak{b} \mid \mathfrak{a}$ if and only if there is an ideal $\mathfrak{c}$ of $\mathcal{O}_K$ with $\mathfrak{a} = \mathfrak{b}\mathfrak{c}$.

Solution. If $\mathfrak{b} \supseteq \mathfrak{a}$, then $\mathfrak{c} := \mathfrak{a}\mathfrak{b}^{-1} \subseteq \mathfrak{b}\mathfrak{b}^{-1} = \mathcal{O}_K$. Thus, $\mathfrak{a} = \mathfrak{b}\mathfrak{c}$, with $\mathfrak{c}$ an ideal of $\mathcal{O}_K$.

If $\mathfrak{a} = \mathfrak{b}\mathfrak{c}$ with $\mathfrak{c} \subseteq \mathcal{O}_K$, then $\mathfrak{a} = \mathfrak{b}\mathfrak{c} \subseteq \mathfrak{b}$.

Exercise 5.3.10 Show that $\gcd(\mathfrak{a}, \mathfrak{b}) = \mathfrak{a} + \mathfrak{b} = \prod_{i=1}^r \wp_i^{\min(e_i, f_i)}$.

Solution. $\mathfrak{a} \subseteq \mathfrak{a} + \mathfrak{b}$, $\mathfrak{b} \subseteq \mathfrak{a} + \mathfrak{b}$, so $\mathfrak{a} + \mathfrak{b} \mid \mathfrak{a}$, $\mathfrak{a} + \mathfrak{b} \mid \mathfrak{b}$.

If $\mathfrak{e} \mid \mathfrak{a}$ and $\mathfrak{e} \mid \mathfrak{b}$, then $\mathfrak{a} \subseteq \mathfrak{e}$, $\mathfrak{b} \subseteq \mathfrak{e}$, so that $\mathfrak{a} + \mathfrak{b} \subseteq \mathfrak{e}$, i.e., $\mathfrak{e} \mid \mathfrak{a} + \mathfrak{b}$. Therefore, $\mathfrak{a} + \mathfrak{b} = \gcd(\mathfrak{a}, \mathfrak{b})$.

Let $\mathfrak{d} = \prod_{i=1}^r \wp_i^{\min(e_i,f_i)}$, and let $\min(e_i, f_i) = a_i$,

$$\begin{aligned} \mathfrak{a} &= \prod_{i=1}^r \wp_i^{e_i} \\ &= \prod_{i=1}^r \wp_i^{e_i - a_i} \prod_{i=1}^r \wp_i^{a_i} \\ &= \prod_{i=1}^r \wp_i^{e_i - a_i} \mathfrak{d}, \quad e_i - a_i \geq 0 \ \forall i. \end{aligned}$$

Thus, $\mathfrak{d} \supseteq \mathfrak{a}$ which implies that $\mathfrak{d} \mid \mathfrak{a}$. Similarly, $\mathfrak{d} \mid \mathfrak{b}$.

Suppose that $\mathfrak{e} \mid \mathfrak{a}$ and $\mathfrak{e} \mid \mathfrak{b}$. Let

$$\mathfrak{e} = \prod_{i=1}^r \wp_i^{k_i}, \quad k_i \geq 0,$$

be the unique factorization of $\mathfrak{e}$ as a product of prime ideals. Suppose $k_i > e_i$ for some $i \in \{1, \ldots, r\}$. We know that $\wp_i^{k_i} \mid \mathfrak{e}$ and $\mathfrak{e} \mid \mathfrak{a}$, so $\wp_i^{k_i} \mid \mathfrak{a}$, i.e., $\wp_i^{k_i} \supseteq \wp_1^{e_1} \cdots \wp_r^{e_r}$. Thus,

$$\begin{aligned} (\wp_i^{-1})^{e_i} \wp_i^{k_i} &\supseteq \wp_1^{e_1} \cdots \wp_{i-1}^{e_{i-1}} \wp_{i+1}^{e_{i+1}} \cdots \wp_r^{e_r}, \\ \wp_i &\supseteq \wp_i^{k_i - e_i} \supseteq \wp_1^{e_1} \cdots \wp_{i-1}^{e_{i-1}} \wp_{i+1}^{e_{i+1}} \cdots \wp_r^{e_r}, \\ \wp_i &\supseteq \wp_j, \end{aligned}$$

for some $j \neq i$, and so $\wp_i = \wp_j$, since $\wp_j$ is maximal. But this is a contradiction, so $k_i \leq e_i$ for all i, and every prime occurring in $\mathfrak{e}$ must occur in $\mathfrak{a}$. Similarly for $\mathfrak{b}$, so $k_i \leq \min(e_i, f_i)$. Thus, $\mathfrak{e} \mid \mathfrak{d}$, so $\mathfrak{d} = \gcd(\mathfrak{a}, \mathfrak{b})$.

Exercise 5.3.11 Show that $\operatorname{lcm}(\mathfrak{a}, \mathfrak{b}) = \mathfrak{a} \cap \mathfrak{b} = \prod_{i=1}^r \wp_i^{\max(e_i,f_i)}$.

Solution. $\mathfrak{a} \supseteq \mathfrak{a} \cap \mathfrak{b}$, $\mathfrak{b} \supseteq \mathfrak{a} \cap \mathfrak{b}$, so $\mathfrak{a} \mid \mathfrak{a} \cap \mathfrak{b}$, $\mathfrak{b} \mid \mathfrak{a} \cap \mathfrak{b}$. Suppose that $\mathfrak{a} \mid \mathfrak{e}$ and $\mathfrak{b} \mid \mathfrak{e}$. Then $\mathfrak{e} \subseteq \mathfrak{a}$, $\mathfrak{e} \subseteq \mathfrak{b}$, so $\mathfrak{e} \subseteq \mathfrak{a} \cap \mathfrak{b}$. Thus, $\mathfrak{a} \cap \mathfrak{b} = \operatorname{lcm}(\mathfrak{a}, \mathfrak{b})$. Let $\mathfrak{m} = \prod_{i=1}^r \wp_i^{\max(e_i,f_i)}$ and let

$$\mathfrak{e} = \prod_{i=1}^r \wp_i^{k_i} \prod_{j=1}^s (\wp_j')^{t_j},$$

where $\wp_1, \ldots, \wp_r, \wp_1', \ldots, \wp_s'$ are distinct prime ideals. Suppose $k_i < e_i$ for some i,

$$\begin{aligned} \mathfrak{e} &\subseteq \mathfrak{a}, \\ \Rightarrow \quad \wp_1^{k_1} \cdots \wp_r^{k_r} (\wp_1')^{t_1} \cdots (\wp_s')^{t_s} &\subseteq \wp_1^{e_1} \cdots \wp_r^{e_r}, \\ \Rightarrow \quad \wp_1^{k_1} \cdots \wp_{i-1}^{k_{i-1}} \wp_{i+1}^{k_{i+1}} \cdots \wp_r^{k_r} (\wp_1')^{t_1} \cdots (\wp_s')^{t_s} &\subseteq \wp_1^{e_1} \cdots \wp_i^{e_i - k_i} \cdots \wp_r^{e_r} \\ &\subseteq \wp_i^{e_i - k_i} \\ &\subseteq \wp_i. \end{aligned}$$

Thus, $\wp_i \supseteq \wp_j$, for some $j \neq i$, or $\wp_i = \wp'_j$, for some j. Neither is true, so $k_i \geq e_i$. Similarly, $k_i \geq f_i$. Thus, $\mathfrak{m} \mid \mathfrak{e}$, so $\mathfrak{m} = \text{lcm}(\mathfrak{a}, \mathfrak{b})$.

Exercise 5.3.12 Suppose $\mathfrak{a}, \mathfrak{b}, \mathfrak{c}$ are ideals of $\mathcal{O}_K$. Show that if $\mathfrak{ab} = \mathfrak{c}^g$ and $\gcd(\mathfrak{a}, \mathfrak{b}) = 1$, then $\mathfrak{a} = \mathfrak{d}^g$ and $\mathfrak{b} = \mathfrak{e}^g$ for some ideals $\mathfrak{d}$ and $\mathfrak{e}$ of $\mathcal{O}_K$. (This generalizes Exercise 1.2.1.)

Solution. We factor uniquely into prime ideals:

$$\mathfrak{a} = \wp_1^{e_1} \cdots \wp_r^{e_r}$$

and

$$\mathfrak{b} = (\wp'_1)^{f_1} \cdots (\wp'_t)^{f_t}$$

where $\wp_1, \ldots, \wp_r, \wp'_1, \ldots, \wp'_t$ are distinct prime ideals since $\gcd(\mathfrak{a}, \mathfrak{b}) = 1$. Now let $\mathfrak{c} = \wp_1^{a_1} \cdots \wp_r^{a_r} (\wp'_1)^{b_1} \cdots (\wp'_t)^{b_t}$. Since $\mathfrak{ab} = \mathfrak{c}^g$, we must have

$$\begin{aligned} e_i &= a_i g, \qquad 1 \leq i \leq r, \\ f_i &= b_i g, \qquad 1 \leq i \leq t, \end{aligned}$$

by unique factorization. Thus $\mathfrak{a} = \mathfrak{d}^g$ and $\mathfrak{b} = \mathfrak{e}^g$ with

$$\begin{aligned} \mathfrak{d} &= \wp_1^{a_1} \cdots \wp_r^{a_r}, \\ \mathfrak{e} &= (\wp'_1)^{a_1} \cdots (\wp'_t)^{a_t}, \end{aligned}$$

as desired.

Exercise 5.3.14 Show that $\text{ord}_\wp(\mathfrak{ab}) = \text{ord}_\wp(\mathfrak{a}) + \text{ord}_\wp(\mathfrak{b})$, where $\wp$ is a prime ideal.

Solution. From a previous exercise, $\mathfrak{a} = \wp^t \mathfrak{a}_1$ and $\mathfrak{b} = \wp^s \mathfrak{b}_1$, where $\wp \not\supseteq \mathfrak{a}_1$, and $\wp \not\supseteq \mathfrak{b}_1$. Thus, $\mathfrak{ab} = \wp^{t+s}\mathfrak{a}_1\mathfrak{b}_1$, so $\wp^{s+t} \mid \mathfrak{ab}$. If $\wp^{s+t+1} \mid \mathfrak{ab}$, then $\mathfrak{ab} = \wp^{s+t+1}\mathfrak{c}$, so $\wp\mathfrak{c} = \mathfrak{a}_1\mathfrak{b}_1$.

Thus, $\wp \supseteq \mathfrak{a}_1\mathfrak{b}_1$, so $\wp \supseteq \mathfrak{a}_1$ or $\wp \supseteq \mathfrak{b}_1$, since $\wp$ is prime. This is a contradiction, so $\text{ord}_\wp(\mathfrak{ab}) = t + s = \text{ord}_\wp(\mathfrak{a}) + \text{ord}_\wp(\mathfrak{b})$.

Exercise 5.3.15 Show that, for $\alpha \neq 0$ in $\mathcal{O}_K$, $N((\alpha)) = |N_K(\alpha)|$.

Solution. Let $\mathcal{O}_K = \mathbb{Z}\omega_1 + \cdots + \mathbb{Z}\omega_n$. There exist $\alpha_i = \sum_{j=1}^n p_{ij}\omega_j$, with $p_{ii} > 0$, $p_{ij} \in \mathbb{Z}$ such that $(\alpha) = \mathbb{Z}\alpha_1 + \cdots + \mathbb{Z}\alpha_n$ and $N((\alpha)) = p_{11} \cdots p_{nn}$, from Theorem 4.2.2.

We also know that $(\alpha) = \mathbb{Z}\alpha\omega_1 + \cdots + \mathbb{Z}\alpha\omega_n$. Now, $N_K(\alpha) = \det(c_{ij})$ where $\alpha\omega_i = \sum_{j=1}^n c_{ij}\omega_j$. And, if $C = (c_{ij}), R = (r_{ij}), P = (p_{ij})$, then

$$\begin{aligned} (\alpha\omega_1, \ldots, \alpha\omega_n)^T &= C(\omega_1, \ldots, \omega_n)^T \\ &= R(\alpha_1, \ldots, \alpha_n)^T \\ &= RP(\omega_1, \ldots, \omega_n)^T. \end{aligned}$$

where $\alpha\omega_i = \sum_{j=i}^{n} r_{ji}\alpha_j$, $r_{ji} \in \mathbb{Z}$. Therefore, by definition,

$$N_K(\alpha) = \det(RP) = \det(R)\det(P).$$

R and R^{-1} have integer entries, since $\{\alpha\omega_i\}$ and $\{\alpha_i\}$ are both $\mathbb{Z}$-bases for (α). Thus, $\det(R) = \pm 1$. So, $\det(C) = \pm\det(P)$, $\det(P) \geq 0$. Thus, $|\det(C)| = \det(P)$.

Thus, $|N_K(\alpha)| = |\det(C)| = \det(P) = N((\alpha))$.

Exercise 5.3.17 If we write $p\mathcal{O}_K$ as its prime factorization,

$$p\mathcal{O}_K = \wp_1^{e_1} \cdots \wp_g^{e_g},$$

show that $N(\wp_i)$ is a power of p and that if $N(\wp_i) = p_i^{f_i}$, $\sum_{i=1}^{g} e_i f_i = n$.

Solution. Since $\wp_i^{e_i} + \wp_j^{e_j} = \mathcal{O}_K$ for $i \neq j$,

$$\mathcal{O}_K/p\mathcal{O}_K \simeq \mathcal{O}_K/\wp_1^{e_1} \oplus \cdots \oplus \mathcal{O}_K/\wp_g^{e_g},$$

by the Chinese Remainder Theorem. Therefore

$$\begin{aligned} p^n &= N(\wp_1^{e_1}) \cdots N(\wp_g^{e_g}) \\ &= N(\wp_1)^{e_1} \cdots N(\wp_g)^{e_g}. \end{aligned}$$

Thus, $N(\wp_i) = p^{f_i}$, for some positive integer f_i, and $n = e_1 f_1 + \cdots + e_g f_g$.

5.4 Dedekind's Theorem

Exercise 5.4.1 Show that $\mathfrak{D}^{-1}$ is a fractional ideal of K and find an integral basis.

Solution. $\mathfrak{D}^{-1}$ is an $\mathcal{O}_K$-module since if $x \in \mathcal{O}_K$ and $y \in \mathfrak{D}^{-1}$, then $xy \in \mathfrak{D}^{-1}$ because

$$\operatorname{Tr}(xy\mathcal{O}_K) \subseteq \operatorname{Tr}(y\mathcal{O}_K) \subseteq \mathbb{Z}.$$

Thus, $\mathcal{O}_K\mathfrak{D}^{-1} \subseteq \mathfrak{D}^{-1}$.

Now, let $\{\omega_1, \ldots, \omega_n\}$ be an integral basis of $\mathcal{O}_K$. There is a dual basis (see Exercise 4.2.1) $\{\omega_1^*, \ldots, \omega_n^*\}$ such that $\operatorname{Tr}(\omega_i\omega_j^*) = \delta_{ij}$, $1 \leq i, j \leq n$. Now $\operatorname{Tr}(\omega_i^*\omega_j) \in \mathbb{Z}$ for all ω_j, ω_i^*, so $\mathbb{Z}\omega_1^* + \cdots + \mathbb{Z}\omega_n^* \subseteq \mathfrak{D}^{-1}$.

We claim that $\mathfrak{D}^{-1} = \mathbb{Z}\omega_1^* + \cdots + \mathbb{Z}\omega_n^*$. Let $x \in \mathfrak{D}^{-1}$. Then

$$x = \sum_{i=1}^{n} a_i\omega_i^*, \qquad a_i \in \mathbb{Q},$$

since $\{\omega_1^*, \ldots, \omega_n^*\}$ is a $\mathbb{Q}$-basis for K. Then

$$\operatorname{Tr}(x\omega_j) = \operatorname{Tr}\left(\sum_{i=1}^{r} a_i\omega_i^*\omega_j\right) = a_j \in \mathbb{Z}.$$

Therefore $x \in \mathbb{Z}\omega_1^* + \cdots + \mathbb{Z}\omega_n^*$. Therefore $\mathcal{D}^{-1} = \mathbb{Z}\omega_1^* + \cdots + \mathbb{Z}\omega_n^*$.

Since for each ω_i^* there is an $a_i \in \mathbb{Z}$ such that $a_i\omega_i^* \in \mathcal{O}_K$, if we let $m = \prod_{i=1}^n a_i$, then $m\mathcal{D}^{-1} \subseteq \mathcal{O}_K$. Thus, $\mathcal{D}^{-1}$ is a fractional ideal.

Exercise 5.4.2 Let $\mathcal{D}$ be the fractional ideal inverse of $\mathcal{D}^{-1}$. We call $\mathcal{D}$ the *different* of K. Show that $\mathcal{D}$ is an ideal of $\mathcal{O}_K$.

Solution. $\mathcal{D}$ is certainly a fractional ideal of $\mathcal{O}_K$ and $\mathcal{D}\mathcal{D}^{-1} = \mathcal{O}_K$. But, from Lemma 4.1.1, $1 \in \mathcal{D}^{-1}$. Thus, $\mathcal{D} \subseteq \mathcal{D}\mathcal{D}^{-1} = \mathcal{O}_K$. Thus, $\mathcal{D}$ is an ideal of $\mathcal{O}_K$.

Exercise 5.4.5 Show that if p is ramified, $p \mid d_K$.

Solution. Since p is ramified, $e_\wp > 1$ for some prime ideal $\wp$ containing p. Thus, $\wp \mid \mathcal{D}$, from the previous theorem, say $\wp\mathfrak{a} = \mathcal{D}$. From the multiplicativity of the norm function, we have $N(\wp)N(\mathfrak{a}) = N(\mathcal{D})$, so $N(\wp) \mid N(\mathcal{D})$. Thus, $p^{f_\wp} \mid d_K$, and so $p \mid d_K$.

5.5 Factorization in $\mathcal{O}_K$

Exercise 5.5.2 If in the previous theorem we do not assume that $\mathcal{O}_K = \mathbb{Z}[\theta]$ but instead that $p \nmid [\mathcal{O}_K : \mathbb{Z}[\theta]]$, show that the same result holds.

Solution. Let m be the index of $\mathbb{Z}[\theta]$ in $\mathcal{O}_K$. Then for $\alpha \in \mathcal{O}_K$, $m\alpha \in \mathbb{Z}[\theta]$. In other words, given any α, we may write $m\alpha = b_0 + b_1\theta + b_2\theta^2 + \cdots + b_{n-1}\theta^{n-1}$. Consider this expression mod p. Since m is coprime to p there is an m' such that $m'm \equiv 1 \pmod{p}$. Then

$$\alpha \equiv b_0m' + b_1m'\theta + \cdots + b_{n-1}m'\theta^{n-1} \pmod{p}.$$

Thus, $\mathcal{O}_K \equiv \mathbb{Z}[\theta] \pmod{p}$.

In the proof of the previous exercise, we only used the fact that $\mathcal{O}_K = \mathbb{Z}[\theta]$ at one point. This was when we wrote that $r_i(\theta) = pa(\theta) + f_i(\theta)b(\theta)$. We now note that we simply need that $r_i(\theta) \equiv pa(\theta) + f_i(\theta) \pmod{p}$. The proof will follow through in the same way, and we deduce that $(p, f_i(\theta))$ is a prime ideal of $\mathbb{Z}[\theta]$. However, since

$$\mathcal{O}_K/(p) \simeq \mathbb{Z}[\theta]/(p),$$

then

$$\mathcal{O}_K/(p, f_i(\theta)) \simeq \mathbb{Z}[\theta]/(p, f_i(\theta)).$$

The rest of the proof will be identical to what was written above.

Exercise 5.5.3 Suppose that $f(x)$ in the previous exercise is Eisensteinian with respect to the prime p. Show that p ramifies totally in K. That is, $p\mathcal{O}_K = (\theta)^n$ where $n = [K : \mathbb{Q}]$.

Solution. By Example 4.3.1, we know that $p \nmid [\mathcal{O}_K : \mathbb{Z}[\theta]]$. Moreover, $f(x) \equiv x^n \pmod{p}$. The result is now immediate from Exercise 5.5.2.

Exercise 5.5.4 Show that $(p) = (1 - \zeta_p)^{p-1}$ when $K = \mathbb{Q}(\zeta_p)$.

Solution. The minimal polynomial of ζ_p is $\Phi_p(x)$, the pth cyclotomic polynomial. Recall that $f(x) = \Phi_p(x+1)$ is p-Eisensteinian, and that $\zeta_p - 1$ is a root. Since $\mathbb{Q}(\zeta_p) = \mathbb{Q}(1 - \zeta_p)$, and $\mathbb{Z}[\zeta_p] = \mathbb{Z}[1 - \zeta_p]$, the previous exercise tells us that $(p) = (1 - \zeta_p)^{p-1}$.

5.6 Supplementary Problems

Exercise 5.6.1 Show that if a ring R is a Dedekind domain and a unique factorization domain, then it is a principal ideal domain.

Solution. Consider an arbitrary prime ideal I of R. Since R is a Dedekind domain, I is finitely generated and we can write $I = (a_1, \ldots, a_n)$ for some set of generators $a_1, \ldots, a_n$. In a unique factorization domain, every pair of elements has a gcd, and so $d = \gcd(a_1, \ldots, a_n)$ exists. Then $(d) = (a_1, \ldots, a_n) = I$, thus proving that R is a principal ideal domain.

Exercise 5.6.2 Using Theorem 5.5.1, find a prime ideal factorization of $5\mathcal{O}_K$ and $7\mathcal{O}_K$ in $\mathbb{Z}[(1+\sqrt{-3})/2]$.

Solution. We now consider $f(x) \pmod{7}$. We have

$$x^2 - x + 1 \equiv x^2 + 6x + 1 \equiv (x+2)(x+4) \pmod{7}$$

so 7 splits and its factorization is

$$(7) = \left(7, \frac{5+\sqrt{-3}}{2}\right)\left(7, \frac{9+\sqrt{-3}}{2}\right).$$

Exercise 5.6.3 Find a prime ideal factorization of $(2), (5), (11)$ in $\mathbb{Z}[i]$.

Solution. The minimal polynomial of i is $x^2 + 1$. We consider it first mod 2.

$$x^2 + 1 \equiv (x+1)^2 \pmod{2}$$

so $(2) = (2, i+1)^2 = (i+1)^2$ since $2i = (i+1)^2$.

$$x^2 + 1 \equiv x^2 - 4 \equiv (x+2)(x-2) \pmod{5}$$

so $(5) = (5, i+2)(5, i-2) = (i+2)(i-2)$ since $(2+i)(2-i) = 5$.

Finally, we consider $f(x) \pmod{11}$. Since if the polynomial reduces it must have an integral root, we check all the possibilities mod 11 and deduce that it is in fact irreducible. Thus, 11 stays prime in $\mathbb{Z}[i]$.

Exercise 5.6.4 Compute the different $\mathcal{D}$ of $K = \mathbb{Q}(\sqrt{-2})$.

Solution. By Theorem 5.4.3, we know that $N(\mathcal{D}) = |d_K| = 8$. Also, we showed in Chapter 2 that $\mathcal{O}_K = \mathbb{Z}[\sqrt{-2}]$ is Euclidean and thus a principal ideal domain. Then $\mathcal{D} = (a + b\sqrt{-2})$ for some $a, b \in \mathbb{Z}$ and $N(\mathcal{D}) = a^2 + 2b^2 = 8$. The only solution in integers is $a = 0, b = \pm 2$, so

$$\mathcal{D} = (2\sqrt{-2}) = (-2\sqrt{-2}).$$

Exercise 5.6.5 Compute the different $\mathcal{D}$ of $K = \mathbb{Q}(\sqrt{-3})$.

Solution. As in the previous exercise, we first observe that $N(\mathcal{D}) = |d_K| = 3$ and since the integers of this ring are the Eisenstein integers, which form a Euclidean ring, $\mathbb{Z}[\rho]$ is a principal ideal domain and $\mathcal{D} = (a+b\rho)$ for some $a, b \in \mathbb{Z}$. We must find all solutions to the equation

$$N(\mathcal{D}) = 3 = a^2 - ab + b^2.$$

Since this is equivalent to $(2a - b)^2 + 3b^2 = 12$, we note that $|b|$ cannot be greater than 2. Checking all possibilities, we find all the elements of $\mathbb{Z}[\rho]$ of norm 3: $2 + \rho$, $-1 + \rho$, $-2 - \rho$, $1 - \rho$, $1 + 2\rho$ and $-1 - 2\rho$. Some further checking reveals that these six elements are all associates, and so they each generate the same principal ideal. Thus, $\mathcal{D} = (2 + \rho)$.

Exercise 5.6.6 Let $K = \mathbb{Q}(\alpha)$ be an algebraic number field of degree n over $\mathbb{Q}$. Suppose $\mathcal{O}_K = \mathbb{Z}[\alpha]$ and that $f(x)$ is the minimal polynomial of α. Write

$$f(x) = (x - \alpha)(b_0 + b_1 x + \cdots b_{n-1}x^{n-1}), \qquad b_i \in \mathcal{O}_K.$$

Prove that the dual basis to $1, \alpha, \ldots, \alpha^{n-1}$ is

$$\frac{b_0}{f'(\alpha)}, \ldots, \frac{b_{n-1}}{f'(\alpha)}.$$

Deduce that

$$\mathcal{D}^{-1} = \frac{1}{f'(\alpha)}(\mathbb{Z}b_0 + \cdots + \mathbb{Z}b_{n-1}).$$

Solution. Let $\alpha_1 = \alpha, \alpha_2, \ldots, \alpha_n$ be the n distinct roots of $f(x)$. We would like to show that

$$\sum_{i=1}^{n} \frac{f(x)}{x - \alpha_i} \cdot \frac{\alpha_i^r}{f'(\alpha_i)} = x^r$$

for $0 \le r \le n - 1$. Define the polynomial

$$g_r(x) = \sum_{i=1}^{n} \frac{f(x)}{x - \alpha_i} \cdot \frac{\alpha_i^r}{f'(\alpha_i)} - x^r.$$

Consider $g_r(\alpha_1)$. Note that $f(\alpha_1)/(\alpha_1-\alpha_i)=0$ for all i except $i=1$. Also,

$$\left(\frac{f(x)}{x-\alpha_1}\right)_{x=\alpha_1} = f'(\alpha_1).$$

Thus, $g_r(\alpha_1) = 0$, and similarly, $g_r(\alpha_i) = 0$ for $1 \leq i \leq n$. Since $\deg(g_r(x)) \leq n-1$, it can have at most $n-1$ roots. As we found n distinct roots, $g_r(x)$ must be identically zero.

For a polynomial $h(x) = c_0 + c_1x + \cdots + c_mx^m \in K[x]$, we define the *trace* of $h(x)$ to be

$$\mathrm{Tr}(h(x)) = \sum_{i=0}^{m} \mathrm{Tr}(c_i)x^i \in \mathbb{Q}[x].$$

Since $\alpha_1, \dots, \alpha_n$ are all the conjugates of α, it is clear that

$$\mathrm{Tr}\left(\frac{f(x)\alpha^r}{(x-\alpha)f'(\alpha)}\right) = \sum_{i=1}^{n} \frac{f(x)\alpha_i^r}{(x-\alpha_i)f'(\alpha_i)} = x^r.$$

But,

$$\mathrm{Tr}\left(\frac{f(x)\alpha^r}{(x-\alpha)f'(\alpha)}\right) = \sum_{i=0}^{n-1} \mathrm{Tr}\left(\frac{b_i\alpha^r}{f'(\alpha)}\right)x^i = x^r.$$

Thus,

$$\mathrm{Tr}\left(\frac{b_i\alpha^r}{f'(\alpha)}\right) = 0$$

unless $i = r$, in which case the trace is 1. Recall that if $\omega_1, \dots, \omega_n$ is a basis, its dual basis $\omega_1^*, \dots, \omega_n^*$ is characterized by $\mathrm{Tr}(\omega_i\omega_j^*) = \delta_{ij}$, the Kronecker delta function. Thus, we have found a dual basis to $1, \alpha, \dots, \alpha^{n-1}$, and it is

$$\frac{b_0}{f'(\alpha)}, \dots, \frac{b_{n-1}}{f'(\alpha)}.$$

By Exercise 5.4.1,

$$\mathcal{D}^{-1} = \frac{1}{f'(\alpha)}(\mathbb{Z}b_0 + \cdots + \mathbb{Z}b_{n-1}).$$

Exercise 5.6.7 Let $K = \mathbb{Q}(\alpha)$ be of degree n over $\mathbb{Q}$. Suppose that $\mathcal{O}_K = \mathbb{Z}[\alpha]$. Prove that $\mathcal{D} = (f'(\alpha))$.

Solution. Let $f(x)$ be the minimal polynomial of α, and let

$$f(x) = (x-\alpha)(b_0 + \cdots + b_{n-1}x^{n-1}).$$

Since $f(x)$ is monic, $b_{n-1} = 1$. Also, $a_{n-1} = b_{n-2} - \alpha b_{n-1}$ which means that $b_{n-2} = a_{n-1} + \alpha$, where a_{n-1} is an integer.

We know from the previous exercise that

$$\mathcal{D}^{-1} = \frac{1}{f'(\alpha)}(\mathbb{Z}b_0 + \cdots + \mathbb{Z}b_{n-1}).$$

Since $b_{n-1} = 1$, $\mathbb{Z} \subseteq \mathbb{Z}b_0 + \cdots + \mathbb{Z}b_{n-1}$. Since $b_{n-2} = a_{n-1} + \alpha$, we can deduce that $\alpha \in \mathbb{Z}b_0 + \cdots + \mathbb{Z}b_{n-1}$, and by considering the expressions for a_i, $0 \leq i \leq n$, we see that in fact $\alpha^i \in \mathbb{Z}b_0 + \cdots + \mathbb{Z}b_{n-1}$ for $1 \leq i \leq n-1$. Thus,

$$\mathbb{Z}[\alpha] \subseteq \mathbb{Z}b_0 + \cdots + \mathbb{Z}b_{n-1} \subseteq \mathbb{Z}[\alpha],$$

and so we have equality. Thus,

$$\mathcal{D}^{-1} = \frac{1}{f'(\alpha)}\mathcal{O}_K$$

and so $\mathcal{D} = (f'(\alpha))$.

Exercise 5.6.8 Compute the different $\mathcal{D}$ of $\mathbb{Q}[\zeta_p]$ where ζ_p is a primitive pth root of unity.

Solution. We can apply the results of the previous exercise to get $\mathcal{D} = (f'(\zeta_p))$.

$$\begin{aligned} f(x) &= \frac{x^p - 1}{x - 1} = x^{p-1} + x^{p-2} + \cdots + x + 1, \\ f'(x) &= \frac{px^{p-1}(x-1) - (x^p - 1)}{(x-1)^2}, \\ f'(\zeta_p) &= \frac{p\zeta_p^{-1}}{\zeta_p - 1}. \end{aligned}$$

Since ζ_p^{-1} is a unit, we find

$$\mathcal{D} = \left(\frac{p}{\zeta_p - 1}\right).$$

From Exercise 5.5.4 we know that $(p) = (1-\zeta_p)^{p-1}$, and so $\mathcal{D} = (1-\zeta_p)^{p-2}$.

Exercise 5.6.9 Let p be a prime, $p \nmid m$, and $a \in \mathbb{Z}$. Show that $p \mid \phi_m(a)$ if and only if the order of $a \pmod p$ is n. (Here $\phi_m(x)$ is the mth cyclotomic polynomial.)

Solution. Since $x^m - 1 = \prod_{d|m} \phi_d(x)$, we have $a^m \equiv 1 \pmod p$. Let k be the order of $a \pmod p$. Then $k \mid m$. If $k < m$, then

$$a^k - 1 = \prod_{d|k} \phi_d(a) \equiv 0 \pmod p$$

so that $\phi_d(a) \equiv 0 \pmod{p}$ for some $d \mid k$. Then

$$a^m - 1 = \phi_m(a)\phi_d(a)(\text{ other factors }) \equiv 0 \pmod{p^2}.$$

Since $\phi_m(a+p) \equiv \phi_m(a) \pmod{p}$ and similarly for $\phi_d(a)$, we also have $(a+p)^m \equiv 1 \pmod{p^2}$. But then $(a+p)^m = a^m + ma^{m-1}p \pmod{p^2}$ so that $ma^{m-1} \equiv 0 \pmod{p}$, a contradiction.

Conversely, suppose that a has order m so that $a^m \equiv 1 \pmod{p}$. Then $\phi_d(a) \equiv 0 \pmod{p}$ for some $d \mid m$. If $d < m$, then the order of $a \pmod{p}$ would be less than m.

Exercise 5.6.10 Suppose $p \nmid m$ is prime. Show that $p \mid \phi_m(a)$ for some $a \in \mathbb{Z}$ if and only if $p \equiv 1 \pmod{m}$. Deduce from Exercise 1.2.5 that there are infinitely many primes congruent to $1 \pmod{m}$.

Solution. If $p \mid \phi_m(a)$, by the previous exercise the order of $a \pmod{p}$ is m so that $m \mid p-1$.

Conversely, if $p \equiv 1 \pmod{m}$, there is an element a of order $m \pmod{p}$ because $(\mathbb{Z}/p\mathbb{Z})^*$ is cyclic. Again by the previous exercise $p \mid \phi_m(a)$.

If there are only finitely many primes $p_1, \dots, p_r$ (say) that are congruent to $1 \pmod{m}$, then setting $a = (p_1 \cdots p_r)m$ we examine the prime divisors of $\phi_m(a)$. Observe that the identity

$$x^m - 1 = \prod_{d \mid m} \phi_d(x)$$

implies that $\phi_m(0) = \pm 1$. Thus, the constant term of $\phi_m(x)$ is ± 1 so that $\phi_m(a)$ is coprime to a and hence coprime to m. (If $\phi_m(a) = \pm 1$, one can replace a by any suitable power of a, so that $|\phi_m(a)| > 1$.)

By what we have proved, any prime divisor p of $\phi_m(a)$ coprime to m must be congruent to $1 \pmod{m}$. The prime p is distinct from $p_1, \dots, p_r$.

Exercise 5.6.11 Show that $p \nmid m$ splits completely in $\mathbb{Q}(\zeta_m)$ if and only if $p \equiv 1 \pmod{m}$.

Solution. Observe that $\phi_m(x)$ has a root mod p if and only if $p \equiv 1 \pmod{m}$ by the previous exercise. But then if it has one root a it has $\varphi(m)$ roots because $m \mid (p-1)$ and so $(\mathbb{Z}/p\mathbb{Z})^*$ has a cyclic subgroup of order m. Thus, $\phi_m(x)$ splits completely if and only if $p \equiv 1 \pmod{m}$.

Exercise 5.6.12 Let p be prime and let a be squarefree and coprime to p. Set $\theta = a^{1/p}$ and consider $K = \mathbb{Q}(\theta)$. Show that $\mathcal{O}_K = \mathbb{Z}[\theta]$ if and only if $a^{p-1} \not\equiv 1 \pmod{p^2}$.

Solution. Assume that $\mathcal{O}_K = \mathbb{Z}[\theta]$. We will show that $a^{p-1} \not\equiv 1 \pmod{p^2}$. By Theorem 5.5.1,

$$p\mathcal{O}_K = \wp^p$$

since $\mathbb{Q}(\theta)$ has degree p over $\mathbb{Q}$. Moreover,

$$\wp = (p, \theta - a).$$

Also, $(\theta - a) \in \wp$ and $(\theta - a) \notin \wp^2$ so that

$$(\theta - a) = \wp\mathfrak{a}$$

for some ideal $\mathfrak{a}$. Taking norms, we find $|N(\theta - a)| = pN\mathfrak{a}$ and $(N\mathfrak{a}, p) = 1$. $(\theta - a)$ is a root of $(x+a)^p - a$ and this polynomial is irreducible since $\mathbb{Q}(\theta - a) = \mathbb{Q}(\theta)$ has degree p over $\mathbb{Q}$. Hence

$$N(\theta - a) = a^p - a = pN\mathfrak{a}$$

so that $a^p \not\equiv a \pmod{p^2}$.

Conversely, suppose that $a^p \not\equiv a \pmod{p^2}$. Then the polynomial

$$(x+a)^p - a$$

is Eisenstein with respect to the prime p. Therefore $p \nmid [\mathcal{O}_K : \mathbb{Z}[\theta - a]]$ by Example 4.3.1. But $\mathbb{Z}[\theta - a] = \mathbb{Z}[\theta]$ so we deduce that $p \nmid [\mathcal{O}_K : \mathbb{Z}[\theta]]$. In addition, $x^p - a$ is Eisenstein with respect to every prime divisor of a. Again, by Example 4.3.1, we deduce that $[\mathcal{O}_K : \mathbb{Z}[\theta]]$ is coprime to a. By Exercises 4.3.3 and 4.2.8,

$$d_{K/\mathbb{Q}}(\theta) = (-1)^{\binom{p}{2}} p^p a^{p-1} = [\mathcal{O}_K : \mathbb{Z}[\theta]]^2 \cdot d_K.$$

Since the index of θ in $\mathcal{O}_K$ is coprime to both p and a, it must equal 1. Thus $\mathcal{O}_K = \mathbb{Z}[\theta]$.

Exercise 5.6.13 Suppose that $K = \mathbb{Q}(\theta)$ and $\mathcal{O}_K = \mathbb{Z}[\theta]$. Show that if $p \mid d_K$, p ramifies.

Solution. We will use the result of Theorem 5.5.1. Let $f(x)$ be the minimal polynomial of $\mathbb{Z}[\theta]$. Suppose that $p \mid d_K$, and

$$f(x) \equiv f_1(x)^{e_1} \cdots f_g(x)^{e_g} \pmod{p}.$$

Since $p \mid d_{K/\mathbb{Q}}(\theta) = \prod(\theta_i - \theta_j)^2$, then $\theta_i \equiv \theta_j$ in $\overline{\mathbb{F}}_p$ for some $i \neq j$. Thus, f has multiple roots in $\overline{\mathbb{F}}_p$. Hence, one of the e_i's is greater than 1.

Exercise 5.6.14 Let $K = \mathbb{Q}(\theta)$ and suppose that $p \mid d_{K/\mathbb{Q}}(\theta)$, $p^2 \nmid d_{K/\mathbb{Q}}(\theta)$. Show that $p \mid d_K$ and p ramifies in K.

Solution. Recall that $d_{K/\mathbb{Q}}(\theta) = m^2 d_K$ where $m = [\mathcal{O}_K : \mathbb{Z}[\theta]]$. Clearly, since $p \mid d_{K/\mathbb{Q}}(\theta)$ but $p^2 \nmid d_{K/\mathbb{Q}}(\theta)$, $p \mid d_K$, and $p \nmid m$. We can now apply the result of Exercise 5.5.2. Using the same argument as in 5.6.13, we deduce that $f(x)$ has a multiple root mod p and so p ramifies in K.

Exercise 5.6.15 Let K be an algebraic number field of discriminant d_K. Show that the normal closure of K contains a quadratic field of the form $\mathbb{Q}(\sqrt{d_K})$.

Solution. Let $\tilde{K}$ be the normal closure of K, $\omega_1, \dots, \omega_n$ an integral basis of K, $\sigma_1, \dots, \sigma_n$ the distinct embeddings of K into $\mathbb{C}$. Then

$$d_K = \det\big(\sigma_i(\omega_j)\big)^2 \in \mathbb{Z}.$$

Thus $\sqrt{d_K} = \det\big(\sigma_i(\omega_j)\big) \in \tilde{K}$.

Exercise 5.6.16 Show that if p ramifies in K, then it ramifies in each of the conjugate fields of K. Deduce that if p ramifies in the normal closure of K, then it ramifies in K.

Solution. Since each embedding $\sigma_i : K \to K^{(i)}$ is an isomorphism of fields, any factorization of

$$(p) = \wp_1^{e_1} \cdots \wp_g^{e_g}$$

takes each prime ideal $\wp_j$ into a conjugate prime ideal $\wp_j^{(i)}$. If some $e_j > 1$ then in each conjugate field, p ramifies. The second part is straightforward upon intersecting with K.

Exercise 5.6.17 Deduce the following special case of Dedekind's theorem: if $p^{2m+1} \| d_K$ show that p ramifies in K.

Solution. By the penultimate exercise, p ramifies in $\mathbb{Q}(\sqrt{d_K})$ and hence in the normal closure. By the previous exercise, p ramifies in K.

Exercise 5.6.18 Determine the prime ideal factorization of (7), (29), and (31) in $K = \mathbb{Q}(\sqrt[3]{2})$.

Solution. By Example 4.3.6, $\mathcal{O}_K = \mathbb{Z}[2^{1/3}]$. We may apply Theorem 5.5.1. Since $x^3 - 2$ is irreducible mod 7, $7\mathcal{O}_K$ is prime in $\mathcal{O}_K$. Since

$$x^3 - 2 \equiv (x+3)(x^2 - 3x + 9) \pmod{29}$$

and the quadratic factor is irreducible, we get

$$29\mathcal{O}_K = \wp_1 \wp_2$$

where $\deg \wp_1 = 1$, $\deg \wp_2 = 2$ and $\wp_1, \wp_2$ are prime ideals. Finally,

$$x^3 - 2 \equiv (x-4)(x-7)(x+11) \pmod{31}$$

so that $31\mathcal{O}_K$ splits completely in K.

Exercise 5.6.19 If L/K is a finite extension of algebraic number field, we can view L as a finite dimensional vector space over K. If $\alpha \in L$, the map $v \mapsto \alpha v$ is a linear mapping and one can define, as before, the *relative norm* $N_{L/K}(\alpha)$ and *relative trace* $Tr_{L/K}(\alpha)$ as the determinant and trace, respectively, of this linear map. If $\alpha \in \mathcal{O}_L$, show that $Tr_{L/K}(\alpha)$ and $N_{L/K}(\alpha)$ lie in $\mathcal{O}_K$.

Solution. By taking a basis $\omega_1, ..., \omega_n$ of L over K and repeating the argument of Lemma 4.1.1, the result follows immediately.

Exercise 5.6.20 If $K \subseteq L \subseteq M$ are finite extensions of algebraic number fields, show that $N_{M/K}(\alpha) = N_{L/K}(N_{M/L}(\alpha))$ and $Tr_{M/K}(\alpha) = Tr_{L/K}(Tr_{M/L}(\alpha))$ for any $\alpha \in M$. (We refer to this as the *transitivity property* of the norm and trace map, respectively.)

Solution. Fix an algebraic closure $\overline{M}$ of M. Let $\sigma_1, ..., \sigma_m$ be the distinct embeddings of L into $\overline{M}$ which are equal to the identity on K. By field theory, we can extend these to embeddings of M into $\overline{M}$. Of these, let $\eta_1, ..., \eta_n$ be the ones trivial on L. If σ is an arbitrary embedding of M into $\overline{M}$ which is trivial on K, then as σ is also an embedding of L into $\overline{M}$, it must be σ_j for some j. Thus, $\sigma_j^{-1}\sigma$ fixes L and so must be an η_i for some i. Thus, every embedding of M is of the form $\sigma_j \circ \eta_i$ so that

$$N_{M/K}(\alpha) = \prod_{i,j} \sigma_j(\eta_i(\alpha)) = \prod_j \sigma_j(N_{M/L}(\alpha)) = N_{L/K}(N_{M/L}(\alpha)),$$

as desired.

Exercise 5.6.21 Let L/K be a finite extension of algebraic number fields. Show that the map

$$Tr_{L/K} : L \times L \to K$$

is non-degenerate.

Solution. This follows from an argument analogous to the proof of Lemma 4.1.4.

Exercise 5.6.22 Let L/K be a finite extension of algebraic number fields. Let $\mathfrak{a}$ be a finitely generated $\mathcal{O}_K$-module contained in L. The set

$$\mathcal{D}_{L/K}^{-1}(\mathfrak{a}) = \{x \in L : \quad Tr_{L/K}(x\mathfrak{a}) \subseteq \mathcal{O}_K\}$$

is called *codifferent* of $\mathfrak{a}$ over K. If $\mathfrak{a} \neq 0$, show that $\mathcal{D}_{L/K}^{-1}(\mathfrak{a})$ is a finitely generated $\mathcal{O}_K$-module. Thus, it is a fractional ideal of L.

Solution. The fact that $\mathcal{D}_{L/K}^{-1}(\mathfrak{a})$ is an $\mathcal{O}_K$-module is clear. To see that it is finitely generated, we take an $\mathcal{O}_K$-basis of $\mathfrak{a}$ and repeat the argument in Exercise 5.4.1 to deduce the result.

Exercise 5.6.23 If in the previous exercise $\mathfrak{a}$ is an ideal of $\mathcal{O}_L$, show that the fractional ideal inverse, denoted $\mathcal{D}_{L/K}(\mathfrak{a})$ of $\mathcal{D}_{L/K}^{-1}(\mathfrak{a})$ is an integral ideal of $\mathcal{O}_L$. (We call $\mathcal{D}_{L/K}(\mathfrak{a})$ the *different* of $\mathfrak{a}$ over K. In the case $\mathfrak{a}$ is $\mathcal{O}_L$, we call it the *relative different* of L/K and denote it by $\mathcal{D}_{L/K}$.)

Solution. We have $\mathcal{D}_{L/K}(\mathfrak{a})\mathcal{D}_{L/K}^{-1}(\mathfrak{a}) = \mathcal{O}_L$ and $1 \in \mathcal{D}_{L/K}^{-1}(\mathfrak{a})$ so that $\mathcal{D}_{L/K}(\mathfrak{a}) \subseteq \mathcal{D}_{L/K}(\mathfrak{a})\mathcal{D}_{L/K}^{-1}(\mathfrak{a}) \subseteq \mathcal{O}_L$ so that $\mathcal{D}_{L/K}(\mathfrak{a})$ is an integral ideal of $\mathcal{O}_L$.

Exercise 5.6.24 Let $K \subseteq L \subseteq M$ be algebraic number fields of finite degree over the rationals. Show that

$$\mathcal{D}_{M/K} = \mathcal{D}_{M/L}(\mathcal{D}_{L/K}\mathcal{O}_M).$$

Solution. We have $x \in \mathcal{D}_{M/L}^{-1}$ if and only if $Tr_{M/L}(x\mathcal{O}_M) \subseteq \mathcal{O}_L$ which is equivalent to

$$\mathcal{D}_{L/K}^{-1}Tr_{M/L}(x\mathcal{O}_M) \subseteq \mathcal{D}_{L/K}^{-1}\mathcal{O}_L \quad \text{iff} \quad Tr_{L/K}(\mathcal{D}_{L/K}^{-1}Tr_{M/L}(x\mathcal{O}_M)) \subseteq \mathcal{O}_K$$

which by transitivity of the trace is equivalent to $Tr_{M/K}(x\mathcal{D}_{L/K}^{-1}) \subseteq \mathcal{O}_K$. That is, we must have

$$x\mathcal{D}_{L/K}^{-1} \subseteq \mathcal{D}_{M/K}^{-1}$$

which is true if and only if

$$x \in \mathcal{D}_{L/K}\mathcal{D}_{M/K}^{-1},$$

which means

$$\mathcal{D}_{M/L}^{-1} = \mathcal{D}_{L/K}\mathcal{D}_{M/K}^{-1},$$

which gives the required result.

Exercise 5.6.25 Let L/K be a finite extension of algebraic number fields. We define the *relative discriminant* of L/K, denoted $d_{L/K}$ as $N_{L/K}(\mathcal{D}_{L/K})$. This is an integral ideal of $\mathcal{O}_K$. If $K \subseteq L \subseteq M$ are as in Exercise 5.6.24, show that

$$d_{M/K} = d_{L/K}^{[M:L]}N_{L/K}(d_{M/L}).$$

Solution. By the transitivity of the norm map and by Exercise 5.6.24, we have

$$N_{M/K}(\mathcal{D}_{M/K}) = N_{M/K}(\mathcal{D}_{M/L}\mathcal{D}_{L/K}) =$$

$$N_{L/K}(N_{M/L}(\mathcal{D}_{L/K}\mathcal{D}_{M/L})) = N_{L/K}(\mathcal{D}_{L/K}^{[M:L]}N_{M/L}(\mathcal{D}_{M/L}))$$

which gives the result. We remark here that Dedekind's theorem concerning ramification extends to relative extensions L/K. More precisely, a prime ideal $\mathfrak{p}$ of $\mathcal{O}_K$ is said to ramify in L if there is a prime ideal $\wp$ of $\mathcal{O}_L$ such that $\wp^2|\mathfrak{p}\mathcal{O}_L$. One can show that $\mathfrak{p}$ ramifies in L if and only if $\mathfrak{p}|d_{L/K}$. The easy part of this assertion that if $\mathfrak{p}$ is ramified then $\mathfrak{p}|d_{L/K}$ can be proved following the argument of Exercise 5.4.5. The converse requires further theory of relative differents. We refer the interested reader to [N].

Exercise 5.6.26 Let L/K be a finite extension of algebraic number fields. Suppose that $\mathcal{O}_L = \mathcal{O}_K[\alpha]$ for some $\alpha \in L$. If $f(x)$ is the minimal polynomial of α over $\mathcal{O}_K$, show that $\mathcal{D}_{L/K} = (f'(\alpha))$.

Solution. This result is identical to Exercises 5.6.6 and 5.6.7. More generally, one can show the following. For each $\theta \in \mathcal{O}_L$ which generates L over K, let $f(x)$ be its minimal polynomial over $\mathcal{O}_K$. Define $\delta_{L/K}(\theta) = f'(\theta)$. Then $\mathcal{D}_{L/K}$ is the ideal generated by the elements $\delta_{L/K}(\theta)$ as θ ranges over such elements. We refer the interested reader to [N].

Exercise 5.6.27 Let K_1, K_2 be algebraic number fields of finite degree over K. If L/K is the compositum of K_1/K and K_2/K, show that the set of prime ideals dividing $d_{L/K}$ and $d_{K_1/K}d_{K_2/K}$ are the same.

Solution. By Exercise 5.6.25, we see that every prime ideal dividing

$$d_{K_1/K}d_{K_2/K}$$

also divides $d_{L/K}$. Suppose now that $\mathfrak{p}$ is a prime ideal of $\mathcal{O}_K$ which divides $d_{L/K}$ but not $d_{K_1/K}$. We have to show that $\mathfrak{p}$ divides $d_{K_2/K}$. By the definition of the relative discriminant, there is a prime ideal $\wp$ of $\mathcal{O}_L$ lying above $\mathfrak{p}$ which divides the different $\mathcal{D}_{L/K}$. This ideal cannot divide $\mathcal{D}_{K_1/K}\mathcal{O}_L$ for this would imply that $\mathfrak{p}$ divides $d_{K_1/K}$, contrary to assumption. Since $\mathcal{D}_{L/K} = \mathcal{D}_{L/K_1}\mathcal{D}_{K_1/K}$, we deduce that $\wp$ divides $\mathcal{D}_{L/K_1}$. Now let $\alpha \in \mathcal{O}_{K_2}$ so that α generates K_2 over K. Let $f(x)$ be its minimal polynomial over K_1 and $g(x)$ its minimal polynomial over K. (We have assumed that we have fixed a common algebraic closure which contains K_1 and K_2.) Then $L = K_1(\alpha)$ and $g(x) = f(x)h(x)$ for some polynomial h over K_1. Hence, $g'(x) = f'(x)h(x)+f(x)h'(x)$ which implies $g'(\alpha) = f'(\alpha)h(\alpha)$. Thus, $g'(\alpha)$ is in the ideal generated by $f'(\alpha)$. By the remark in the solution of Exercise 5.6.26, we deduce that $f'(\alpha) \in \mathcal{D}_{L/K} \subseteq \wp$. Therefore, $g'(\alpha) \in \wp$. The same remark enables us to deduce that $g'(\alpha) \in \mathcal{D}_{K_2/K}$ implying that $\mathfrak{p}$ divides $d_{K_2/K}$.

Exercise 5.6.28 Let L/K be a finite extension of algebraic number fields. If $\tilde{L}$ denotes the normal closure, show that a prime $\mathfrak{p}$ of $\mathcal{O}_K$ is unramified in L if and only if it is unramified in $\tilde{L}$.

Solution. If we apply the preceding exercise to the compositum of the conjugate fields of L, the result is immediate.

Chapter 6

The Ideal Class Group

6.1 Elementary Results

Exercise 6.1.2 Show that given $\alpha, \beta \in \mathcal{O}_K$, there exist $t \in \mathbb{Z}, |t| \leq H_K$, and $w \in \mathcal{O}_K$ so that $|N(\alpha t - \beta w)| < |N(\beta)|$.

Solution. If we apply Lemma 6.1.1 with α replaced by α/β, we conclude that there exist $t \in \mathbb{Z}, |t| \leq H_K$, and $w \in \mathcal{O}_K$ such that

$$|N(t\alpha/\beta - w)| < 1.$$

This implies $|N(t\alpha - w\beta)| < |N(\beta)|$.

6.2 Finiteness of the Ideal Class Group

Exercise 6.2.1 Show that the relation $\sim$ defined above is an equivalence relation.

Solution. It is trivial that $\mathcal{A} \sim \mathcal{A}$, and if $\mathcal{A} \sim \mathcal{B}$ then $\mathcal{B} \sim \mathcal{A}$, for any ideals $\mathcal{A}$ and $\mathcal{B}$. Suppose now that $\mathcal{A} \sim \mathcal{B}$, and $\mathcal{B} \sim \mathcal{C}$. That is, there exist $\alpha, \beta, \gamma, \theta \in \mathcal{O}_K$ such that $(\alpha)\mathcal{A} = (\beta)\mathcal{B}$, and $(\gamma)\mathcal{B} = (\theta)\mathcal{C}$. It is now easily seen that $(\alpha\gamma)\mathcal{A} = (\beta\theta)\mathcal{C}$. Thus, $\mathcal{A} \sim \mathcal{B}$ and $\mathcal{B} \sim \mathcal{C}$ imply $\mathcal{A} \sim \mathcal{C}$.

Hence, $\sim$ is an equivalence relation.

Exercise 6.2.3 Show that each equivalence class of ideals has an integral ideal representative.

Solution. Suppose $\mathcal{A}$ is a fractional ideal in K. Let $\mathcal{A} = \mathfrak{b}/\mathfrak{c}$, with $\mathfrak{b}, \mathfrak{c} \subseteq \mathcal{O}_K$.

We know from Exercise 4.4.1 that $\mathfrak{c} \cap \mathbb{Z} \neq \{0\}$, so there exists $0 \neq t \in \mathbb{Z}$ such that $t \in \mathfrak{c}$. Thus, $\mathfrak{c} \supseteq (t) = t\mathcal{O}_K$, and so $\mathfrak{c}$ divides (t). This implies that there exists an integral ideal $\mathfrak{e} \subseteq \mathcal{O}_K$ such that

$$\mathfrak{c}\mathfrak{e} = (t). \tag{6.1}$$

We now have

$$(t)\mathcal{A} = (t)\frac{\mathfrak{b}}{\mathfrak{c}} = \frac{\mathfrak{ceb}}{\mathfrak{c}} = \mathfrak{eb} \subseteq \mathcal{O}_K.$$

Thus, $\mathcal{A} \sim \mathfrak{be} \subseteq \mathcal{O}_K$, and the result is proved.

Exercise 6.2.4 Prove that for any integer $x > 0$, the number of integral ideals $\mathfrak{a} \subseteq \mathcal{O}_K$ for which $N(\mathfrak{a}) \leq x$ is finite.

Solution. Since the norm is multiplicative and takes values > 1 on prime ideals, and since integral ideals have unique factorization, it is sufficient to prove that there are only a finite number of prime ideals $\wp$ with $N(\wp) \leq x$.

Now, any prime $\wp$ contains exactly one prime $p \in \mathbb{Z}$, as shown in Exercise 4.4.4. Thus, $\wp$ occurs in the factorization of $(p) \subseteq \mathcal{O}_K$ into prime ideals. Since $N(\wp) \geq 2$, we have $N(\wp) = p^t$ for some $t \geq 1$. This implies there are at most n possibilities for such $\wp$, since the factorization $(p) = \prod_{i=1}^{s} \wp_i^{a_i}$ implies that $p^n = N((p)) = \prod_{i=1}^{s} N(\wp_i)^{a_i}$ leading to $s \leq n$. Moreover, $p \leq N(\wp) \leq x$. This proves the exercise.

Exercise 6.2.6 Show that the product defined above is well-defined, and that $\mathcal{H}$ together with this product form a group, of which the equivalence class containing the principal ideals is the identity element.

Solution. To show that the product defined above is well-defined we only need to show that if $\mathcal{A}_1 \sim \mathcal{B}_1$ and $\mathcal{A}_2 \sim \mathcal{B}_2$, then $\mathcal{A}_1\mathcal{A}_2 \sim \mathcal{B}_1\mathcal{B}_2$. Indeed, by definition, there exist $\alpha_1, \alpha_2, \beta_1, \beta_2 \in \mathcal{O}_K$ such that $(\alpha_1)\mathcal{A}_1 = (\beta_1)\mathcal{B}_1$ and $(\alpha_2)\mathcal{A}_2 = (\beta_2)\mathcal{B}_2$. Therefore

$$(\alpha_1\alpha_2)\mathcal{A}_1\mathcal{A}_2 = (\beta_1\beta_2)\mathcal{B}_1\mathcal{B}_2.$$

Thus, $\mathcal{A}_1\mathcal{A}_2 \sim \mathcal{B}_1\mathcal{B}_2$.

Now, it is easy to check that $\mathcal{H}$ with the product defined above is closed, associative, commutative, and has the class of principal ideals as the identity element. Thus, to finish the exercise, we need to show that each element of $\mathcal{H}$ does have an inverse. Suppose $\mathcal{C}$ is an arbitrary element of $\mathcal{H}$. Let $\mathfrak{a} \subseteq \mathcal{O}_K$ be a representative of $\mathcal{C}$ (we showed in Exercise 6.2.3 that every equivalence class of ideals contains an integral representative). If we proceed as we did when deriving equation (6.1), we conclude that there exists an integral ideal $\mathfrak{b}$ such that $\mathfrak{ab}$ is principal. It then follows immediately that the class containing $\mathfrak{b}$ is the inverse of $\mathcal{C}$.

Exercise 6.2.7 Show that the constant C_K in Theorem 6.2.2 could be taken to be the greatest integer less than or equal to H_K, the Hurwitz constant.

Solution. As in Lemma 6.1.1, let $\{\omega_1, \omega_2, \dots, \omega_n\}$ be an integral basis of $\mathcal{O}_K$. Let $\mathcal{C}$ be a given class of ideals. We denote by $\mathcal{C}^{-1}$ the inverse class of

$\mathcal{C}$ in $\mathcal{H}$. Let $\mathfrak{a}$ be an integral representative of $\mathcal{C}^{-1}$. Consider the following set

$$S = \left\{ s \in \mathcal{O}_K | s = \sum_{i=1}^{n} m_i \omega_i,\ m_i \in \mathbb{Z},\ 0 \leq m_i < (N(\mathfrak{a}))^{1/n} + 1 \right\}.$$

Then $|S| \geq N(\mathfrak{a}) + 1$. Since $N(\mathfrak{a}) = [\mathcal{O}_K : \mathfrak{a}]$, we can find distinct $a, b \in S$ such that $a \equiv b \pmod{\mathfrak{a}}$. Thus, $(a - b) \subseteq \mathfrak{a}$. This implies that there exists an integral ideal $\mathfrak{b}$ such that $(a - b) = \mathfrak{ab}$. It is easy to observe that $\mathfrak{b} \in \mathcal{C}$.

We may write $a - b = \sum_{i=1}^{n} p_i \omega_i$. Since $a, b \in S$, $|p_i| \leq (N(\mathfrak{a}))^{1/n} + 1$, and so we have

$$\begin{aligned} |N(a-b)| &= \left| \prod_{j=1}^{n} \left(\sum_{i=1}^{n} p_i \omega_i^{(j)} \right) \right| \\ &\leq \prod_{j=1}^{n} \left(\sum_{i=1}^{n} |p_i| |\omega_i^{(j)}| \right) \\ &\leq \left[(N(\mathfrak{a}))^{1/n} + 1 \right]^n \prod_{j=1}^{n} \left(\sum_{i=1}^{n} |\omega_i^{(j)}| \right) \\ &\leq \left[(N(\mathfrak{a}))^{1/n} + 1 \right]^n H_K. \end{aligned}$$

We also know that since $(a - b) = \mathfrak{ab}$, $|N(a - b)| = N(\mathfrak{a})N(\mathfrak{b})$. Thus,

$$N(\mathfrak{b}) \leq \left[1 + (N(\mathfrak{a}))^{-1/n} \right]^n H_K.$$

However, observe that we can always replace $\mathfrak{a}$ by the ideal $c\mathfrak{a}$, in the same equivalence class, for any $c \in \mathcal{O}_K \backslash \{0\}$, and with $|N(c)|$ arbitrarily large; we can therefore make $\left[1 + (N(\mathfrak{a}))^{-1/n}\right]^n$ arbitrarily close to 1.

Thus, every equivalence class $\mathcal{C}$ has an integral representative $\mathfrak{b}$ with $N(\mathfrak{b}) \leq H_K$. This implies that every ideal is equivalent to another integral ideal with norm less than or equal to H_K.

6.3 Diophantine Equations

Exercise 6.3.2 Let $k > 0$ be a squarefree positive integer. Suppose that $k \equiv 1, 2 \pmod 4$, and k does not have the form $k = 3a^2 \pm 1$ for an integer a. Consider the equation

$$x^2 + k = y^3. \tag{6.4}$$

Show that if 3 does not divide the class number of $\mathbb{Q}(\sqrt{-k})$, then this equation has no integral solution.

Solution. Similar to what was done in Example 6.3.1, y must be odd (consider congruences modulo 4). Also, if a prime $p \mid (x, y)$, then $p \mid k$;

and hence, since k is squarefree (so, in particular, k is not divisible by p^2), by dividing both sides of the equation (6.4) by p, we end up having a contradiction modulo p. Thus, x and y are coprime.

Suppose now that (x, y) is an integral solution to equation (6.4). As given, $k \equiv 1, 2 \pmod 4$, so $-k \equiv 3, 2 \pmod 4$. Thus, the integers in $K = \mathbb{Q}(\sqrt{-k})$ are $\mathbb{Z}[\sqrt{-k}]$. We consider the factorization

$$(x + \sqrt{-k})(x - \sqrt{-k}) = y^3, \tag{6.5}$$

in the ring of integers $\mathbb{Z}[\sqrt{-k}]$.

As in Example 6.3.1, suppose a prime $\wp$ divides the gcd of the ideals $(x + \sqrt{-k})$ and $(x - \sqrt{-k})$ (which implies $\wp$ divides (y)). Then $\wp$ divides $(2x)$. Since y is odd, $\wp$ does not divide (2). Thus, $\wp$ divides (x). This contradicts the fact that x and y are coprime. Hence, $(x + \sqrt{-k})$ and $(x - \sqrt{-k})$ are coprime. Equation (6.5) now implies that

$$(x + \sqrt{-k}) = \mathfrak{a}^3 \quad \text{and} \quad (x - \sqrt{-k}) = \mathfrak{b}^3,$$

for some ideals $\mathfrak{a}$ and $\mathfrak{b}$.

Let $h(K)$ be the class number of the field K, then $\mathfrak{c}^{h(K)}$ is principal for any ideal $\mathfrak{c}$. As given, $3 \nmid h(K)$, so $(3, h(K)) = 1$. Thus, since $\mathfrak{a}^3$ and $\mathfrak{b}^3$ are principal, $\mathfrak{a}$ and $\mathfrak{b}$ are also principal. We must have

$$(x + \sqrt{-k}) = \varepsilon(a + b\sqrt{-k})^3, \tag{6.6}$$

for some integers a, b, and a unit $\varepsilon \in \mathbb{Z}[\sqrt{-k}]$.

Let $\varepsilon = x_1 + x_2\sqrt{-k}$. Then, since $\alpha \in \mathbb{Z}[\sqrt{-k}]$ is a unit if and only if $N(\alpha) = \pm 1$, we have

$$x_1^2 + kx_2^2 = \pm 1. \tag{6.7}$$

As given $k > 0$ and k is square-free, so $k > 1$. Thus, equation (6.7) implies $x_2 = 0$ and $x_1 = \pm 1$. Hence, $\varepsilon = \pm 1$, and in equation (6.6) it could be absorbed into the cube. We have

$$(x + \sqrt{-k}) = (a + b\sqrt{-k})^3.$$

This implies $1 = b(3a^2 - kb^2)$. It is clear that $b \mid 1$, so $b = \pm 1$. Both cases lead to either $k = 3a^2 + 1$ or $k = 3a^2 - 1$, which violates the hypothesis.

Hence, we conclude that equation (6.4) does not have an integral solution.

6.4 Exponents of Ideal Class Groups

Exercise 6.4.1 Fix a positive integer $g > 1$. Suppose that n is odd, greater than 1 and $n^g - 1 = d$ is squarefree. Show that the ideal class group of $\mathbb{Q}(\sqrt{-d})$ has an element of order g.

Solution. Since d is even and squarefree, $d \equiv 2 \pmod 4$. The ring of integers of $\mathbb{Q}(\sqrt{-d})$ is $\mathbb{Z}[\sqrt{-d}]$. We have the ideal factorization:

$$(n)^g = (n^g) = (1+d) = (1+\sqrt{-d})(1-\sqrt{-d}).$$

The ideals $(1+\sqrt{-d})$ and $(1-\sqrt{-d})$ are coprime since n is odd. Thus by Theorem 5.3.13, each of the ideals $(1+\sqrt{-d}),(1-\sqrt{-d})$ must be gth powers. Thus

$$\begin{aligned} \mathfrak{a}^g &= (1+\sqrt{-d}), \\ (\mathfrak{a}')^g &= (1-\sqrt{-d}), \end{aligned}$$

with $\mathfrak{a}\mathfrak{a}' = (n)$. Hence $\mathfrak{a}$ has order dividing g in the class group.

Suppose $\mathfrak{a}^m = (u+v\sqrt{-d})$ for some $u, v \in \mathbb{Z}$. Note that v cannot be zero for otherwise $\mathfrak{a}^m = (u)$ implies that $(\mathfrak{a}')^m = (u)$ so that $(u) = \gcd(\mathfrak{a}^m, (\mathfrak{a}')^m)$, contrary to $\gcd(\mathfrak{a}, \mathfrak{a}') = 1$. Therefore $v \neq 0$.

Now take norms of the equation $\mathfrak{a}^m = (u+v\sqrt{-d})$ to obtain

$$n^m = u^2 + v^2 d \geq d = n^g - 1.$$

If $m \leq g-1$, we get $n^{g-1} \geq n^g - 1$ which implies that $1 \geq n^{g-1}(n-1) \geq 2$, a contradiction.

Therefore $\mathfrak{a}^g = (1+\sqrt{-d})$ and $\mathfrak{a}^m$ is not principal for any $m < g$. Thus there is an element of order g in the ideal class group of $\mathbb{Q}(\sqrt{-d})$.

Exercise 6.4.2 Let g be odd and greater than 1. If $d = 3^g - x^2$ is squarefree with x odd and satisfying $x^2 < 3^g/2$, show that $\mathbb{Q}(\sqrt{-d})$ has an element of order g in the class group.

Solution. Observe that $d \equiv 2 \pmod 4$ so the ring of integers of $\mathbb{Q}(\sqrt{-d})$ is $\mathbb{Z}[\sqrt{-d}]$. The factorization

$$3^g = (x+\sqrt{-d})(x-\sqrt{-d})$$

shows that 3 splits in $\mathbb{Q}(\sqrt{-d})$, as the ideals $(x+\sqrt{-d})$ and $(x-\sqrt{-d})$ are coprime. Thus

$$(3) = \wp_1 \wp_1'.$$

We must have

$$(x+\sqrt{-d}) = \wp_1^g.$$

Therefore, the order of $\wp_1$ in the ideal class group is a divisor of g. If $\wp_1^m = (u+v\sqrt{-d})$, then $3^m = u^2 + v^2 d$. If $v \neq 0$, we deduce $3^m \geq d > 3^g/2$ which is a contradiction if $m \leq g-1$. Either $\wp_1$ has order g or $v = 0$. In the latter case, we get $u^2 = 3^m$, a contradiction since m is odd.

Exercise 6.4.3 Let g be odd. Let N be the number of squarefree integers of the form $3^g - x^2$, x odd, $0 < x^2 < 3^g/2$. For g sufficiently large, show that $N \gg 3^{g/2}$. Deduce that there are infinitely many imaginary quadratic fields whose class number is divisible by g.

Solution. The number of integers under consideration is

$$\frac{1}{2\sqrt{2}}3^{g/2} + O(1).$$

From these, we will remove any number divisible by the square of a prime. Since g is odd, $x^2 \equiv 3^g \pmod 4$ implies that $x^2 \equiv -1 \pmod 4$ has a solution. This is a contradiction. Therefore $4 \nmid 3^g - x^2$. If $3 \mid x$, then $3 \mid 3^g - x^2$ so we remove such numbers. Their count is

$$\frac{1}{6\sqrt{2}}3^{g/2} + O(1).$$

If p is odd and greater than 3, the number of $3^g - x^2$ divisible by p^2 is at most

$$\frac{3^{g/2}}{p^2\sqrt{2}} + O(1).$$

Thus,

$$N \geq \frac{3^{g/2}}{\sqrt{2}}\left\{\frac{1}{2} - \frac{1}{6} - \sum_{\substack{p^2<3^g \\ p\geq 5}} \frac{1}{p^2} + O\left(\frac{3^{g/2}}{g}\right)\right\}$$

by using Exercise 1.1.26. Now

$$\begin{aligned}\sum_{p\geq 5}\frac{1}{p^2} &\leq \sum_{n=5}^{\infty}\frac{1}{n(n-1)} = \left(\tfrac{1}{4} - \tfrac{1}{5}\right) + \left(\tfrac{1}{5} - \tfrac{1}{6}\right) + \cdots \\ &= \tfrac{1}{4}.\end{aligned}$$

Since $\frac{1}{2} - \frac{1}{4} - \frac{1}{6} = \frac{1}{12}$, we see $N \gg 3^{g/2}$. By the previous exercise, each of these values gives rise to a distinct quadratic field whose class group has exponent divisible by g. By applying this result for powers of g we deduce that there are infinitely many imaginary quadratic fields of class number divisible by g.

(This argument is due to Ankeny and Chowla.)

6.5 Supplementary Problems

Exercise 6.5.1 Show that the class number of $K = \mathbb{Q}(\sqrt{-19})$ is 1.

Solution. We know that $1, (1+\sqrt{-19})/2$ forms an integral basis. We then write

$$\omega_1^{(1)} = 1, \qquad \omega_2^{(1)} = \frac{1+\sqrt{-19}}{2},$$

$$\omega_1^{(2)} = 1, \qquad \omega_2^{(2)} = \frac{1-\sqrt{-19}}{2},$$

and use this to find the Hurwitz constant

$$\begin{aligned} H_K &= \prod_{j=1}^{2}\left(\sum_{i=1}^{2}|\omega_i^{(j)}|\right) \\ &= \left(1+\left|\frac{1+\sqrt{-19}}{2}\right|\right)\left(1+\left|\frac{1-\sqrt{-19}}{2}\right|\right) \\ &= 13.53\cdots. \end{aligned}$$

Just as in Example 6.2.8, we examine all the primes $p \leq 13$ to determine the prime ideals with $N(\wp) \leq 13$. The primes in question are $2, 3, 5, 7, 11$, and 13. They factor in $\mathbb{Z}[(1+\sqrt{-19})/2]$ as follows: $2, 3$, and 13 stay prime, and

$$\begin{aligned} 5 &= \left(\frac{1+\sqrt{-19}}{2}\right)\left(\frac{1-\sqrt{-19}}{2}\right), \\ 7 &= \left(\frac{3+\sqrt{-19}}{2}\right)\left(\frac{3-\sqrt{-19}}{2}\right), \\ 11 &= \left(\frac{5+\sqrt{-19}}{2}\right)\left(\frac{5-\sqrt{-19}}{2}\right). \end{aligned}$$

These are all principal ideals and thus are all equivalent. This shows that the class number of $K = \mathbb{Q}(\sqrt{-19})$ is 1.

Exercise 6.5.2 (Siegel) Let C be a symmetric, bounded domain in $\mathbb{R}^n$. (That is, C is bounded and if $x \in C$ so is $-x$.) If $\mathrm{vol}(C) > 1$, then there are two distinct points $P, Q \in C$ such that $P - Q$ is a lattice point.

Solution. Let $\varphi(x) = 1$ or 0 according as $x \in C$ or not. Then set

$$\psi(x) = \sum_{\gamma \in \mathbb{Z}^n} \varphi(x+\gamma).$$

Clearly, $\psi(x)$ is bounded and integrable. thus

$$\begin{aligned} \int_{\mathbb{R}^n/\mathbb{Z}^n} \psi(x)\,dx &= \int_{\mathbb{R}^n/\mathbb{Z}^n} \sum_{\gamma\in\mathbb{Z}^n} \varphi(x+\gamma)\,dx \\ &= \sum_{\gamma\in\mathbb{Z}^n} \int_{\mathbb{R}^n/\mathbb{Z}^n} \varphi(x+\gamma)\,dx \\ &= \sum_{\gamma\in\mathbb{Z}^n} \int_{\gamma+\mathbb{R}^n/\mathbb{Z}^n} \varphi(x)\,dx \\ &= \int_{\mathbb{R}^n} \varphi(x)\,dx \\ &= \mathrm{vol}(C) > 1. \end{aligned}$$

Since $\psi(x)$ takes only integer values, we must have $\psi(x) \geq 2$ for some x. Therefore, there are two distinct points $P+\gamma, P+\gamma'$ in C so their difference is a lattice point.

Exercise 6.5.3 If C is any convex, bounded, symmetric domain of volume $> 2^n$, show that C contains a non-zero lattice point. (C is said to be *convex* if $x, y \in C$ implies $\lambda x + (1-\lambda)y \in C$ for $0 \leq \lambda \leq 1$.)

Solution. By the previous question, the bounded symmetric domain $\frac{1}{2}C$ contains two distinct points $\frac{1}{2}P$ and $\frac{1}{2}Q$ such that $\frac{1}{2}P - \frac{1}{2}Q$ is a lattice point, because

$$\mathrm{vol}\left(\tfrac{1}{2}C\right) = \frac{\mathrm{vol}(C)}{2^n} > 1.$$

Since C is convex,

$$0 \neq \gamma = \tfrac{1}{2}P - \tfrac{1}{2}Q \in C$$

as $P, Q \in C$. This is a nonzero lattice point in C.

Exercise 6.5.4 Show in the previous question if the volume $\geq 2^n$, the result is still valid, if C is closed.

Solution. We can enlarge our domain by ε to create C_ε of volume $> 2^n$. For each ε, C_ε contains a lattice point. Since

$$\lim_{\varepsilon \to 0} C_\varepsilon = C,$$

C also contains a lattice point (perhaps on the boundary).

Exercise 6.5.5 Show that there exist bounded, symmetric convex domains with volume $< 2^n$ that do not contain a lattice point.

Solution. Consider $-1 < x_i < 1, 1 \leq i \leq n$. This hypercube has volume 2^n and the only lattice point it contains is 0.

Exercise 6.5.6 (Minkowski) For $x = (x_1, \ldots, x_n)$, let

$$L_i(x) = \sum_{j=1}^{n} a_{ij}x_j, \qquad 1 \leq i \leq n,$$

be n linear forms with real coefficients. Let C be the domain defined by

$$|L_i(x)| \leq \lambda_i, \qquad 1 \leq i \leq n.$$

Show that if $\lambda_1 \cdots \lambda_n \geq |\det A|$ where $A = (a_{ij})$, then C contains a nonzero lattice point.

Solution. Clearly C is convex, bounded, and symmetric. We want to compute the volume of $C/2$,

$$\underset{\substack{|L_i(x)|\leq\lambda_i/2\\ 1\leq i\leq n}}{\int\cdots\int} dx_1\cdots dx_n.$$

We make a linear change of variables: $y = Ax$. Then

$$dy_1 \cdots dy_n = (\det A)\, dx_1 \cdots dx_n,$$

so we get

$$\mathrm{vol}(C/2) = \frac{\lambda_1 \cdots \lambda_n}{|\det A|} \geq 1$$

from which the result follows because C is closed.

Exercise 6.5.7 Suppose that among the n linear forms above, $L_i(x), 1 \leq i \leq r_1$ are real (i.e., $a_{ij} \in \mathbb{R}$), and $2r_2$ are not real (i.e., some a_{ij} may be nonreal). Further assume that

$$L_{r_1+r_2+j} = \overline{L_{r_1+j}}, \qquad 1 \leq j \leq r_2.$$

That is,

$$L_{r_1+r_2+j}(x) = \sum_{k=1}^{n} \overline{a}_{r_1+j,k} x_k, \qquad 1 \leq j \leq r_2.$$

Now let C be the convex, bounded symmetric domain defined by

$$|L_i(x)| \leq \lambda_i, \qquad 1 \leq i \leq n,$$

with $\lambda_{r_1+j} = \lambda_{r_1+r_2+j}$, $1 \leq j \leq r_2$. Show that if $\lambda_1 \cdots \lambda_n \geq |\det A|$, then C contains a nonzero lattice point.

Solution. We replace the nonreal linear forms by real ones and apply the previous result. Set

$$L'_{r_1+j} = \frac{L_{r_1+j} + L_{r_1+r_2+j}}{2}$$

and

$$L''_{r_1+j} = \frac{L_{r_1+j} - L_{r_1+r_2+j}}{2}.$$

Then L'_{r_1+j}, L''_{r_1+j} are linear forms. Clearly, if

$$\begin{aligned} |L'_{r_1+j}| &\leq \frac{\lambda_{r_1+j}}{\sqrt{2}}, \\ |L''_{r_1+j}| &\leq \frac{\lambda_{r_1+j}}{\sqrt{2}}, \end{aligned}$$

then

$$|L_{r_1+j}| \leq \lambda_{r_1+j},$$

so we replace $|L_{r_1+j}| \leq \lambda_{r_1+j}$, $|L_{r_1+r_2+j}| \leq \lambda_{r_1+j}$ with L'_{r_1+j} and L''_{r_1+j} satisfying the inequalities above. We deduce by the results established in the previous questions, that this domain contains a nonzero lattice point provided

$$\frac{\lambda_1 \cdots \lambda_n 2^{-r_2}}{|\det A'|} > 1,$$

where A' is the appropriately modified matrix. A simple linear algebra computation shows $\det A' = 2^{-r_2} \det A$.

Exercise 6.5.8 Using the previous result, deduce that if K is an algebraic number field with discriminant d_K, then every ideal class contains an ideal $\mathfrak{b}$ satisfying $N\mathfrak{b} \leq \sqrt{|d_K|}$.

Solution. Let $\mathfrak{a}$ be any integral ideal and $\alpha_1, \ldots, \alpha_n$ an integral basis of $\mathfrak{a}$. Consider the linear forms

$$L_i(x) = \sum_{j=1}^{n} \alpha_j^{(i)} x_j$$

and the bounded symmetric convex domain defined by

$$|L_i(x)| \leq |\Delta|^{1/n},$$

where $|\Delta| = |\det(\alpha_j^{(i)})|$. By the previous question, the system has a nontrivial integral solution, $(x_1, \ldots, x_n)$. Let

$$\omega = x_1\alpha_1 + \cdots + x_n\alpha_n \in \mathfrak{a}.$$

Then $(\omega) \subseteq \mathfrak{a}$ so that for some ideal $\mathfrak{b}$, $\mathfrak{ab} = (\omega)$. But

$$|N(\omega)| = |N\mathfrak{a}||N\mathfrak{b}| \leq |\Delta|$$

by construction. Also,

$$|\Delta|^2 = (N\mathfrak{a})^2|d_K|$$

by Exercise 4.4.5. Hence, $|N\mathfrak{b}| \leq \sqrt{|d_K|}$. Given any ideal $\mathfrak{a}$ we have found an ideal $\mathfrak{b}$ in the inverse class whose norm is less than or equal to $\sqrt{|d_K|}$.

Exercise 6.5.9 Let X_t consist of points

$$(x_1, \ldots, x_r, y_1, z_1, \ldots, y_s, z_s)$$

in $\mathbb{R}^{r+2s}$ where the coordinates satisfy

$$|x_1| + \cdots + |x_r| + 2\sqrt{y_1^2 + z_1^2} + \cdots + 2\sqrt{y_s^2 + z_s^2} < t.$$

Show that X_t is a bounded, convex, symmetric domain.

Solution. The fact that X_t is bounded and symmetric is clear. To see convexity, let

$$P = (a_1, \ldots, a_r, b_1, c_1, \ldots, b_s, c_s)$$

and

$$Q = (d_1, \ldots, d_r, e_1, f_1, \ldots, e_s, f_s)$$

be points of X_t. We must show that $\lambda P + \mu Q \in X_t$ whenever $\lambda, \mu \geq 0$ and $\lambda + \mu = 1$. Clearly

$$|\lambda a_i + \mu d_i| \leq \lambda |a_i| + \mu |d_i|.$$

Also

$$\sqrt{(\lambda b_i + \mu e_i)^2 + (\lambda c_i + \mu f_i)^2} \leq \lambda \sqrt{b_i^2 + c_i^2} + \mu \sqrt{e_i^2 + f_i^2},$$

as is easily verified. From these inequalities, it follows that $\lambda P + \mu Q \in X_t$ so that X_t is convex.

Exercise 6.5.10 In the previous question, show that the volume of X_t is

$$\frac{2^{r-s}\pi^s t^n}{n!},$$

where $n = r + 2s$.

Solution. We begin by making a change of variables to polar coordinates: $2y_j = \rho_j \cos\theta_j$, $2x_j = \rho_j \sin\theta_j$, $4\,dy_j\,dz_j = \rho_j\,d\rho_j\,d\theta_j$ so that integrating over $x_i \geq 0$ for $1 \leq i \leq r$ gives

$$\begin{aligned} \mathrm{vol}(X_t) &= 2^r \cdot 2^{-2s} \int \rho_1 \cdots \rho_s \, dx_1 \cdots dx_r \, d\rho_1 \cdots d\rho_s \, d\theta_1 \cdots d\theta_s \\ &= 2^r 2^{-2s} (2\pi)^s \int_{Y_t} \rho_1 \cdots \rho_s \, dx_1 \cdots dx_r \, d\rho_1 \cdots d\rho_s, \end{aligned}$$

where

$$Y_t = \{(x_1, \ldots, x_r, \rho_1, \ldots, \rho_s) : x_i, \rho_j \geq 0, x_1 + \cdots + x_r + \rho_1 + \cdots + \rho_s \leq t\}.$$

Let $f_{r,s}(t)$ denote the value of the above integral. By changing variables, it is clear that

$$\begin{aligned} f_{r,s}(1) &= \int_0^1 f_{r-1,s}(1 - x_1)\,dx_1 \\ &= f_{r-1,s}(1) \int_0^1 x_1^{r-1+2s}\,dx_1 \\ &= \frac{1}{r+2s} f_{r-1,s}(1). \end{aligned}$$

Proceeding inductively, we get

$$f_{r,s}(1) = \frac{(2s)!}{(r+2s)!} f_{0,s}(1).$$

Now we evaluate $f_{0,s}(1)$ by integrating with respect to ρ_s first:

$$\begin{aligned} f_{0,s}(1) &= \int_0^1 \rho_s f_{0,s-1}(1-\rho_s)\, d\rho_s \\ &= f_{0,s-1}(1) \int_0^1 \rho_s(1-\rho_s)^{2s-2}\, d\rho_s \\ &= \frac{f_{0,s-1}(1)}{2s(2s-1)}. \end{aligned}$$

Again, proceeding inductively, we find $f_{0,s}(1) = 1/(2s)!$ so that

$$f_{r,s}(1) = \frac{1}{(r+2s)!} = \frac{1}{n!}.$$

This completes the proof.

Exercise 6.5.11 Let C be a bounded, symmetric, convex domain in $\mathbb{R}^n$. Let $a_1, \ldots, a_n$ be linearly independent vectors in $\mathbb{R}^n$. Let A be the $n \times n$ matrix whose rows are the a_i's. If

$$\mathrm{vol}(C) > 2^n |\det A|,$$

show that there exist rational integers $x_1, \ldots, x_n$ (not all zero) such that

$$x_1 a_1 + \cdots + x_n a_n \in C.$$

Solution. Consider the set D of all $(x_1, \ldots, x_n) \in \mathbb{R}^n$ such that

$$x_1 a_1 + \cdots + x_n a_n \in C.$$

It is easily seen that D is bounded, symmetric, and convex because C is. Moreover, $D = A^{-1}C$ so that by linear algebra,

$$\mathrm{vol}(D) = \mathrm{vol}(C)(|\det A|)^{-1}.$$

Thus, if $\mathrm{vol}(D) > 2^n$, then D contains a lattice point $(x_1, \ldots, x_n) \neq 0$ such that $x_1 a_1 + \cdots + x_n a_n \in C$. But $\mathrm{vol}(D) > 2^n$ is equivalent to

$$\mathrm{vol}(C) > 2^n |\det A|,$$

as desired.

Exercise 6.5.12 (Minkowski's Bound) Let K be an algebraic number field of degree n over $\mathbb{Q}$. Show that each ideal class contains an ideal $\mathfrak{a}$ satisfying

$$N\mathfrak{a} \leq \frac{n!}{n^n} \left(\frac{4}{\pi}\right)^{r_2} |d_K|^{1/2},$$

where r_2 is the number of pairs of complex embeddings of K, and d_K is the discriminant.

Solution. Given any ideal $\mathfrak{b}$, let $\omega_1, \ldots, \omega_n$ be a basis of $\mathfrak{b}$. Let

$$\begin{aligned} a_i &= (\sigma_1(\omega_i), \ldots, \sigma_{r_1}(\omega_i), \operatorname{Re}(\sigma_{r_1+1}(\omega_i)), \\ &\quad \operatorname{Im}(\sigma_{r_1+1}(\omega_i)), \ldots, \operatorname{Re}(\sigma_{r_1+r_2}(\omega_i)), \operatorname{Im}(\sigma_{r_1+r_2}(\omega_i))) \in \mathbb{R}^n. \end{aligned}$$

Then the a_i are linearly independent vectors in $\mathbb{R}^n$. Consider the bounded symmetric convex domain X_t defined in Exercise 6.5.9 above (with $r = r_1, s = r_2$). By Exercise 6.5.10 above, the volume of X_t is

$$\frac{2^{r_1 - r_2}\pi^{r_2}t^n}{n!}.$$

If t is chosen so that this volume is greater than $2^n|\det A|$, then X_t contains a lattice point $(x_1, \ldots, x_n)$ so that

$$0 \neq x_1 a_1 + \cdots + x_n a_n \in X_t.$$

Let us set $\alpha = x_1\omega_1 + \cdots + x_n\omega_n \in \mathfrak{b}$. By the arithmetic mean – geometric mean inequality, we find

$$|N(\alpha)|^{1/n} < \frac{t}{n}$$

so that

$$|N(\alpha)| < \left(\frac{t}{n}\right)^n.$$

Moreover, $\det A = 2^{-r_2}|N(\mathfrak{b})||d_K|^{1/2}$ and

$$\frac{t^n}{n!} = \frac{2^n|\det A|}{2^{r_1 - r_2}\pi^{r_2}} = \left(\frac{4}{\pi}\right)^{r_2} |N\mathfrak{b}||d_K|^{1/2}.$$

Thus, there is an $\alpha \in \mathfrak{b}$, $\alpha \neq 0$ such that

$$|N(\alpha)| \leq \frac{n!}{n^n}\left(\frac{4}{\pi}\right)^{r_2} |N\mathfrak{b}||d_K|^{1/2}.$$

Write $(\alpha) = \mathfrak{a}\mathfrak{b}$ for some ideal $\mathfrak{a}$. Then

$$N\mathfrak{a} \leq \frac{n!}{n^n}\left(\frac{4}{\pi}\right)^{r_2} |d_K|^{1/2},$$

as desired.

Exercise 6.5.13 Show that if $K \neq \mathbb{Q}$, then $|d_K| > 1$. Thus, by Dedekind's theorem, in any nontrivial extension of K, some prime ramifies.

Solution. Since $N\mathfrak{a} \geq 1$, we have by the Minkowski bound,

$$1 \leq \frac{n!}{n^n}\left(\frac{4}{\pi}\right)^{r_2} |d_K|^{1/2}$$

so that

$$|d_K|^{1/2} \geq \frac{n^n}{n!}\left(\frac{\pi}{4}\right)^{r_2} \geq \frac{n^n}{n!}\left(\frac{\pi}{4}\right)^{n/2} = c_n,$$

say. Then

$$\frac{c_{n+1}}{c_n} = \left(\frac{\pi}{4}\right)^{1/2}\left(1+\frac{1}{n}\right)^n,$$

which is greater than 1 for every positive n. Hence $c_{n+1} > c_n$. We have $c_2 > 1$ so that $|d_K| > 1$, if $n \geq 2$.

Exercise 6.5.14 If K and L are algebraic number fields such that d_K and d_L are coprime, show that $K \cap L = \mathbb{Q}$. Deduce that

$$[KL : \mathbb{Q}] = [K : \mathbb{Q}][L : \mathbb{Q}].$$

Solution. If $M = K \cap L$, then by a result of Chapter 4, $d_M \mid d_K$ and $d_M \mid d_L$. Since d_K and d_L are coprime, $d_M = 1$. But then, by the previous exercise, $M = \mathbb{Q}$. We have

$$[KL : \mathbb{Q}] = [KL : K][K : \mathbb{Q}].$$

Let $L = \mathbb{Q}(\theta)$ and g its minimal polynomial over $\mathbb{Q}$. If $[KL : K] < [L : \mathbb{Q}]$, then the minimal polynomial h of θ over K divides g and has degree smaller than that of g. Thus the coefficients of h generate a proper extension T (say) of $\mathbb{Q}$ which is necessarily contained in K. Hence, $d_T | d_K$. If we let $\tilde{L}$ be the normal closure of L over $\mathbb{Q}$, then $h \in \tilde{L}[x]$. We now need to use the fact that primes which ramify in L are the same as the ones that ramify in $\tilde{L}$ (see Exercise 5.6.28). Since T is contained in $\tilde{L}$, we see that $d_T | d_{\tilde{L}}$ and by the quoted fact, we deduce that d_L and d_K have a common prime factor if $d_T > 1$, which is contrary to hypothesis. Thus, $d_T = 1$ and by 6.5.13 we deduce $T = \mathbb{Q}$, a contradiction.

Exercise 6.5.15 Using Minkowski's bound, show that $\mathbb{Q}(\sqrt{5})$ has class number 1.

Solution. The discriminant of $\mathbb{Q}(\sqrt{5})$ is 5 and the Minkowski bound is

$$\frac{2!}{2^2}\sqrt{5} = \frac{\sqrt{5}}{2} = 1.11\ldots.$$

The only ideal of norm less than $\sqrt{5}/2$ is the trivial ideal which is principal.

Exercise 6.5.16 Using Minkowski's bound, show that $\mathbb{Q}(\sqrt{-5})$ has class number 2.

Solution. The discriminant of $\mathbb{Q}(\sqrt{-5})$ is -20 and the Minkowski bound is

$$\frac{2}{\pi}\sqrt{20} = \frac{4}{\pi}(2.236\ldots) = 2.84\ldots.$$

We need to look at ideals of norm 2. There is only one ideal of norm 2 and by Exercise 5.2.5 we know that $\mathbb{Z}[\sqrt{-5}]$ is not a principal ideal domain. Hence the class number must be 2.

Exercise 6.5.17 Compute the class numbers of the fields $\mathbb{Q}(\sqrt{2})$, $\mathbb{Q}(\sqrt{3})$, and $\mathbb{Q}(\sqrt{13})$.

Solution. The discriminants of these fields are $8, 12$, and 13 respectively. The Minkowski bound is

$$\tfrac{1}{2}\sqrt{|d_K|} < \tfrac{1}{2}\sqrt{13} = 1.802\ldots.$$

The only ideal of norm less than 1.8 is the trivial ideal, which is principal, so the class number is 1. (Recall that in Exercises 2.4.5 and 2.5.4 we showed that the ring of integers of $\mathbb{Q}(\sqrt{2})$ and $\mathbb{Q}(\sqrt{3})$ are Euclidean and hence PIDs. So that the class number is 1 for each of these was already known to us from Chapter 2.)

Exercise 6.5.18 Compute the class number of $\mathbb{Q}(\sqrt{17})$.

Solution. The discriminant of $\mathbb{Q}(\sqrt{17})$ is 17 and the Minkowski bound is

$$\tfrac{1}{2}\sqrt{17} = 2.06\ldots.$$

We need to consider ideals of norm 2. Since

$$-2 = \frac{9-17}{4} = \frac{3-\sqrt{17}}{2}\cdot\frac{3+\sqrt{17}}{2},$$

2 splits and the principal ideals $((3+\sqrt{17})/2)$ and $((3-\sqrt{17})/2)$ are the only ones of norm 2. Therefore, the class number is 1.

Exercise 6.5.19 Compute the class number of $\mathbb{Q}(\sqrt{6})$.

Solution. The discriminant is 24 and the Minkowski bound is

$$\tfrac{1}{2}\sqrt{24} = \sqrt{6} = 2.44\ldots,$$

2 ramifies in $\mathbb{Q}(\sqrt{6})$. Moreover,

$$-2 = (2-\sqrt{6})(2+\sqrt{6})$$

so that the ideal $(2-\sqrt{6})$ is the only one of norm 2 since $(2+\sqrt{6})/(2-\sqrt{6})$ is a unit. Thus, the class number is 1.

Exercise 6.5.20 Show that the fields $\mathbb{Q}(\sqrt{-1})$, $\mathbb{Q}(\sqrt{-2})$, $\mathbb{Q}(\sqrt{-3})$, and $\mathbb{Q}(\sqrt{-7})$ each have class number 1.

Solution. The Minkowski bound for an imaginary quadratic field K is

$$\frac{2}{\pi}\sqrt{|d_K|}.$$

The given fields have discriminants equal to $-4, -8, -3, -7$, respectively. Since

$$\frac{2}{\pi}\sqrt{8} = \frac{4\sqrt{2}}{\pi} = 1.80\ldots,$$

we deduce that every ideal is principal.

Exercise 6.5.21 Let K be an algebraic number field of degree n over $\mathbb{Q}$. Prove that

$$|d_K| \geq \left(\frac{\pi}{4}\right)^n \left(\frac{n^n}{n!}\right)^2.$$

Solution. This follows directly from Minkowski's bound.

Exercise 6.5.22 Show that $|d_K| \to \infty$ as $n \to \infty$ in the preceding question.

Solution. From integral calculus,

$$\log n! = n\log n - n + O(\log n),$$

so that

$$\log|d_K| \geq \left(2 + \log\frac{\pi}{4}\right)n + O(\log n).$$

Exercise 6.5.23 (Hermite) Show that there are only finitely many algebraic number fields with a given discriminant.

Solution. We give a brief hint of the proof. From the preceding question, the degree n of K is bounded. By Minkowski's theorem, we can find an element $\alpha \neq 0$ in $\mathcal{O}_K$ so that

$$|\alpha^{(1)}| \leq \sqrt{|d_K|}, \quad |\alpha^{(i)}| < 1, \quad i = 2, \ldots, r.$$

We must show α generates K, but this is not difficult. With these inequalities, the coefficients of the minimal polynomial of α are bounded. Since the coefficients are integers, there are only finitely many such polynomials.

Exercise 6.5.24 Let p be a prime $\equiv 11 \pmod{12}$. If $p > 3^n$, show that the ideal class group of $\mathbb{Q}(\sqrt{-p})$ has an element of order greater than n.

Solution. Since $-p \equiv 1 \pmod 3$, $x^2 \equiv -p \pmod 3$ has a solution and so 3 splits in $\mathbb{Q}(\sqrt{-3})$. Write

$$(3) = \wp_1\wp_1'.$$

We claim the order of $\wp_1$ in the ideal class group is at least n. If not, $\wp_1^m$ is principal and equals $(u + v\sqrt{-p})$ (say) for some $m < n$. Taking norms, we see

$$3^m \equiv u^2 + pv^2$$

has a solution. If $v \neq 0$, this is a contradiction since $p > 3^n$. If $v = 0$, we get m is even, so that

$$\wp_1^m = (u) = (3^{m/2}) = \wp_1^{m/2}(\wp_1')^{m/2},$$

contradicting unique factorization.

Exercise 6.5.25 Let $K = \mathbb{Q}(\alpha)$ where α is a root of the polynomial $f(x) = x^5 - x + 1$. Prove that $\mathbb{Q}(\alpha)$ has class number 1.

Solution. Since $f(x)$ is a polynomial of the type described in Exercise 4.5.4, we deduce immediately that $d_{K/\mathbb{Q}}(\alpha) = 5^5 - 4^4 = 2869 = 19 \cdot 151$. Since $d_{K/\mathbb{Q}}(\alpha)$ is squarefree, Exercise 4.5.26 tells us that $\mathcal{O}_K = \mathbb{Z}[\alpha]$ and $d_K = 2869$. A quick look at the graph of $f(x) = x^5 - x + 1$ shows that $r_1 = 1$ and $r_2 = 2$.

Using Minkowski's bound, we find that every ideal class must contain an ideal $\mathfrak{a}$ of norm strictly less than 4. Therefore we must look at the numbers 2 and 3 to see how they factor in this ring. Let $\wp$ be an ideal such that $N\wp = 2$ or 3. Then $\wp$ is prime, because $\mathcal{O}_K/\wp$ is a field. Recall that to find $\wp$ we consider $f(x) \mod p$. If

$$f(x) \equiv f_1^{e_1}(x) \cdots f_g^{e_g}(x) \pmod{p},$$

then $p\mathcal{O}_K$ factors as $\wp_1^{e_1} \cdots \wp_g^{e_g}$ with $N\wp_i = p^{f_i}$ where f_i is the degree of $f_i(x)$.

First, suppose we have an ideal $\wp$ with $N\wp = 2$. Then $\wp$ must appear in the prime factorization of $2\mathcal{O}_K$, and so in the factorization of $f(x) \pmod 2$, there is a linear factor. However, it is easy to see that $x^5 - x + 1$ has no linear factor mod 2, and so there are no ideals of norm 2. By a similar argument, we see that $x^5 - x + 1$ must have a linear factor mod 3, if there exists an ideal $\wp$ with norm 3. Again, it is clear that $x^5 - x + 1$ can have no linear factor, and we conclude that there are no ideals of norm 3 in $\mathcal{O}_K$. Thus, the only ideal of norm less than 4 is the trivial ideal, and we conclude that $\mathcal{O}_K$ has class number 1.

Exercise 6.5.26 Determine the class number of $\mathbb{Q}(\sqrt{14})$.

Solution. The Minkowski bound in this case is

$$\frac{2!}{2^2}\sqrt{14 \cdot 4} = \sqrt{14} = 3.74\ldots.$$

Then we must check ideals of norm less than or equal to 3. Therefore we must look at $2\mathcal{O}_K$ and $3\mathcal{O}_K$ to see how they factor in this ring. Since 3 is inert, there are no ideals of norm 3. However, 2 ramifies as

$$(2) = (4 + \sqrt{14})(4 - \sqrt{14}) = \wp^2,$$

where $\wp = (4 + \sqrt{14})$. This is a principal ideal, and we conclude that all ideals of $\mathbb{Q}(\sqrt{14})$ are principal and so the class number is 1.

Exercise 6.5.27 If K is an algebraic number field of finite degree over $\mathbb{Q}$ with d_K squarefree, show that K has no non-trivial subfields.

Solution. By Exercise 6.5.13 and 5.6.25, any proper subfield would introduce a power into the discriminant of K.

Chapter 7

Quadratic Reciprocity

7.1 Preliminaries

Exercise 7.1.1 Let p be a prime and $a \neq 0$. Show that $x^2 \equiv a \pmod{p}$ has a solution if and only if $a^{(p-1)/2} \equiv 1 \pmod{p}$.

Solution. $\Rightarrow$ Suppose that $x^2 \equiv a \pmod{p}$ has a solution. Let x_0 be this solution, i.e., $x_0^2 \equiv a \pmod{p}$. But then,

$$a^{(p-1)/2} \equiv (x_0^2)^{(p-1)/2} \equiv x_0^{p-1} \equiv 1 \pmod{p}.$$

The last congruence follows from Fermat's Little Theorem.

$\Leftarrow$ We begin by noting that $a \not\equiv 0 \pmod{p}$. So $a \pmod{p}$ can be viewed as an element of $(\mathbb{Z}/p\mathbb{Z})^\times$, the units of $(\mathbb{Z}/p\mathbb{Z})$. Since $(\mathbb{Z}/p\mathbb{Z})^\times$ is a cyclic group, there exists some generator g such that $\langle g \rangle = (\mathbb{Z}/p\mathbb{Z})^\times$. So, $a = g^k$, where $1 \leq k \leq p-1$. From our hypothesis,

$$a^{(p-1)/2} \equiv g^{k(p-1)/2} \equiv 1 \pmod{p}.$$

Because the order of g is $p-1$, $p-1 | k(p-1)/2$. But this implies that $2|k$. So $k = 2k'$. So, we can write $a \pmod{p}$ as

$$a \equiv g^k \equiv g^{2k'} \equiv (g^{k'})^2 \pmod{p}.$$

Hence a is a square mod p, completing the proof.

Exercise 7.1.2 Using Wilson's theorem and the congruence

$$k(p-k) \equiv -k^2 \pmod{p}$$

compute $(-1/p)$ for all primes p.

Solution. The case when $p = 2$ is trivial since every odd number is congruent to 1 (mod 2). Then we will assume that p is an odd prime. To begin, we recall Wilson's theorem (proved in Exercise 1.4.10) which states that for any prime p, we have $(p-1)! \equiv -1 \pmod{p}$. We note that $(p-1)!$ can be expressed as

$$(p-1)! = 1 \cdot 2 \cdot 3 \cdots \frac{p-1}{2} \cdot \left(p - \frac{p-1}{2}\right) \cdot \left(p - \left(\frac{p-1}{2} - 1\right)\right) \cdots (p-1).$$

Thus, when we mod out by p on both sides, we get

$$\begin{aligned} -1 &\equiv 1 \cdot 2 \cdot 3 \cdots \frac{p-1}{2} \cdot \left(-\frac{p-1}{2}\right) \cdot \left(-\left(\frac{p-1}{2} - 1\right)\right) \cdots (-1) \\ &\equiv (-1)^{(p-1)/2} \left[\left(\frac{p-1}{2}\right)!\right]^2 \pmod{p}. \end{aligned}$$

If $p \equiv 1 \pmod 4$, then it follows that $(p-1)/2 = 2a$ for some integer a. Hence, from the above identity,

$$(-1)^{(p-1)/2} \left[\left(\frac{p-1}{2}\right)!\right]^2 \equiv \left[\left(\frac{p-1}{2}\right)!\right]^2 \equiv -1 \pmod{p}.$$

So, -1 is a quadratic residue mod p if $p \equiv 1 \pmod 4$.

If $p \equiv 3 \pmod 4$, we find that $(p-1)/2$ is odd. If $x^2 \equiv -1 \pmod p$, then by Exercise 7.1.1, $(-1)^{(p-1)/2} \equiv 1 \pmod p$. But since $p \equiv 3 \pmod 4$, we know that $(-1)^{(p-1)/2} \equiv -1 \pmod p$. So, there can be no solutions to $x^2 \equiv -1 \pmod p$ if $p \equiv 3 \pmod 4$.

Thus

$$\left(\frac{-1}{p}\right) = \begin{cases} 1 & \text{if } p \equiv 1 \pmod 4, \\ -1 & \text{if } p \equiv 3 \pmod 4. \end{cases}$$

Finally, we observe that we can encode this information more compactly as

$$\left(\frac{-1}{p}\right) = (-1)^{(p-1)/2}.$$

Exercise 7.1.3 Show that

$$a^{(p-1)/2} \equiv \left(\frac{a}{p}\right) \pmod{p}.$$

Solution. If $p \mid a$, then the conclusion is trivial. So, suppose p does not divide a. By Fermat's Little Theorem, $a^{p-1} \equiv 1 \pmod p$. We can factor this statement as

$$a^{p-1} - 1 \equiv (a^{(p-1)/2} - 1)(a^{(p-1)/2} + 1) \equiv 0 \pmod{p}.$$

Thus, $a^{(p-1)/2} \equiv \pm 1 \pmod p$. We will consider each case separately.

If $a^{(p-1)/2} \equiv 1 \pmod{p}$, then by Exercise 7.1.1, there exists a solution to the equation $x^2 \equiv a \pmod{p}$. But this implies $(a/p) = 1$.

If $a^{(p-1)/2} \equiv -1 \pmod{p}$, then Exercise 7.1.1 tells us there is no solution to the equation $x^2 \equiv a \pmod{p}$. So, $(a/p) = -1$. We conclude that

$$a^{(p-1)/2} \equiv \left(\frac{a}{p}\right) \pmod{p}.$$

Exercise 7.1.4 Show that

$$\left(\frac{ab}{p}\right) = \left(\frac{a}{p}\right)\left(\frac{b}{p}\right).$$

Solution. We will use Exercise 7.1.3 to prove this result. Thus,

$$\left(\frac{ab}{p}\right) \equiv (ab)^{(p-1)/2} \pmod{p}.$$

Similarly,

$$\left(\frac{a}{p}\right) \equiv a^{(p-1)/2} \pmod{p},$$

and

$$\left(\frac{b}{p}\right) \equiv b^{(p-1)/2} \pmod{p}.$$

But then

$$\left(\frac{ab}{p}\right) \equiv (ab)^{(p-1)/2} = a^{(p-1)/2}b^{(p-1)/2} \equiv \left(\frac{a}{p}\right)\left(\frac{b}{p}\right) \pmod{p}.$$

Thus

$$\left(\frac{ab}{p}\right) \equiv \left(\frac{a}{p}\right)\left(\frac{b}{p}\right) \pmod{p}.$$

But because the Legendre symbol only takes on the values ± 1, we can rewrite this statement as

$$\left(\frac{ab}{p}\right) = \left(\frac{a}{p}\right)\left(\frac{b}{p}\right),$$

which is what we wished to show.

Exercise 7.1.5 If $a \equiv b \pmod{p}$, then

$$\left(\frac{a}{p}\right) = \left(\frac{b}{p}\right).$$

Solution. This is clear from the definition of the Legendre symbol. If a is a quadratic residue mod p then so is b. The same is true if a is a quadratic nonresidue.

Exercise 7.1.7 Show that the number of quadratic residues mod p is equal to the number of quadratic nonresidues mod p.

Solution. The equation $a^{(p-1)/2} \equiv 1 \pmod{p}$ will have $(p-1)/2$ solutions. This can be deduced from the fact that $(\mathbb{Z}/p\mathbb{Z})^\times$ is a cyclic group with some generator g, where g is a $(p-1)$st root of unity. So, for any even power of g, i.e., g^{2k}, we will have

$$(g^{2k})^{(p-1)/2} \equiv (g^{p-1})^k \equiv 1 \pmod{p}.$$

For any odd power,

$$(g^{2k+1})^{(p-1)/2} \equiv (g^{(p-1)/2}) \not\equiv 1 \pmod{p}.$$

The last congruence holds since g is a $(p-1)$st root of unity. Thus, half of the elements of $(\mathbb{Z}/p\mathbb{Z})^\times$ will correspond to some even power of g, and hence, $a^{(p-1)/2} \equiv 1 \pmod{p}$. But this in turn implies that there are $(p-1)/2$ elements such that $(a/p) = 1$. Since there are $(p-1)$ residues mod p, there are $(p-1) - (p-1)/2 = (p-1)/2$ residues that are not squares. But now we have that the number of quadratic residues mod p and the number of quadratic nonresidues mod p are equal.

Exercise 7.1.8 Show that

$$\sum_{a=1}^{p-1} \left(\frac{a}{p}\right) = 0$$

for any fixed prime p.

Solution. From Exercise 7.1.7, the number of residues equals the number of nonresidues. So, there are $(p-1)/2$ residues, and $(p-1)/2$ nonresidues. Thus

$$\sum_{a=1}^{p-1} \left(\frac{a}{p}\right) = \frac{p-1}{2}(1) + \frac{p-1}{2}(-1) = 0.$$

7.2 Gauss Sums

Exercise 7.2.2 Show that

$$S^q \equiv \left(\frac{q}{p}\right) S \pmod{q},$$

where q and p are odd primes.

Solution. Let $K = \mathbb{Q}(\zeta_q)$, where ζ_q is a primitive qth root of unity. Let $R = \mathcal{O}_K$ be its ring of integers. So

$$\begin{aligned} S^q &= \left[\sum_{a \bmod p} \left(\frac{a}{p}\right) \zeta_p^a\right]^q \\ &\equiv \sum_{a \bmod p} \left(\frac{a}{p}\right)^q \zeta_p^{aq} \pmod{qR}. \end{aligned}$$

This follows from the fact that $(x_1 + \cdots + x_n)^q \equiv x_1^q + \cdots + x_n^q \pmod{q}$. Also, because q is an odd prime, and because (a/p) only takes on the values ± 1, we have $(a/p)^q = (a/p)$. Hence

$$S^q \equiv \sum_{a \bmod p} \left(\frac{a}{p}\right) \zeta_p^{aq} \pmod{qR}.$$

However,

$$\left(\frac{a}{p}\right) = \left(\frac{a}{p}\right)\left(\frac{q^2}{p}\right) = \left(\frac{aq^2}{p}\right).$$

So, it follows that

$$\begin{aligned} S^q &\equiv \sum_{a \bmod p} \left(\frac{aq^2}{p}\right) \zeta_p^{aq} \pmod{qR} \\ &\equiv \left(\frac{q}{p}\right) \sum_{a \bmod p} \left(\frac{aq}{p}\right) \zeta_p^{aq} \pmod{qR}. \end{aligned}$$

But as a runs through all the residue classes mod p, so will aq. Thus,

$$S^q \equiv \left(\frac{q}{p}\right) S \pmod{qR}.$$

From this, it follows that

$$S^q \equiv \left(\frac{q}{p}\right) S \pmod{q},$$

completing the proof.

7.3 The Law of Quadratic Reciprocity

Exercise 7.3.2 Let q be an odd prime. Prove:

(a) If $q \equiv 1 \pmod 4$, then q is a quadratic residue mod p if and only if $p \equiv r \pmod q$, where r is a quadratic residue mod q.

(b) If $q \equiv 3 \pmod 4$, then q is a quadratic residue mod p if and only if $p \equiv \pm b^2 \pmod{4q}$, where b is an odd integer prime to q.

Solution. (a) We begin by rewriting the result of Theorem 7.3.1 in the following equivalent form:

$$\left(\frac{p}{q}\right)\left(\frac{q}{p}\right) = (-1)^{\frac{p-1}{2}\cdot\frac{q-1}{2}}.$$

Since $q \equiv 1 \pmod 4$, $(q-1)/2$ is even, so we will have

$$\left(\frac{p}{q}\right)\left(\frac{q}{p}\right) = 1.$$

From this it follows that either $(p/q) = 1 = (q/p)$, or $(p/q) = -1 = (q/p)$. We can now prove the statement.

$\Rightarrow$ If $(q/p) = 1$, then $(p/q) = 1$. So, by Exercise 7.1.5, $p \equiv r \pmod{q}$, where r is a quadratic residue mod q.

$\Leftarrow$ Suppose $p \equiv r \pmod{q}$, and $(r/q) = 1$. But then $(p/q) = (r/q)$, and thus, $(p/q) = 1$. Since $(q/p) = (p/q)$ by quadratic reciprocity, we must also have that $(q/p) = 1$, which implies that q is a quadratic residue mod p. This completes the proof of (a).

(b) Suppose $q \equiv 3 \pmod{4}$. Then, quadratic reciprocity gives us

$$\left(\frac{q}{p}\right) = (-1)^{\frac{p-1}{2}} \left(\frac{p}{q}\right).$$

$\Rightarrow$ Suppose $(q/p) = 1$. Then, we have two cases:

Case 1. $(-1)^{(p-1)/2} = -1$ and $(p/q) = -1$.

Case 2. $(-1)^{(p-1)/2} = 1$ and $(p/q) = 1$.

Case 1. First we note that $(-1)^{(p-1)/2} = -1$ implies that $p \equiv 3 \pmod{4}$. Because $q \equiv 3 \pmod{4}$, we have from an earlier exercise that

$$\left(\frac{-1}{q}\right) = \left(\frac{p}{q}\right) = -1.$$

But then, we find that

$$\left(\frac{-1}{q}\right) = \left(\frac{-1}{q}\right)\left(\frac{b^2}{q}\right) = \left(\frac{-b^2}{q}\right).$$

Hence $(p/q) = (-b^2/q)$, and $p \equiv -b^2 \pmod{q}$. We can suppose that b is odd. If not, we can replace it with $b' = b + q$, which is odd. Since b is odd, $b = 2n + 1$. Thus $b^2 = 4n^2 + 4n + 1$, and so $-b^2 \equiv 3 \pmod{4}$. Since we already deduced that $p \equiv 3 \pmod{4}$, we have $p \equiv -b^2 \pmod{4}$. Because $p \equiv -b^2 \pmod{q}$, we conclude that $p \equiv -b^2 \pmod{4q}$.

Case 2. $(-1)^{(p-1)/2} = 1$ implies that $p \equiv 1 \pmod{4}$. Also, $(p/q) = (b^2/q)$. Assume b is odd for the same reason given above. So $p \equiv b^2 \pmod{q}$. Because b is odd, we have $b = 2n + 1$, which means that $b^2 \equiv 1 \pmod{4}$. But $p \equiv 1 \pmod{4}$, so clearly $p \equiv b^2 \pmod{4}$. Since we also know that $p \equiv b^2 \pmod{q}$, we conclude that $p \equiv b^2 \pmod{4q}$.

$\Leftarrow$ Suppose we have $p \equiv \pm b^2 \pmod{4q}$, where b is coprime to q. We will examine each case, $p \equiv b^2 \pmod{4q}$ and $p \equiv -b^2 \pmod{4q}$ separately.

If $p \equiv b^2 \pmod{4q}$, we have $p \equiv b^2 \equiv 1 \pmod{4}$, and $p \equiv b^2 \pmod{q}$. But $p \equiv 1 \pmod{4}$ implies that $(-1)^{(p-1)/2} = 1$. Since $p \equiv b^2 \pmod{q}$ it follows that

$$\left(\frac{p}{q}\right) = 1 = \left(\frac{b^2}{q}\right).$$

Hence

$$\left(\frac{q}{p}\right) = (-1)^{\frac{p-1}{2}} \left(\frac{p}{q}\right) = 1 \cdot 1 = 1,$$

so q is a quadratic residue mod p.

If $p \equiv -b^2 \pmod{4q}$, we have both that $p \equiv -b^2 \equiv 3 \pmod 4$, and $p \equiv -b^2 \pmod q$. So, $(-1)^{(p-1)/2} = -1$ from the fact that $p \equiv 3 \pmod 4$. From $p \equiv -b^2 \pmod q$, we deduce that

$$\left(\frac{p}{q}\right) = \left(\frac{-b^2}{q}\right) = \left(\frac{-1}{q}\right)\left(\frac{b^2}{q}\right) = \left(\frac{-1}{q}\right).$$

Since $q \equiv 3 \pmod 4$, we know that $(-1/q) = -1$. So, $(p/q) = -1$. Thus,

$$\left(\frac{q}{p}\right) = (-1)^{(p-1)/2}\left(\frac{p}{q}\right) = (-1)(-1) = 1.$$

It now follows that q is a quadratic residue mod p. This completes the proof.

Exercise 7.3.3 Compute $(5/p)$ and $(7/p)$.

Solution. We will first compute $(5/p)$. Since $5 \equiv 1 \pmod 4$, we can use part (a) of Exercise 7.3.2. So $(5/p) = 1$ if and only if $p \equiv r \pmod 5$, where r is a quadratic residue mod 5. It is easy to determine which r are quadratic residues mod 5; $1^2 \equiv 1, 2^2 \equiv 4, 3^2 \equiv 4, 4^2 \equiv 1$. So, 1 and 4 are quadratic residues mod 5, while 2 and 3 are not. Thus

$$\left(\frac{5}{p}\right) = \begin{cases} 1 & \text{if } p \equiv 1, 4 \pmod 5, \\ -1 & \text{if } p \equiv 2, 3 \pmod 5, \\ 0 & \text{if } p = 5. \end{cases}$$

Now, we will find $(7/p)$. Since $7 \equiv 3 \pmod 4$, we must use part (b) of Exercise 7.3.2. So, we have to compute all the residues mod 28 of all the squares of odd integers prime to 7. Some calculation reveals that

$$1^2, 13^2, 15^2, 27^2 \equiv 1 \pmod{28},$$

$$3^2, 11^2, 17^2, 25^2 \equiv 9 \pmod{28},$$

and

$$5^2, 9^2, 17^2, 23^2 \equiv 25 \pmod{28}.$$

Thus

$$\left(\frac{7}{p}\right) = \begin{cases} 1 & \text{if } p \equiv \pm 1, \pm 9, \pm 25 \pmod{28}, \\ -1 & \text{if } p \equiv \pm 5, \pm 11, \pm 13 \pmod{28}, \\ 0 & \text{if } p = 7. \end{cases}$$

7.4 Quadratic Fields

Exercise 7.4.1 Find the discriminant of $K = \mathbb{Q}(\sqrt{d})$ when:

(a) $d \equiv 2, 3 \pmod 4$; and

(b) $d \equiv 1 \pmod 4$.

Solution. (a) If $d \equiv 2, 3 \pmod 4$, then $\omega_1 = 1, \omega_2 = \sqrt{d}$ forms an integral basis for $\mathcal{O}_K$. Then an easy calculation shows that $d_K = 4d$.

(b) For $d \equiv 1 \pmod 4$, then $\omega_1 = 1, \omega_2 = (1 + \sqrt{d})/2$ forms an integral basis. Then $d_K = d$.

(See also Example 4.3.4.)

Exercise 7.4.3 Assume that p is an odd prime. Show that $(d/p) = 0$ if and only if $p\mathcal{O}_K = \wp^2$, where $\wp$ is prime.

Solution. $\Rightarrow$ We claim that $p\mathcal{O}_K = (p, \sqrt{d})^2$. Notice that

$$(p, \sqrt{d})^2 = (p^2, p\sqrt{d}, d) = (p)(p, \sqrt{d}, d/p).$$

Because d is squarefree, d/p and p are relatively prime. So, $(p, \sqrt{d}, d/p) = 1$. Since $(p, \sqrt{d})$ is a prime ideal (for the same reason given above), we have shown that $p\mathcal{O}_K$ ramifies.

$\Leftarrow$ Once again, let m be the discriminant of $K = \mathbb{Q}(\sqrt{d})$. Since $p\mathcal{O}_K$ ramifies we can find some $a \in \wp$, but $a \notin p\mathcal{O}_K$. So, $a = x + y(m + \sqrt{m})/2$. Since $a^2 \in p\mathcal{O}_K$, we get

$$\frac{(2x + ym)^2 + my^2}{4} + 2y\frac{(2x + ym)}{4}\sqrt{m} \in p\mathcal{O}_K.$$

Thus, $p \mid (2x + ym)^2 + my^2$ and $p \mid y(2x + ym)$. If $p \mid y$, then $p \mid 2x$. Since p is odd, $p \mid x$. But then $a \in p\mathcal{O}_K$. This is a contradiction. So, $p \mid (2x + ym)$ and $p \mid my^2$. Thus, $p \mid m$. But this means that $(d/p) = 0$, since p is an odd prime.

Exercise 7.4.4 Assume p is an odd prime. Then $(d/p) = -1$ if and only if $p\mathcal{O}_K = \wp$, where $\wp$ is prime.

Solution. This follows immediately from Theorem 7.4.3 and Exercise 7.4.3. If $(d/p) = -1$, then we know that $p\mathcal{O}_K$ does not split, nor does it ramify. So $p\mathcal{O}_K$ must stay inert. Conversely, if $p\mathcal{O}_K$ is inert, the only possible value for (d/p) is -1.

7.5 Primes in Special Progressions

Exercise 7.5.1 Show that there are infinitely many primes of the form $4k + 1$.

Solution. Suppose there are only a finite number of primes of this form. Let $p_1, p_2, \dots, p_n$ be these primes. Let $d = (2p_1p_2\cdots p_n)^2 + 1$. Let p be a prime that divides this number. So $d \equiv 0 \pmod{p}$, which implies that $(-1/p) = 1$.

From Exercise 7.1.2, we know that this only occurs if $p \equiv 1 \pmod{4}$. So, $p \in \{p_1, p_2, \dots, p_n\}$. But for any $p_i \in \{p_1, p_2, \dots, p_n\}$, p_i does not divide d by construction. So $p \notin \{p_1, p_2, \dots, p_n\}$. Thus, our assumption is incorrect, and so, there must exist an infinite number of primes of the form $4k + 1$.

Exercise 7.5.2 Show that there are infinitely many primes of the form $8k + 7$.

Solution. Suppose the statement is false, that is, there exist only a finite number of primes of the form $8k + 7$. Let $p_1, p_2, \dots, p_n$ be these primes. Construct the following integer, $d = (4p_1p_2\cdots p_n)^2 - 2$. Let p be any prime that divides this number. But then $(4p_1p_2\cdots p_n)^2 \equiv 2 \pmod{p}$, so $(2/p) = 1$. From Theorem 7.1.6, we can deduce that $p \equiv \pm 1 \pmod{8}$.

We claim that all the odd primes that divide d cannot have the form $8k + 1$. We observe that $2 \mid (4p_1p_2\cdots p_n)^2 - 2$. So, any odd prime that divides d must divide $8(p_1p_2\cdots p_n)^2 - 1$. If all primes were of the form $8k + 1$, we would have

$$8(p_1p_2\cdots p_n)^2 - 1 = (8k_1 + 1)^{e_1} \cdot (8k_2 + 1)^{e_2} \cdots (8k_m + 1)^{e_m}.$$

But now consider this equation mod 8. We find that $-1 \equiv 1 \pmod{8}$, which is clearly false. So, there must be at least one odd prime p of the form $8k + 7$ that divides d.

So, $p \in \{p_1, p_2, \dots, p_n\}$. But p cannot be in $\{p_1, p_2, \dots, p_n\}$ since every p_i leaves a remainder of -2 when dividing d. So, $\{p_1, p_2, \dots, p_n\}$ does not contain all the primes of the form $8k + 7$. But we assumed that it did. We have arrived at a contradiction. Therefore, there must be an infinite number of primes of the form $8k + 7$.

Exercise 7.5.3 Show that $p \equiv 4 \pmod{5}$ for infinitely many primes p.

Solution. From the preceding comments, we know that we can use a Euclid-type proof to prove this assertion since $4^2 \equiv 1 \pmod{5}$. So, suppose there exists only a finite number of such primes, say $p_1, p_2, \dots, p_n$. Consider the integer $25(p_1p_2\cdots p_n)^2 - 5$. Then for any prime divisor not equal to 5, we will have $(5/p) = 1$. From Exercise 7.3.3, we know that this will only occur if $p \equiv 1, 4 \pmod{5}$. Since $p \neq 5$, then $p \mid 5(p_1\cdots p_n)^2 - 1$. So, all the prime divisors cannot be congruent to 1 (mod 5), because if this was true, $-1 \equiv 1 \pmod{5}$ which is clearly false. So, there is some $p \equiv 4 \pmod{5}$ that divides $5(p_1\cdots p_n)^2 - 1$. But p is not any of the $p_1, \dots, p_n$ since none of these numbers divide $5(p_1\cdots p_n)^2 - 1$. This gives us a contradiction.

7.6 Supplementary Problems

Exercise 7.6.1 Compute $(11/p)$.

Solution. Since $11 \equiv 3 \pmod 4$, we use Exercise 7.3.2 (b) which says that 11 is a quadratic residue mod p if and only if $p \equiv \pm b^2 \pmod{44}$ where b is an odd integer prime to p. If we compute $b^2 \pmod{44}$ for all possible b, we get

$$\begin{aligned} 1^2, 21^2, 23^2, 43^2 &\equiv 1 \pmod{44}, \\ 3^2, 19^2, 25^2, 41^2 &\equiv 9 \pmod{44}, \\ 5^2, 17^2, 27^2, 39^2 &\equiv 25 \pmod{44}, \\ 7^2, 15^2, 29^2, 37^2 &\equiv 5 \pmod{44}, \\ 9^2, 13^2, 31^2, 35^2 &\equiv 37 \pmod{44}. \end{aligned}$$

It is now easy to determine the primes for which 11 is a quadratic residue by applying Exercise 7.3.2.

Exercise 7.6.3 If $p \equiv 1 \pmod 3$, prove that there are integers a, b such that $p = a^2 - ab + b^2$.

Solution. Let $\rho = (1+\sqrt{-3})/2$ and consider $\mathbb{Q}(\sqrt{-3})$. The ring of integers of $\mathbb{Q}(\sqrt{-3})$ is $\mathbb{Z}[\rho]$. Since $p \equiv 1 \pmod 3$, $x^2+3 \equiv 0 \pmod p$ has a solution. Hence p splits in $\mathbb{Z}[\rho]$. Now use the fact that $\mathbb{Z}[\rho]$ is Euclidean.

Exercise 7.6.4 If $p \equiv \pm 1 \pmod 8$, show that there are integers a, b such that $a^2 - 2b^2 = \pm p$.

Solution. Consider the Euclidean ring $\mathbb{Z}[\sqrt{2}]$.

Exercise 7.6.5 If $p \equiv \pm 1 \pmod 5$, show that there are integers a, b such that $a^2 + ab - b^2 = \pm p$.

Solution. Let $\omega = (1+\sqrt{5})/2$ and consider $\mathbb{Z}[\omega]$.

Exercise 7.6.10 Show that the number of solutions of the congruence

$$x^2 + y^2 \equiv 1 \pmod p,$$

$0 < x < p$, $0 < y < p$ (p an odd prime), is even if and only if $p \equiv \pm 3 \pmod 8$.

Solution. Pair up the solutions, (x, y) with (y, x). The number of solutions is even unless (x, x) is a solution which means that 2 is a square mod p.

Exercise 7.6.11 If p is a prime such that $p - 1 = 4q$ with q prime, show that 2 is a primitive root mod p.

Solution. The only possible orders of 2 (mod p) are $1, 2, 4, q, 2q$, or $4q$. Since $p > q$, the orders 1, 2, 4 are impossible. If the order is q or $2q$, then 2 is a quadratic residue mod p. However, $p \equiv 5 \pmod 8$.

Exercise 7.6.12 (The Jacobi Symbol) Let Q be a positive odd number. We can write $Q = q_1 q_2 \cdots q_s$ where the q_i are odd primes, not necessarily distinct. Define the Jacobi symbol

$$\left(\frac{a}{Q}\right) = \prod_{j=1}^{s} \left(\frac{a}{q_i}\right).$$

If Q and Q' are odd and positive, show that:

(a) $(a/Q)(a/Q') = (a/QQ')$.

(b) $(a/Q)(a'/Q) = (aa'/Q)$.

(c) $(a/Q) = (a'/Q)$ if $a \equiv a' \pmod Q$.

Solution. All of these are evident from the properties of the Legendre symbol. For (c), note that $a \equiv a' \pmod Q$ implies $a \equiv a' \pmod{q_i}$ for $i = 1, 2, \dots, s$.

Exercise 7.6.13 If Q is odd and positive, show that

$$\left(\frac{-1}{Q}\right) = (-1)^{(Q-1)/2}.$$

Solution.

$$\left(\frac{-1}{Q}\right) = \prod_{j=1}^{s} \left(\frac{-1}{q_j}\right) = \prod_{j=1}^{s} (-1)^{(q_j-1)/2} = (-1)^{\sum_{j=1}^{s}(q_j-1)/2}.$$

Now, if a and b are odd, observe that

$$\frac{ab-1}{2} - \left(\frac{a-1}{2} + \frac{b-1}{2}\right) = \frac{(a-1)(b-1)}{2} \equiv 0 \pmod 2.$$

Hence

$$\frac{a-1}{2} + \frac{b-1}{2} \equiv \frac{ab-1}{2} \pmod 2.$$

Applying this observation repeatedly in our context gives

$$\left(\frac{-1}{Q}\right) = (-1)^{(Q-1)/2}.$$

Exercise 7.6.14 If Q is odd and positive, show that $(2/Q) = (-1)^{(Q^2-1)/8}$.

Solution. If a and b are odd, then

$$\frac{a^2b^2-1}{8} - \left(\frac{a^2-1}{8} + \frac{b^2-1}{8}\right) = \frac{(a^2-1)(b^2-1)}{8} \equiv 0 \pmod 8$$

so we have

$$\frac{a^2-1}{8}+\frac{b^2-1}{8}\equiv\frac{a^2b^2-1}{8} \pmod 8.$$

Again applying this repeatedly in our context gives the result.

Exercise 7.6.15 (Reciprocity Law for the Jacobi Symbol) If P and Q are odd, positive, and coprime, show that

$$\left(\frac{P}{Q}\right)\left(\frac{Q}{P}\right)=(-1)^{\frac{P-1}{2}\cdot\frac{Q-1}{2}}.$$

Solution. Write $P=\prod_{i=1}^r p_i$ and $Q=\prod_{j=1}^s q_j$. Then

$$\begin{aligned}\left(\frac{P}{Q}\right) &= \prod_{j=1}^{s}\left(\frac{P}{q_j}\right)=\prod_{j=1}^{s}\prod_{i=1}^{r}\left(\frac{p_i}{q_j}\right)\\ &= \prod_{j=1}^{s}\prod_{i=1}^{r}\left(\frac{q_j}{p_i}\right)(-1)^{\frac{p_i-1}{2}\cdot\frac{q_j-1}{2}}\\ &= \left(\frac{Q}{P}\right)(-1)^{\sum_{i,j}\frac{p_i-1}{2}\cdot\frac{q_j-1}{2}}\end{aligned}$$

by the reciprocity law for the Legendre symbol. But, as noted in the previous exercises,

$$\sum_{i=1}^{r}\frac{p_i-1}{2}\equiv\frac{P-1}{2} \pmod 2$$

and

$$\sum_{j=1}^{s}\frac{q_j-1}{2}\equiv\frac{Q-1}{2} \pmod 2,$$

which completes the proof.

Exercise 7.6.16 (The Kronecker Symbol) We can define (a/n) for any integer $a\equiv 0$ or $1 \pmod 4$, as follows. Define

$$\left(\frac{a}{2}\right)=\left(\frac{a}{-2}\right)=\begin{cases}0 & \text{if } a\equiv 0 \pmod 4,\\ 1 & \text{if } a\equiv 1 \pmod 8,\\ -1 & \text{if } a\equiv 5 \pmod 8.\end{cases}$$

For general n, write $n=2^c n_1$, with n_1 odd, and define

$$\left(\frac{a}{n}\right)=\left(\frac{a}{2}\right)^c\left(\frac{a}{n_1}\right),$$

where $(a/2)$ is defined as above and (a/n_1) is the Jacobi symbol.

Show that if d is the discriminant of a quadratic field, and n,m are positive integers, then

$$\left(\frac{d}{n}\right)=\left(\frac{d}{m}\right) \quad \text{for } n\equiv m \pmod d$$

and
$$\left(\frac{d}{n}\right)=\left(\frac{d}{m}\right)\operatorname{sgn} d \quad \text{for } n\equiv -m \pmod d.$$

Solution. Let $d=2^a d', n=2^b n', m=2^c m'$ with d',n',m' odd. If $a>0$, the case $b>0$ is trivial since then $c>0$ and both the symbols are zero. So suppose $b=c=0$. Then

$$\left(\frac{d}{n}\right)=\left(\frac{2^a d'}{n}\right)=\left(\frac{2}{n}\right)^a\left(\frac{d'}{n}\right)=(-1)^{a(n^2-1)/8}\left(\frac{n}{d'}\right)(-1)^{(n-1)(d'-1)/4}$$

and similarly

$$\left(\frac{d}{m}\right)=\left(\frac{2^a d'}{m}\right)=(-1)^{a(m^2-1)/8}\left(\frac{m}{d'}\right)(-1)^{(m-1)(d'-1)/2}.$$

Since $4\mid d$, the first factors coincide for m and n. The same is true for the other factors in the case $n\equiv m \pmod d$. But if $n\equiv -m \pmod d$, they differ by $\operatorname{sgn} d'$ which is $\operatorname{sgn} d$.

In the case $a=0$, we note $d\equiv 1 \pmod 4$. Then

$$\left(\frac{d}{n}\right)=\left(\frac{d}{2^b n'}\right)=\left(\frac{d}{2}\right)^b\left(\frac{d}{n'}\right)=\left(\frac{2}{d}\right)^b\left(\frac{d}{n'}\right)$$

since $(d/2)=(2/d)$ for $d\equiv 1 \pmod 4$. Thus $(d/n)=(n/d)$ and $(-1/d)=\operatorname{sgn} d$. Therefore

$$\left(\frac{d}{m}\right)=\left(\frac{d}{n}\right)\quad \text{for } m,n>0,\ \ m\equiv n \pmod d$$

and
$$\left(\frac{d}{n}\right)=\left(\frac{n}{d}\right)=\left(\frac{-m}{d}\right)=\operatorname{sgn} d\left(\frac{m}{d}\right)=\left(\frac{d}{m}\right)\operatorname{sgn} d$$

for $n\equiv -m \pmod d$.

Exercise 7.6.17 If p is an odd prime show that the least positive quadratic nonresidue is less than $\sqrt{p}+1$.

(It is a famous conjecture of Vinogradov that the least quadratic nonresidue mod p is $O(p^\varepsilon)$ for any $\varepsilon>0$.)

Solution. Let n be the least positive quadratic nonresidue and m the least such that $mn>p$, so that $n(m-1)\le p$. Since p is prime, $n(m-1)<p<mn$. Now $mn-p<n$ so that

$$1=\left(\frac{mn-p}{p}\right)=\left(\frac{mn}{p}\right)=-\left(\frac{m}{p}\right).$$

Therefore $m\ge n$, so that $(n-1)^2<n(n-1)<p$.

Exercise 7.6.18 Show that $x^4 \equiv 25 \pmod{1013}$ has no solution.

Solution. First observe that 1013 is prime. If $x^4 \equiv 25 \pmod{1013}$ had a solution x_0, then $x_0^2 \equiv \pm 5 \pmod{1013}$. However,

$$\left(\frac{\pm 5}{1013}\right) = \left(\frac{5}{1013}\right) = \left(\frac{1013}{5}\right) = \left(\frac{3}{5}\right) = -1$$

so the congruence has no solutions.

Exercise 7.6.19 Show that $x^4 \equiv 25 \pmod{p}$ has no solution if p is a prime congruent to 13 or 17 (mod 20).

Solution. If the congruence has a solution, then

$$1 = \left(\frac{\pm 5}{p}\right) = \left(\frac{5}{p}\right) = \left(\frac{p}{5}\right) = \left(\frac{\pm 3}{5}\right) = -1,$$

a contradiction.

Exercise 7.6.20 If p is a prime congruent to 13 or 17 (mod 20), show that $x^4 + py^4 = 25z^4$ has no solutions in integers.

Solution. We may suppose that $\gcd(x, y, z) = 1$, because otherwise we can cancel the common factor. Also $\gcd(p, z) = 1$ for otherwise $p \mid x$ and $p \mid y$. Now reduce the equation mod p. We have a solution to $x^4 \equiv 25z^4 \pmod{p}$ so that by the previous question,

$$1 = \left(\frac{\pm 5z^2}{p}\right) = \left(\frac{\pm 5}{p}\right) = -1,$$

a contradiction.

Exercise 7.6.21 Compute the class number of $\mathbb{Q}(\sqrt{33})$.

Solution. The discriminant is 33 and the Minkowski bound is

$$\tfrac{1}{2}\sqrt{33} = 2.87\ldots.$$

Since $33 \equiv 1 \pmod{8}$, 2 splits as a product of two ideals each of norm 2. Moreover,

$$2 = \frac{\sqrt{33}-5}{2} \cdot \frac{\sqrt{33}+5}{2}$$

and so the principal ideals

$$\left(\frac{\sqrt{33}-5}{2}\right) \quad \text{and} \quad \left(\frac{\sqrt{33}+5}{2}\right)$$

are the only ones of norm 2. Hence the class number is 1.

Exercise 7.6.22 Compute the class number of $\mathbb{Q}(\sqrt{21})$.

Solution. The discriminant is 21 and the Minkowski bound is 2.29.... However, $21 \equiv 5 \pmod 8$ so that 2 is inert in $\mathbb{Q}(\sqrt{21})$. Therefore, there are no ideals of norm 2 and the class number is 1.

Exercise 7.6.23 Show that $\mathbb{Q}(\sqrt{-11})$ has class number 1.

Solution. The field has discriminant -11 and Minkowski's bound is

$$\frac{2}{\pi}\sqrt{11} = 2.11\ldots.$$

We must examine ideals of norm 2. Since $-11 \equiv 5 \pmod 8$, by Theorem 7.4.5, 2 is inert in $\mathbb{Q}(\sqrt{-11})$, so that there are no ideals of norm 2. Hence the class number is 1.

Exercise 7.6.24 Show that $\mathbb{Q}(\sqrt{-15})$ has class number 2.

Solution. The field has discriminant -15 and Minkowski's bound is

$$\frac{2}{\pi}\sqrt{15} = 2.26\ldots.$$

Since $-15 \equiv 1 \pmod 8$, by Theorem 7.4.5, 2 splits as a product of two ideals $\wp, \wp'$ each of norm 2. If $\wp$ were principal, then

$$\wp = \left(\frac{u + v\sqrt{-15}}{2}\right)$$

for integers u, v. However, $8 = u^2 + 15v^2$ has no solution. Thus $\wp$ is not principal and the class number is 2.

Exercise 7.6.25 Show that $\mathbb{Q}(\sqrt{-31})$ has class number 3.

Solution. The discriminant is -31 and the Minkowski bound is

$$\frac{2}{\pi}\sqrt{31} = 3.26\ldots.$$

So we must consider ideals of norm less than or equal to 3.

Since $-31 \equiv 1 \pmod 8$, 2 splits as $\wp_2 \cdot \wp_2'$ (say). Since

$$\left(\frac{-31}{3}\right) = -1,$$

3 is inert so there are no ideals of norm 3. Moreover, neither $\wp_2$ or $\wp_2'$ are principal since we cannot solve

$$8 = u^2 + 31v^2.$$

Since

$$8 = \frac{1^2 + (31) \cdot 1^2}{4} = \left(\frac{1 - \sqrt{-31}}{2}\right)\left(\frac{1 + \sqrt{-31}}{2}\right)$$

and $((1-\sqrt{-31})/2), ((1+\sqrt{-31})/2)$ are coprime, $\wp_2^3$ and $(\wp_2')^3$ are principal. $\wp_2$ cannot be principal since $16 = u^2 + 31v^2$ has no solution except $(u, v) = (\pm 4, 0)$ in which case

$$\wp_2^2 = (4) = (2)^2 = \wp_2^2(\wp_2')^2,$$

which implies $(\wp_2')^2 = 1$, a contradiction. Thus, as each ideal class contains either $(1), \wp_2$ or $\wp_2'$ and because $\wp_2^2$ is inequivalent to (1) or $\wp_2$ we must have $\wp_2^2 \sim \wp_2'$. Thus, $\wp_2^3 \sim (1)$. Thus, the class number is 3.

Chapter 8

The Structure of Units

8.1 Dirichlet's Unit Theorem

Exercise 8.1.1 (a) Show that there are only finitely many roots of unity in K.
(b) Show, similarly, that for any positive constant c, there are only finitely many $\alpha \in \mathcal{O}_K$ for which $|\alpha^{(i)}| \leq c$ for all i.

Solution. (a) Suppose that $\alpha^m = 1$. Then $\alpha \in \mathcal{O}_K$, $|\alpha|^m = 1 \Rightarrow |\alpha| = 1$ and, if $\sigma_1, \ldots, \sigma_n$ are the distinct embeddings of K in $\mathbb{C}$, then, for each $\alpha^{(i)} = \sigma_i(\alpha)$, we have that $\sigma_i(\alpha)^m = \sigma_i(\alpha^m) = 1 \Rightarrow |\alpha^{(i)}| = 1$ for $i = 1, \ldots, n$.

The characteristic polynomial of α is

$$f_\alpha(x) = \prod_{i=1}^{n}(x - \alpha^{(i)}) = x^n + a_{n-1}x^{n-1} + \cdots + a_0 \in \mathbb{Z}[x].$$

Now,

$$a_{n-j} = (-1)^j s_j(\alpha^{(1)}, \ldots, \alpha^{(n)}),$$

where $s_j(\alpha^{(1)}, \ldots, \alpha^{(n)})$ is the jth symmetric function in the $\alpha^{(i)}$, i.e., the sum of all products of the $\alpha^{(i)}$, taken j at a time. This implies that

$$|a_{n-j}| \leq \binom{n}{j} \leq n!.$$

Thus, since the a_j's are bounded, there are only finitely many choices for the coefficients of the characteristic polynomial of a root of unity $\alpha \in K$ and, hence, only finitely many such roots of unity.

(b) Suppose that $\alpha \in \mathcal{O}_K$ such that $|\alpha^{(i)}| \leq c$ for all $i = 1, \ldots, n$. As in (a), let

$$f_\alpha(x) = \prod_{i=1}^{n}(x - \alpha^{(i)}) = x^n + a_{n-1}x^{n-1} + \cdots + a_0 \in \mathbb{Z}[x]$$

be the characteristic polynomial of α. Then

$$|a_{n-j}| = |s_j(\alpha^{(1)}, \ldots, \alpha^{(n)})| \leq c^j \binom{n}{j} \leq c^n n!.$$

Thus, the coefficients of the characteristic polynomial of such an α in $\mathcal{O}_K$ are bounded. There are, therefore, only finitely many such α.

Exercise 8.1.2 Show that W_K, the group of roots of unity in K, is cyclic, of even order.

Solution. Let $\alpha_1, \ldots, \alpha_l$ be the roots of unity in K. For $j = 1, \ldots, l$, $\alpha_j^{q_j} = 1$ for some q_j which implies that $\alpha_j = e^{2\pi i p_j/q_j}$, for some $0 \leq p_j \leq q_j - 1$. Let $q_0 = \prod_{i=1}^{l} q_j$. Then, clearly, each $\alpha_i \in \langle e^{2\pi i/q_0} \rangle$ so W_K is a subgroup of the cyclic group $\langle e^{2\pi i/q_0} \rangle$ and is, thus, cyclic. Moreover, since $\{\pm 1\} \subseteq W_K$, W_K has even order.

Exercise 8.1.7 (a) Let Γ be a lattice of dimension n in $\mathbb{R}^n$ and suppose that $\{v_1, \ldots, v_n\}$ and $\{w_1, \ldots, w_n\}$ are two bases for Γ over $\mathbb{Z}$. Let V and W be the $n \times n$ matrices with rows consisting of the v_i's and w_i's, respectively. Show that $|\det V| = |\det W|$. Thus, we can unambiguously define the *volume of the lattice* Γ, $\text{vol}(\Gamma) =$ the absolute value of the determinant of the matrix formed by taking, as its rows, any basis for Γ over $\mathbb{Z}$.

(b) Let $\varepsilon_1, \ldots, \varepsilon_r$ be a fundamental system of units for a number field K. Show that the *regulator of* K, $R_K = |\det(\log|\varepsilon_j^{(i)}|)|$, is independent of the choice of $\varepsilon_1, \ldots, \varepsilon_r$.

Solution. (a) Since $\{w_1, \ldots, w_n\}$ is a $\mathbb{Z}$-basis for Γ, we can express each v_i as a $\mathbb{Z}$-linear combination of the w_j's, say $v_i = \sum_{j=1}^{n} a_{ij} w_j$. Setting $A = (a_{ij})$, we have that $V = AW$. Since $\{v_1, \ldots, v_n\}$ is also an integral basis for Γ, the matrix A is invertible, $A \in GL_n(\mathbb{Z}) \Rightarrow \det A \in \mathbb{Z}^* = \{\pm 1\}$. Thus, $|\det V| = |\det A||\det W| = |\det W|$.

(b) As before, define

$$\begin{aligned} f : U_K &\rightarrow \mathbb{R}^r, \\ \varepsilon &\mapsto (\log|\varepsilon^{(1)}|, \ldots, \log|\varepsilon^{(r)}|). \end{aligned}$$

Im f is a lattice of dimension r in $\mathbb{R}^r$, with $\mathbb{Z}$-basis $\{f(\varepsilon_1), \ldots, f(\varepsilon_r)\}$, for any system of fundamental units, $\varepsilon_1, \ldots, \varepsilon_r$. By definition, R_K is the absolute value of the determinant of the matrix formed by taking, as its rows, the $f(\varepsilon^{(i)})$'s. By (a) then, R_K is independent of the particular system of fundamental units.

Exercise 8.1.8 (a) Show that, for any real quadratic field $K = \mathbb{Q}(\sqrt{d})$, where d is a positive squarefree integer, $U_K \simeq \mathbb{Z}/2\mathbb{Z} \times \mathbb{Z}$. That is, there is a fundamental unit $\varepsilon \in U_K$ such that $U_K = \{\pm\varepsilon^k : k \in \mathbb{Z}\}$. Conclude that the equation $x^2 - dy^2 = 1$ (erroneously dubbed *Pell's equation*) has infinitely many integer solutions for $d \equiv 2, 3$ *mod* 4 and that the equation $x^2 - dy^2 = 4$ has infinitely many integer solutions for $d \equiv 1$ *mod* 4.

(b) Let $d \equiv 2, 3 \pmod 4$. Let b be the smallest positive integer such that one of $db^2 \pm 1$ is a square, say a^2, $a > 0$. Then $a + b\sqrt{d}$ is a unit. Show that it is the fundamental unit. Using this algorithm, determine the fundamental units of $\mathbb{Q}(\sqrt{2})$, $\mathbb{Q}(\sqrt{3})$.

(c) Devise a similar algorithm to compute the fundamental unit in $\mathbb{Q}(\sqrt{d})$, for $d \equiv 1 \pmod 4$. Determine the fundamental unit of $\mathbb{Q}(\sqrt{5})$.

Solution. (a) Since $K \subseteq \mathbb{R}$, the only roots of unity in K are $\{\pm 1\}$, so $W_K = \{\pm 1\}$. Moreover, since there are $r_1 = 2$ real and $2r_2 = 0$ nonreal embeddings of K in $\mathbb{C}$, by Dirichlet's theorem, we have that $U_K \simeq W_K \times \mathbb{Z} \simeq \mathbb{Z}/2\mathbb{Z} \times \mathbb{Z}$.

Suppose that $d \equiv 2, 3 \pmod 4$, so that $\mathcal{O}_K = \mathbb{Z}[\sqrt{d}]$. If $\varepsilon = a + b\sqrt{d} \in U_K^2$, then

$$N_K(\varepsilon) = (a + b\sqrt{d})(a - b\sqrt{d}) = a^2 - db^2 = 1.$$

i.e., each $\varepsilon = a + b\sqrt{d} \in U_K^2$ yields a solution $(a, b) \in \mathbb{Z}^2$ to the equation $x^2 - dy^2 = 1$. Since $U_K^2 \simeq \mathbb{Z}$ is infinite, there are infinitely many such solutions.

Suppose, now, that $d \equiv 1 \pmod 4$, so that

$$\mathcal{O}_K = \left\{ \frac{a + b\sqrt{d}}{2} : a, b \in \mathbb{Z}, a \equiv b \pmod 2 \right\}.$$

If $\varepsilon = (a + b\sqrt{d})/2 \in U_K^2$, then $N_K(\varepsilon) = (a^2 - db^2)/4 = 1 \Rightarrow a^2 - db^2 = 4$. Since U_K^2 is infinite and each of its elements yields an integral solution to $x^2 - dy^2 = 4$, this equation has infinitely many solutions $x, y \in \mathbb{Z}$.

(b) We have that $db^2 \pm 1 = a^2$ so $N_K(a + b\sqrt{d}) = \pm 1$ and $a + b\sqrt{d} \in U_K$. Also, $a, b > 0$ means that $a + b\sqrt{d} > 1$ so $a + b\sqrt{d} = \varepsilon^k$, for some $k \geq 1$, where ε is the fundamental unit in $\mathbb{Q}(\sqrt{d})$. If $k > 1$, then write $\varepsilon = \alpha + \beta\sqrt{d}$, $\alpha, \beta > 0$. It is easy to see that $a + b\sqrt{d} = (\alpha + \beta\sqrt{d})^k$ implies that $\alpha < a$ and $\beta < b$. But $d\beta^2 \pm 1 = \alpha^2$, contradicting the minimality of b. Therefore, $a + b\sqrt{d}$ is, in fact, the fundamental unit.

In $\mathbb{Q}(\sqrt{2})$, $2(1)^2 - 1 = 1^2 \Rightarrow 1 + \sqrt{2}$ is the fundamental unit. In $\mathbb{Q}(\sqrt{3})$, $3(1)^2 + 1 = 2^2 \Rightarrow 2 + \sqrt{3}$ is the fundamental unit.

(c) Let $d \equiv 1 \pmod 4$. The same argument as in (b) shows that, if b is the smallest positive integer such that one of $db^2 \pm 4$ is a square, say $a^2, a > 0$, then $(a + b\sqrt{d})/2$ is the fundamental unit in $\mathbb{Q}(\sqrt{d})$. In $\mathbb{Q}(\sqrt{5})$, $5(1)^2 - 4 = 1^2 \Rightarrow (1 + \sqrt{5})/2$ is the fundamental unit.

Exercise 8.1.9 (a) For an imaginary quadratic field $K = \mathbb{Q}(\sqrt{-d})$ (d a positive, squarefree integer), show that

$$U_K \simeq \begin{cases} \mathbb{Z}/4\mathbb{Z} & \text{for } d = 1, \\ \mathbb{Z}/6\mathbb{Z} & \text{for } d = 3, \\ \mathbb{Z}/2\mathbb{Z} & \text{otherwise.} \end{cases}$$

(b) Show that U_K is finite $\Leftrightarrow K = \mathbb{Q}$ or K is an imaginary quadratic field.

(c) Show that, if there exists an embedding of K in $\mathbb{R}$, then $W_K \simeq \{\pm 1\} \simeq \mathbb{Z}/2\mathbb{Z}$. Conclude that, in particular, this is the case if $[K : \mathbb{Q}]$ is odd.

Solution. (a) Suppose that $-d \equiv 2, 3 \pmod 4$ so that $\mathcal{O}_K = \mathbb{Z}[\sqrt{-d}]$. Let $a + b\sqrt{-d} \in U_K$. Then $a^2 + db^2 = 1$ (since $a^2, b^2, d \geq 0$, $a^2 + db^2 \neq -1$). If $b = 0$, then $a = \pm 1$. If $b \neq 0$, then $a^2 + db^2 \geq d$; thus,

$$d > 1 \quad \Rightarrow \quad U_K = \{\pm 1\} \simeq \mathbb{Z}/2\mathbb{Z}.$$

For $d = 1$, $a^2 + db^2 = a^2 + b^2 = 1 \Rightarrow (a, b) \in \{(\pm 1, 0), (0, \pm 1)\}$

$$\Rightarrow \quad U_K = \{\pm 1, \pm i\} = \langle i \rangle \simeq \mathbb{Z}/4\mathbb{Z}.$$

Suppose now that $-d \equiv 1 \pmod 4$. Then $(a + b\sqrt{-d})/2 \in U_K \Rightarrow a^2 + db^2 = 4$. If $d > 4$, then the only solutions to this equation are $(a, b) = (\pm 2, 0)$

$$\Rightarrow \quad U_K \simeq \mathbb{Z}/2\mathbb{Z}.$$

If $d = 3$, then $a^2 + db^2 = a^2 + 3b^2 = 4$. Then $(a, b) \in \{(\pm 2, 0), (\pm 1, \pm 1)\}$, so that

$$U_K = \{\pm\zeta_3, \pm\zeta_3^2, \pm 1\} = \langle -\zeta_3 \rangle \simeq \mathbb{Z}/6\mathbb{Z}$$

(where $\zeta_3 = (-1 + \sqrt{-3})/2$).

(b) We have already shown that for K a quadratic imaginary field, U_K is finite. Now, suppose that U_K is finite. Then $r_1 + r_2 = 1$ which implies that either $r_1 = 1, r_2 = 0$ so that $[K : \mathbb{Q}] = r_1 + 2r_2 = 1$ and $K = \mathbb{Q}$ or $r_1 = 0, r_2 = 1$ and $[K : \mathbb{Q}] = 2$ and $K \not\subseteq \mathbb{R}$, so that K is quadratic imaginary.

(c) If there exists an embedding $\sigma : K \hookrightarrow \mathbb{R}$, then $K \simeq \sigma(K)$ and in particular $W_K \simeq W_{\sigma(K)} \subseteq \mathbb{R}$. Since the only real roots of unity are $\{\pm 1\}$, we must have $W_K = \sigma^{-1}(\{\pm 1\}) = \{\pm 1\}$. In particular, if $[K : \mathbb{Q}] = r_1 + 2r_2$ is odd, then r_1 is odd and, therefore, ≥ 1.

Exercise 8.1.11 Let $[K : \mathbb{Q}] = 3$ and suppose that K has only one real embedding. Then, by Exercise 8.1.8 (c), $W_K = \{\pm 1\}$ implies that $U_K = \{\pm u^k : k \in \mathbb{Z}\}$, where $u > 1$ is the fundamental unit in K.

(a) Let $u, \rho e^{i\theta}, \rho e^{-i\theta}$ be the $\mathbb{Q}$-conjugates of u. Show that $u = \rho^{-2}$ and that $d_{K/\mathbb{Q}}(u) = -4\sin^2\theta(\rho^3 + \rho^{-3} - 2\cos\theta)^2$.

(b) Show that $|d_{K/\mathbb{Q}}(u)| < 4(u^3 + u^{-3} + 6)$.

(c) Conclude that $u^3 > d/4 - 6 - u^{-3} > d/4 - 7$, where $d = |d_K|$.

Solution. (a) $N_K(u) = u\rho^2 = \pm 1$, so $u = \pm\rho^{-2}$, but $u > 1$ implies that $u = \rho^{-2}$,

$$\begin{aligned} d_{K/\mathbb{Q}}(u) &= \prod_{1\le r<s\le 3} (u^{(r)} - u^{(s)})^2 \\ &= (\rho^{-2} - \rho e^{i\theta})^2(\rho^{-2} - \rho e^{-i\theta})^2(\rho e^{i\theta} - \rho e^{-i\theta})^2 \\ &= \left(\rho^{-4} - \rho(e^{i\theta} + e^{-i\theta}) + \rho^2\right)^2\left(\rho(e^{i\theta} - e^{-i\theta})\right)^2 \\ &= \left(\rho^{-4} - \rho(2\cos\theta) + \rho^2\right)^2\left(\rho(2i\sin\theta)\right)^2 \\ &= -4\sin^2\theta(\rho^3 + \rho^{-3} - 2\cos\theta)^2. \end{aligned}$$

(b)

$$|d_{K/\mathbb{Q}}(u)| = 4\sin^2\theta(\rho^3 + \rho^{-3} - 2\cos\theta)^2.$$

Now set $x = \rho^3 + \rho^{-3}$, $c = \cos\theta$ and consider

$$f(x) = (1 - c^2)(x - 2c)^2 - x^2.$$

This function attains a maximum when $x = -2(1-c^2)/c$ and this maximum is $4(1 - c^2) \le 4$. Therefore

$$\begin{aligned} |d_{K/\mathbb{Q}}(u)| &\le 4(\rho^3 + \rho^{-3})^2 + 16 \\ &= 4(u^3 + u^{-3} + 6). \end{aligned}$$

(c) Since $d = |d_K| < |d_{K/\mathbb{Q}}(u)|$, we have

$$u^3 > \frac{d}{4} - 6 - u^{-3} > \frac{d}{4} - 7.$$

Exercise 8.1.12 Let $\alpha = \sqrt[3]{2}, K = \mathbb{Q}(\alpha)$. Given that $d_K = -108$:

(a) Show that, if u is the fundamental unit in K, $u^3 > 20$.

(b) Show that $\beta = (\alpha - 1)^{-1} = \alpha^2 + \alpha + 1$ is a unit, $1 < \beta < u^2$. Conclude that $\beta = u$.

Solution. (a) By Exercise 8.1.11, $u^3 > 108/4 - 7 = 20$ so $u^2 > 20^{2/3} > 7$.

(b) Computation shows that $\frac{1}{7} < \alpha - 1 < 1$ and therefore $1 < (\alpha-1)^{-1} < 7 < u^2$. Since β is a power of u, this power must be 1. Therefore $\beta = u$ is the fundamental unit in K.

Exercise 8.1.13 (a) Show that, if $\alpha \in K$ is a root of a monic polynomial $f \in \mathbb{Z}[x]$ and $f(r) = \pm 1$, for some $r \in \mathbb{Z}$, then $\alpha - r$ is a unit in K.

(b) Using the fact that if $K = \mathbb{Q}(\sqrt[3]{m})$, then $d_K = -27m^2$, for any cubefree integer m, determine the fundamental unit in $K = \mathbb{Q}(\sqrt[3]{7})$.

(c) Determine the fundamental unit in $K = \mathbb{Q}(\sqrt[3]{3})$.

Solution. (a) Let $g(x) = f(x+r)$. Then $g(x)$ is a monic polynomial in $\mathbb{Z}[x]$ with constant term $g(0) = f(r) = \pm 1$. Since $g(\alpha - r) = 0$, the minimal polynomial of $\alpha - r \in \mathcal{O}_K$ divides $g(x)$ and, thus, has constant term ± 1. Therefore $N_K(\alpha - r) = \pm 1$ so $\alpha - r \in U_K$.

(b) $d_K = -27 \cdot 7^2$ implies that $u^3 > 27 \cdot 7^2/4 - 7 > 323$ and so $u^2 > (323)^{2/3} > 47$ by Exercise 8.1.11.

Let $f(x) = x^3 - 7$ and note that $f(2) = 1$. Then $\sqrt[3]{7} - 2 \in U_K$. Also, $1/u^2 < 1/47 < 2 - \sqrt[3]{7} < 1$ implies that $1 < (2 - \sqrt[3]{7})^{-1} < u^2$. Therefore, $(2 - \sqrt[3]{7})^{-1}$ is the fundamental unit of $\mathbb{Q}(\sqrt[3]{7})$.

(c) Let $\alpha = \sqrt[3]{3}$, $K = \mathbb{Q}(\alpha)$. $d_K = -27 \cdot 3^2 = -3^5$. We observe that α^2 is a root of $f(x) = x^3 - 9$ and $f(2) = -1$ so $\alpha^2 - 2 \in U_K$. We have $u^3 > 3^5/4 - 7 > 53$ and thus $u^2 > 14$. Since $\frac{1}{14} < \alpha^2 - 1 < 1, 1 < (\alpha^2 - 2)^{-1} < 14 < u^2$. Thus, $(\alpha^2 - 2)^{-1}$ is the fundamental unit of $K = \mathbb{Q}(\alpha)$.

8.2 Units in Real Quadratic Fields

Exercise 8.2.1 (a) Consider the continued fraction $[a_0, \ldots, a_n]$. Define the sequences $p_0, \ldots, p_n$ and $q_0, \ldots, q_n$ recursively as follows:

$$\begin{aligned} p_0 &= a_0, & q_0 &= 1, \\ p_1 &= a_0 a_1 + 1, & q_1 &= a_1, \\ p_k &= a_k p_{k-1} + p_{k-2}, & q_k &= a_k q_{k-1} + q_{k-2}, \end{aligned}$$

for $k \geq 2$. Show that the kth convergent $C_k = p_k/q_k$.

(b) Show that $p_k q_{k-1} - p_{k-1} q_k = (-1)^{k-1}$, for $k \geq 1$.

(c) Derive the identities

$$C_k - C_{k-1} = \frac{(-1)^{k-1}}{q_k q_{k-1}},$$

for $1 \leq k \leq n$, and

$$C_k - C_{k-2} = \frac{a_k (-1)^k}{q_k q_{k-2}},$$

for $2 \leq k \leq n$.

(d) Show that

$$C_1 > C_3 > C_5 > \cdots,$$
$$C_0 < C_2 < C_4 < \cdots,$$

and that every odd-numbered convergent C_{2j+1}, $j \geq 0$, is greater than every even-numbered convergent C_{2k}, $k \geq 0$.

Solution. (a) We prove this by induction on k.

For $k = 0$, $C_0 = [a_0] = p_0/q_0$. For $k = 1$,

$$C_1 = [a_0, a_1] = a_0 + \frac{1}{a_1} = \frac{a_0 a_1 + 1}{a_1} = \frac{p_1}{q_1}.$$

For $k > 1$, suppose that

$$C_k = \frac{p_k}{q_k} = \frac{a_k p_{k-1} + p_{k-2}}{a_k q_{k-1} + q_{k-2}}.$$

Since $p_{k-1}, p_{k-2}, q_{k-1}, q_{k-2}$ depend only on $a_0, \ldots, a_{k-1}$,

$$\begin{aligned} C_{k+1} &= \left[a_0, a_1, \ldots, a_{k-1}, a_k + \frac{1}{a_{k+1}}\right] \\ &= \frac{(a_k + \frac{1}{a_{k+1}})p_{k-1} + p_{k-2}}{(a_k + \frac{1}{a_{k+1}})q_{k-1} + q_{k-2}} \\ &= \frac{a_{k+1}(a_k p_{k-1} + p_{k-2}) + p_{k-1}}{a_{k+1}(a_k q_{k-1} + q_{k-2}) + q_{k-1}} \\ &= \frac{a_{k+1} p_k + p_{k-1}}{a_{k+1} q_k + q_{k-1}} = \frac{p_{k+1}}{q_{k+1}}. \end{aligned}$$

(b) Again, we apply induction on k. For $k = 1$,

$$p_1 q_0 - p_0 q_1 = (a_0 a_1 + 1) \cdot 1 - a_0 a_1 = 1.$$

For $k \geq 1$,

$$\begin{aligned} p_{k+1} q_k - p_k q_{k+1} &= (a_{k+1} p_k + p_{k-1}) q_k - p_k (a_{k+1} q_k + q_{k-1}) \\ &= p_{k-1} q_k - p_k q_{k-1} = -(-1)^{k-1} = (-1)^k, \end{aligned}$$

by our induction hypothesis.

(c) By (b),

$$p_k q_{k-1} - q_k p_{k-1} = (-1)^{k-1}.$$

Dividing by $q_k q_{k-1}$, we obtain the first identity. Now,

$$C_k - C_{k-2} = \frac{p_k}{q_k} - \frac{p_{k-2}}{q_{k-2}} = \frac{p_k q_{k-2} - p_{k-2} q_k}{q_k q_{k-2}}.$$

But

$$\begin{aligned} p_k q_{k-2} - p_{k-2} q_k &= (a_k p_{k-1} + p_{k-2}) q_{k-2} - p_{k-2}(a_k q_{k-1} + q_{k-2}) \\ &= a_k (p_{k-1} q_{k-2} - p_{k-2} q_{k-1}) = a_k (-1)^{k-2}, \end{aligned}$$

establishing the second identity.

(d) By (c),

$$C_k - C_{k-2} = \frac{a_k (-1)^k}{q_k q_{k-2}}.$$

Thus, $C_k < C_{k-2}$, for k odd and $C_k > C_{k-2}$, for k even. In addition,

$$C_{2m} - C_{2m-1} = \frac{(-1)^{2m-1}}{q_{2m} q_{2m-1}} < 0 \quad \Rightarrow \quad C_{2m-1} > C_{2m},$$

$$\Rightarrow \quad C_{2k} < C_{2(j+k+1)} < C_{2(j+k)+1} < C_{2j+1}.$$

Exercise 8.2.2 Let $\{a_i\}_{i\geq 0}$ be an infinite sequence of integers with $a_i \geq 0$ for $i \geq 1$ and let $C_k = [a_0, \ldots, a_k]$. Show that the sequence $\{C_k\}$ converges.

Solution. By Exercise 8.2.1 (d), we have

$$C_1 > C_3 > C_5 > \cdots .$$

Moreover, each $C_{2j+1} > a_0$ so that the sequence $\{C_{2j+1}\}_{j\geq 0}$ is decreasing and bounded from below and is, thus, convergent, say $\lim_{j\to\infty} C_{2j+1} = \alpha_1$. Also,

$$C_0 < C_2 < C_4 < \cdots$$

and $C_{2j} < C_{2k+1}$ for all $j, k \geq 0$. In particular, each $C_{2j} < C_1$. The sequence $\{C_{2j}\}_{j\geq 0}$ is increasing and bounded from above and, therefore, also converges, say $\lim_{j\to\infty} C_{2j} = \alpha_2$. We will show that $\alpha_1 = \alpha_2$.

Since each $a_i \geq 1, q_0, q_1 \geq 1$, we easily see, by induction on k, that $q_k = a_k q_{k-1} + q_{k-2} \geq 2k - 3$. By Exercise 8.2.1 (c),

$$C_{2j+1} - C_{2j} = \frac{1}{q_{2j+1}q_{2j}} \leq \frac{1}{(4j-1)(4j-3)} \to 0,$$

as $j \to \infty$. Thus, both sequences converge to the same limit $\alpha = \alpha_1 = \alpha_2$ and

$$\lim_{j\to\infty} C_j = \alpha.$$

Exercise 8.2.3 Let $\alpha = \alpha_0$ be an irrational real number greater than 0. Define the sequence $\{a_i\}_{i\geq 0}$ recursively as follows:

$$a_k = [\alpha_k], \qquad \alpha_{k+1} = \frac{1}{\alpha_k - a_k}.$$

Show that $\alpha = [a_0, a_1, \ldots]$ is a representation of α as a simple continued fraction.

Solution. By induction on k, we easily see that each α_k is irrational. Therefore $\alpha_{k+1} > 1$ which means that $a_{k+1} \geq 1$ so that $[a_0, a_1, \ldots]$ is a simple continued fraction. Also,

$$\begin{aligned}\alpha = \alpha_0 &= [\alpha_0] + (\alpha_0 - [\alpha_0]) = a_0 + \frac{1}{\alpha_1} \\ &= [a_0, \alpha_1] = [a_0, a_1, \alpha_2] = \cdots = [a_0, a_1, \ldots, a_k, \alpha_{k+1}],\end{aligned}$$

for all k. By Exercise 8.2.1 (a),

$$\alpha = \frac{\alpha_{k+1}p_k + p_{k-1}}{\alpha_{k+1}q_k + q_{k-1}}$$

so that

$$\begin{aligned}|\alpha - C_k| &= \left|\frac{\alpha_{k+1}p_k + p_{k-1}}{\alpha_{k+1}q_k + q_{k-1}} - \frac{p_k}{q_k}\right| \\ &= \left|\frac{-(p_k q_{k-1} - p_{k-1}q_k)}{(\alpha_{k+1}q_k + q_{k-1})q_k}\right| \\ &= \left|\frac{1}{(\alpha_{k+1}q_k + q_{k-1})q_k}\right| \\ &< \frac{1}{q_k^2} \leq \frac{1}{(2k-3)^2} \to 0\end{aligned}$$

as $k \to \infty$. Thus,

$$\alpha = \lim_{k\to\infty} C_k = [a_0, a_1, \ldots].$$

Exercise 8.2.5 Let d be a positive integer, not a perfect square. Show that, if $|x^2 - dy^2| < \sqrt{d}$ for positive integers x, y, then x/y is a convergent of the continued fraction of $\sqrt{d}$.

Solution. Suppose first that $0 < x^2 - dy^2 < \sqrt{d}$. Then

$$(x + y\sqrt{d})(x - y\sqrt{d}) > 0 \quad \Rightarrow \quad x > y\sqrt{d},$$

$$\begin{aligned}\Rightarrow \quad \left|\sqrt{d} - \frac{x}{y}\right| &= \frac{x}{y} - \sqrt{d} = \frac{x - y\sqrt{d}}{y} \\ &= \frac{x^2 - dy^2}{y(x + y\sqrt{d})} < \frac{x^2 - dy^2}{y(2y\sqrt{d})} < \frac{1}{2y^2}.\end{aligned}$$

Thus, by Theorem 8.2.4 (b), x/y is a convergent of the continued fraction of $\sqrt{d}$.

Similarly, if $-\sqrt{d} < x^2 - dy^2 < 0$, then

$$0 < y^2 - \frac{1}{d}x^2 < \frac{1}{\sqrt{d}} \quad \Rightarrow \quad y > \frac{x}{\sqrt{d}},$$

$$\begin{aligned}\Rightarrow \quad \left|\frac{1}{\sqrt{d}} - \frac{y}{x}\right| &= \frac{y}{x} - \frac{1}{\sqrt{d}} = \frac{y - x/\sqrt{d}}{x} \\ &= \frac{y^2 - x^2/d}{x(y + x/\sqrt{d})} < \frac{y^2 - x^2/d}{2x^2/\sqrt{d}} < \frac{1}{2x^2}.\end{aligned}$$

Thus y/x is a convergent of the continued fraction of $1/\sqrt{d}$.

Let α be any irrational number. Then $\alpha = [a_0, a_1, \ldots]$ implies $1/\alpha = [0, a_0, a_1, \ldots]$. We therefore have that the $(k+1)$th convergent of the continued fraction of $1/\alpha$ is the reciprocal of the kth convergent of α, for all $k \geq 0$.

Using this fact, we find that, as before, x/y is a convergent of the continued fraction of $\sqrt{d}$.

Exercise 8.2.6 Let α be a quadratic irrational (i.e, the minimal polynomial of the real number α over $\mathbb{Q}$ has degree 2). Show that there are integers P_0, Q_0, d such that

$$\alpha = \frac{P_0 + \sqrt{d}}{Q_0}$$

with $Q_0 | (d - P_0^2)$. Recursively define

$$\begin{aligned} \alpha_k &= \frac{P_k + \sqrt{d}}{Q_k}, \\ a_k &= [\alpha_k], \\ P_{k+1} &= a_k Q_k - P_k, \\ Q_{k+1} &= \frac{d - P_{k+1}^2}{Q_k}, \end{aligned}$$

for $k = 0, 1, 2, \ldots$. Show that $[a_0, a_1, a_2, \ldots]$ is the simple continued fraction of α.

Solution. There exist $a, b, e, f \in \mathbb{Z}$, $e, f > 0$, e not a perfect square, such that

$$\alpha = \frac{a + b\sqrt{e}}{f} = \frac{af + \sqrt{eb^2 f^2}}{f^2}$$

and, evidently, $f^2 | (a^2 f^2 - eb^2 f^2)$. Set $P_0 = af, Q_0 = f^2, d = eb^2 f^2$. This sequence is well-defined, since d is not a perfect square $\Rightarrow Q_k \neq 0$ for all k. By Exercise 8.2.3, it will suffice to show that $\alpha_{k+1} = \frac{1}{\alpha_k - a_k}$ for all k.

$$\begin{aligned} \alpha_k - a_k &= \frac{P_k + \sqrt{d}}{Q_k} - a_k \\ &= \frac{\sqrt{d} - (a_k Q_k - P_k)}{Q_k} \\ &= \frac{\sqrt{d} - P_{k+1}}{Q_k} \\ &= \frac{d - P_{k+1}^2}{Q_k(\sqrt{d} + P_{k+1})} \\ &= \frac{Q_k Q_{k+1}}{Q_k(\sqrt{d} + P_{k+1})} \\ &= \frac{1}{\alpha_{k+1}}. \end{aligned}$$

Exercise 8.2.7 Show that the simple continued fraction expansion of a quadratic irrational α is periodic.

Solution. By Exercise 8.2.6, we may write

$$\alpha = \frac{P_0 + \sqrt{d}}{Q_0}$$

with $Q_0|(P_0^2 - d)$. Setting

$$\begin{aligned} \alpha_k &= \frac{P_k + \sqrt{d}}{Q_k}, \\ a_k &= [\alpha_k], \\ P_{k+1} &= a_k Q_k - P_k, \\ Q_{k+1} &= \frac{d - P_{k+1}^2}{Q_k}, \end{aligned}$$

we have

$$\begin{aligned} &P_0 = 0, && P_1 = d, \\ &Q_0 = 1, && Q_1 = 1, \\ &\alpha_0 = \sqrt{d^2+1}, && \alpha_1 = d + \sqrt{d^2+1}, \\ &a_0 = d, && a_1 = 2d. \end{aligned}$$

and $\alpha = [a_0, a_1, \ldots]$. Now,

$$\alpha = \frac{\alpha_k p_{k-1} + p_{k-2}}{\alpha_k q_{k-1} + q_{k-2}}$$

and if α' denotes the $\mathbb{Q}$-conjugate of α,

$$\alpha' = \frac{\alpha_k' p_{k-1} + p_{k-2}}{\alpha_k' q_{k-1} + q_{k-2}} \quad \Rightarrow \quad \alpha_k' = -\frac{q_{k-2}}{q_{k-1}} \left(\frac{\alpha' - C_{k-2}}{\alpha' - C_{k-1}} \right).$$

Since $C_{k-1}, C_{k-2} \to \alpha$ as $k \to \infty$,

$$\frac{\alpha' - C_{k-2}}{\alpha' - C_{k-1}} \to 1.$$

Therefore, $\alpha_k' < 0$, and, since $\alpha_k > 0$, $\alpha_k - \alpha_k' = 2\sqrt{d}/Q_k > 0$, for all sufficiently large k. We also have $Q_k Q_{k+1} = d - P_{k+1}^2$ so

$$Q_k \leq Q_k Q_{k+1} = d - P_{k+1}^2 \leq d$$

and

$$P_{k+1}^2 \leq d - Q_k \leq d$$

for sufficiently large k. Thus there are only finitely many possible values for P_k, Q_k and we conclude that there exist integers $i < j$ such that $P_i = P_j, Q_i = Q_j$. Then $a_i = a_j$ and, since the a_i are defined recursively, we have

$$\alpha = [a_0, a_1, \ldots, a_{i-1}, \overline{a_i, \ldots, a_{j-1}}].$$

Exercise 8.2.8 Show that, if d is a positive integer but not a perfect square, and $\alpha = \alpha_0 = \sqrt{d}$, then

$$p_{k-1}^2 - dq_{k-1}^2 = (-1)^k Q_k,$$

for all $k \geq 1$, where p_k/q_k is the kth convergent of the continued fraction of α and Q_k is as defined in Exercise 8.2.6.

Solution. By inspection, $p_0^2 - dq_0^2 = [\sqrt{d}]^2 - d = -Q_1$. Now, suppose that $k \geq 2$. Writing

$$\sqrt{d} = \alpha_0 = [a_0, a_1, \ldots, a_{k-1}, \alpha_k] = \frac{\alpha_k p_{k-1} + p_{k-2}}{\alpha_k q_{k-1} + q_{k-2}}$$

and recalling that $\alpha_k = (P_k + \sqrt{d})/Q_k$, we have

$$\sqrt{d} = \frac{(P_k + \sqrt{d})p_{k-1} + Q_k p_{k-2}}{(P_k + \sqrt{d})q_{k-1} + Q_k q_{k-2}},$$

$$\Rightarrow \quad dq_{k-1} + (P_k q_{k-1} + Q_k q_{k-2})\sqrt{d} = P_k p_{k-1} + Q_k p_{k-2} + p_{k-1}\sqrt{d}.$$

Equating coefficients in $\mathbb{Q}(\sqrt{d})$, we have

$$dq_{k-1} = P_k p_{k-1} + Q_k p_{k-2}$$

and

$$p_{k-1} = P_k q_{k-1} + Q_k q_{k-2}.$$

Computation yields

$$p_{k-1}^2 - dq_{k-1}^2 = (p_{k-1}q_{k-2} - p_{k-2}q_{k-1})Q_k = (-1)^k Q_k.$$

Exercise 8.2.10 (a) Find the simple continued fractions of $\sqrt{6}, \sqrt{23}$.

(b) Using Theorem 8.2.9 (c), compute the fundamental unit in both $\mathbb{Q}(\sqrt{6})$ and $\mathbb{Q}(\sqrt{23})$.

Solution. (a) Using notation of previous exercises, setting $\alpha = \alpha_0 = \sqrt{6}$, we have

$$\begin{array}{lll} P_0 = 0, & P_1 = 2, & P_2 = 2, \\ Q_0 = 1, & Q_1 = 2, & Q_2 = 1, \\ \alpha_0 = \sqrt{6}, & \alpha_1 = \dfrac{2+\sqrt{6}}{2} & \alpha_2 = 2 + \sqrt{6}, \\ a_0 = 2, & a_1 = 2, & a_2 = 4. \end{array}$$

Thus, the period of the continued fraction of α is $2 \Rightarrow \sqrt{6} = [a_0, \overline{a_1, a_2}] = [2, \overline{2, 4}]$. Applying the same procedure, we find $\sqrt{23} = [4, \overline{1, 3, 1, 8}]$.

(b) For $\sqrt{6}$, $C_1 = p_1/q_1 = [a_0, a_1] = a_0 + 1/a_1 = 2 + 1/2 = 5/2$. Thus, the fundamental unit in $\mathbb{Q}(\sqrt{6})$ is $5 + 2\sqrt{6}$.

For $\sqrt{23}$, $C_3 = [4, 1, 3, 1] = 24/5$. Therefore the fundamental unit in $\mathbb{Q}(\sqrt{23})$ is $24 + 5\sqrt{23}$.

Exercise 8.2.11 (a) Show that $[d, \overline{2d}]$ is the continued fraction of $\sqrt{d^2+1}$.

(b) Conclude that, if d^2+1 is squarefree, $d \equiv 1, 3 \pmod 4$, then the fundamental unit of $\mathbb{Q}(\sqrt{d^2+1})$ is $d + \sqrt{d^2+1}$. Compute the fundamental unit of $\mathbb{Q}(\sqrt{2}), \mathbb{Q}(\sqrt{10}), \mathbb{Q}(\sqrt{26})$.

(c) Show that the continued fraction of $\sqrt{d^2+2}$ is $[d, \overline{d, 2d}]$.

(d) Conclude that, if d^2+2 is squarefree, then the fundamental unit of $\mathbb{Q}(\sqrt{d^2+2})$ is $d^2+1+d\sqrt{d^2+2}$. Compute the fundamental unit in $\mathbb{Q}(\sqrt{3})$, $\mathbb{Q}(\sqrt{11})$, $\mathbb{Q}(\sqrt{51})$, and $\mathbb{Q}(\sqrt{66})$.

Solution. (a) Observing that $d^2 < d^2+1 < (d+1)^2$ for all $d > 0$, we see that $[\sqrt{d^2+1}] = d$ and setting $\alpha = \alpha_0 = \sqrt{d^2+1}$, we have

$$\begin{aligned} P_0 &= 0, & P_1 &= d, \\ Q_0 &= 1, & Q_1 &= 1, \\ \alpha_0 &= \sqrt{d^2+1}, & \alpha_1 &= d + \sqrt{d^2+1}, \\ a_0 &= d, & a_1 &= 2d. \end{aligned}$$

This implies that the period of the continued fraction of $\sqrt{d^2+1}$ is 1. Therefore $\sqrt{d^2+1} = [a_0, \overline{a_1}] = [d, \overline{2d}]$.

(b) $d \equiv 1, 3 \pmod 4$ and thus $d^2+1 \equiv 2 \pmod 4$. Thus, if d^2+1 is squarefree, then the fundamental unit of $\mathbb{Q}(\sqrt{d^2+1})$ is $p_0 + q_0\sqrt{d^2+1} = d + \sqrt{d^2+1}$.

(c) Observing that $d^2 < d^2+2 < (d+1)^2$ for all $d \geq 1$ we get $[\sqrt{d^2+2}] = d$ and setting $\alpha = \alpha_0 = \sqrt{d^2+2}$, we have

$$\begin{aligned} P_0 &= 0, & P_1 &= d, & P_2 &= d, \\ Q_0 &= 1, & Q_1 &= 2, & Q_2 &= 1, \\ \alpha_0 &= \sqrt{d^2+2}, & \alpha_1 &= \frac{d+\sqrt{d^2+2}}{2}, & \alpha_2 &= d + \sqrt{d^2+2}, \\ a_0 &= d, & a_1 &= d, & a_2 &= 2d. \end{aligned}$$

Therefore the period of the continued fraction of $\sqrt{d^2+2}$ is 2, so

$$\sqrt{d^2+2} = [a_0, \overline{a_1, a_2}] = [d, \overline{d, 2d}]$$

and thus

$$\frac{p_1}{q_1} = d + \frac{1}{d} = \frac{d^2+1}{d}.$$

(d) For all d, $d^2+2 \equiv 2, 3 \pmod 4$ so, if d is squarefree, the fundamental unit in $\mathbb{Q}(\sqrt{d^2+2})$ is $p_1 + q_1\sqrt{d^2+2} = d^2+1+d\sqrt{d^2+2}$.

8.3 Supplementary Problems

Exercise 8.3.1 If n^2-1 is squarefree, show that $n+\sqrt{n^2-1}$ is the fundamental unit of $\mathbb{Q}(\sqrt{n^2-1})$.

Solution. The continued fraction of $\sqrt{n^2-1}$ is $[n-1, \overline{1, 2(n-1)}]$, so that the first two convergents are $(n-1)/1$ and $n/1$. Now apply Theorem 8.2.9.

Exercise 8.3.2 Determine the units of an imaginary quadratic field from first principles.

Solution. We have already determined the units of $\mathbb{Q}(i)$ and $\mathbb{Q}(\sqrt{-3})$. They are $\pm 1, \pm i$ and $\pm 1, \pm\rho, \pm\rho^2$, respectively. Now we determine the units of an arbitrary field $\mathbb{Q}(\sqrt{-d})$ where $-d \equiv 2, 3 \pmod 4$. All units are of the form $a + b\sqrt{-d}$ with $a^2 + db^2 = 1$. It is easy to see that since $d \geq 2$, this has no solution except for $a = \pm 1, b = 0$. If $-d \equiv 1 \pmod 4$, then units will be of the form $(a + b\sqrt{-d})/2$ where $a \equiv b \pmod 2$ and $a^2 + db^2 = 4$. Since we already know the units for $d = 3$, then $d \geq 7$ and once again, the only solution is $a = \pm 1, b = 0$.

Exercise 8.3.3 Suppose that $2^{2n} + 1 = dy^2$ with d squarefree. Show that $2^n + y\sqrt{d}$ is the fundamental unit of $\mathbb{Q}(\sqrt{d})$, whenever $\mathbb{Q}(\sqrt{d}) \neq \mathbb{Q}(\sqrt{5})$.

Solution. Suppose not. Let $(a + b\sqrt{d})/2$ be the fundamental unit. Then

$$2^n + y\sqrt{d} = \left(\frac{a + b\sqrt{d}}{2}\right)^j$$

for some integer j. If an odd prime p divides j, we can write

$$2^n + y\sqrt{d} = \left(\frac{u + v\sqrt{d}}{2}\right)^p,$$

where $(u + v\sqrt{d})/2$ is again a unit. Hence

$$2^{n+p} = \sum_k \binom{p}{2k} v^{2k} d^k u^{p-2k} = uA,$$

say. Clearly, $(u, A) = 1$ so we must have either $u = 1$, $A = 2^{n+p}$ or $u = 2^{n+p}, A = 1$. The latter case is impossible since

$$A = pv^{p-1} d^{(p-1)/2} > 1.$$

Hence $u = 1$. But then $dv^2 = 5$ or -3. The former case is ruled out by the hypothesis and the latter is impossible. Thus, if $2^n + y\sqrt{d}$ is not the fundamental unit, it must be the square of a unit. But then, this means that

$$2^{n+4} + 4y\sqrt{d} = (u + v\sqrt{d})^2 = (u^2 + dv^2) + 2uv\sqrt{d}$$

and this case is also easily ruled out.

Exercise 8.3.4 (a) Determine the continued fraction expansion of $\sqrt{51}$ and use it to obtain the fundamental unit ε of $\mathbb{Q}(\sqrt{51})$.

(b) Prove from first principles that all units of $\mathbb{Q}(\sqrt{51})$ are given by $\varepsilon^n, n \in \mathbb{Z}$.

Solution. See Exercise 8.2.11.

Exercise 8.3.5 Determine a unit $\neq \pm 1$ in the ring of integers of $\mathbb{Q}(\theta)$ where $\theta^3 + 6\theta + 8 = 0$.

Solution. Consider $\eta = 1 + \theta$.

Exercise 8.3.6 Let p be an odd prime > 3 and suppose that it does not divide the class number of $\mathbb{Q}(\zeta_p)$. Show that

$$x^p + y^p + z^p = 0$$

is impossible for integers x, y, z such that $p \nmid xyz$.

Solution. We factor

$$x^p + y^p = \prod_{i=0}^{p-1} (x + \zeta_p^i y) = -z^p.$$

Since $p \nmid xyz$, the terms $(x + \zeta_p^i y)$ are mutually coprime for $0 \leq i \leq p-1$. Moreover, viewing the equation as an ideal equation gives

$$(x + \zeta_p^i y) = \mathfrak{a}_i^p$$

for some ideal $\mathfrak{a}_i$. Since p does not divide the class number, $\mathfrak{a}_i$ itself must be principal. Therefore

$$(x + \zeta_p^i y) = \varepsilon \alpha^p,$$

where ε is a unit in $\mathbb{Q}(\zeta_p)$ and α is an integer of $\mathbb{Q}(\zeta_p)$.

By Exercise 4.3.7, $\mathbb{Z}[\zeta_p]$ is the ring of integers of $\mathbb{Q}(\zeta_p)$ and

$$1, \zeta_p, \zeta_p^2, \ldots, \zeta_p^{p-2}$$

is an integral basis. Therefore, we can express

$$\alpha = a_0 + a_1 \zeta_p + \cdots + a_{p-2} \zeta_p^{p-2},$$

so that

$$\alpha^p \equiv a_0^p + \cdots + a_{p-2}^p \equiv a \pmod{p},$$

where $a = a_0 + \cdots + a_{p-2}$, by a simple application of Fermat's little Theorem (Exercise 1.1.13). Also, we may write $\varepsilon = \zeta_p^s \eta$ where $0 \leq s < p$ and η is a real unit, by Theorem 8.1.10. Hence, for $i = 1$,

$$x + \zeta_p y \equiv \zeta_p^s \beta \pmod{p},$$

where β is a real integer of $\mathbb{Q}(\zeta_p)$. Also, by complex conjugation,

$$x + \zeta_p^{-1} y \equiv \zeta_p^{-s} \beta \pmod{p}.$$

Then

$$\zeta_p^{-s}(x+\zeta_p y) - \zeta_p^s(x+\zeta_p^{-1}y) \equiv 0 \pmod p.$$

Since $1, \zeta_p, \ldots, \zeta_p^{p-2}$ is an integral basis for $\mathbb{Q}(\zeta_p)$,

$$c_0 + c_1\zeta_p + \cdots + c_{p-2}\zeta_p^{p-2} \equiv 0 \pmod p$$

holds if and only if

$$c_0 \equiv c_1 \equiv \cdots \equiv c_{p-2} \equiv 0 \pmod p.$$

Write $\zeta_p^{p-s}(x+\zeta_p y) - (\zeta_p^s x + \zeta_p^{s-1}y) \equiv 0 \pmod p$. If $2 < s \leq p-2$ and $s \neq (p+1)/2$, the powers of ζ_p are all different and less than $p-1$ in the above congruence. Hence $x \equiv y \equiv 0 \pmod p$, a contradiction.

If $s = (p+1)/2$, then $x + \zeta_p y \equiv \zeta_p x + y \pmod p$ so that $x \equiv y \pmod p$. Similarly, $x \equiv z \pmod p$ so that

$$x^p + y^p + z^p \equiv 3x^p \equiv 0 \pmod p.$$

Since $p > 3$, we get $p \mid x$, a contradiction.

If $s = 0$, then $x + \zeta_p y - (x + \zeta_p^{-1}y) \equiv 0 \pmod p$ so that $y \equiv 0 \pmod p$, a contradiction.

If $s = 1$, $\zeta_p^{-1}x + y \equiv \zeta_p x + y \pmod p$ and so $p \mid x$, a contradiction.

If $s = 2$, $x + \zeta_p y \equiv \zeta_p^4 x + \zeta_p^3 y$ and so $x \equiv y \equiv 0 \pmod p$, again a contradiction.

Finally, if $s = p-1$,

$$\zeta_p(x+\zeta_p y) \equiv \zeta_p^{-1}x + \zeta_p^{-2}y \pmod p,$$

that is,

$$\zeta_p^3(x+\zeta_p y) \equiv \zeta_p x + y \pmod p,$$

which gives $p \mid x$ (since $p \geq 5$), again a contradiction.

Exercise 8.3.7 Let K be a quadratic field of discriminant d. Let P_0 denote the group of principal fractional ideals $\alpha\mathcal{O}_K$ with $\alpha \in K$ satisfying $N_K(\alpha) > 0$. The quotient group H_0 of all nonzero fractional ideals modulo P_0 is called the *restricted class group* of K. Show that H_0 is a subgroup of the ideal class group H of K and $[H : H_0] \leq 2$.

Solution. If $d < 0$, the norm of any nonzero element is greater than 0 and so the notions of restricted class group and ideal class group coincide. If $d > 0$, then $H = H_0$ if and only if there exists a unit in K of norm -1. This is because $\sqrt{d}\mathcal{O}_K$ is in the same coset of $\mathcal{O}_K \pmod{P_0}$ if and only if $\sqrt{d} = \varepsilon\alpha$ with $N_K(\alpha) > 0$, and ε is a unit. Since $N_K(\sqrt{d}) < 0$, this can happen if and only if ε is a unit of norm -1.

Exercise 8.3.8 Given an ideal $\mathfrak{a}$ of a quadratic field K, let $\mathfrak{a}'$ denote the conjugate ideal. If K has discriminant d, write

$$|d| = p_1^{\alpha_1} p_2 \cdots p_t,$$

where $p_1 = 2$, $\alpha_1 = 0, 2$, or 3 and $p_2, \ldots, p_t$ are distinct odd primes. If we write $p_i \mathcal{O}_K = \wp_i^2$ show that for any ideal $\mathfrak{a}$ of $\mathcal{O}_K$ satisfying $\mathfrak{a} = \mathfrak{a}'$ we can write

$$\mathfrak{a} = r\wp_1^{a_1} \cdots \wp_t^{a_t},$$

$r > 0$, $a_i = 0, 1$ uniquely.

Solution. We first factor $\mathfrak{a}$ as a product of prime ideals. The fact that $\mathfrak{a} = \mathfrak{a}'$ implies that if a prime ideal $\wp$ divides $\mathfrak{a}$, so does $\wp'$. If $\wp$ is inert then it is principal and generated by a rational prime. If $\wp$ splits, then $\wp$ and $\wp'$ (which both occur in $\mathfrak{a}$ with the same multiplicity) can be paired to give again a principal ideal generated by a rational prime. Only the ramified prime ideals cannot be so paired. Since $\wp_i^2 = p_i \mathcal{O}_K$, we see immediately that $\mathfrak{a}$ has a factorization of the form described above. If we have another factorization

$$\mathfrak{a} = r_1 \wp^{b_1} \cdots \wp^{b_t}, \qquad r_1 > 0, \quad b_i = 0, 1,$$

then taking norms we obtain $r^2 p_1^{a_1} \cdots p_t^{a_t} = r_1^2 p_1^{b_1} \cdots p_t^{b_t}$ so that $a_i \equiv b_i \pmod 2$. Since $a_i = 0$ or 1, and $b_i = 0$ or 1, this means $a_i = b_i$. Hence $r = r_1$.

Exercise 8.3.9 An ideal class C of H_0 is said to be *ambiguous* if $C^2 = 1$ in H_0. Show that any ambiguous ideal class is equivalent (in the restricted sense) to one of the at most 2^t ideal classes

$$\wp_1^{a_1} \cdots \wp_t^{a_t}, \qquad a_i = 0, 1.$$

Solution. Let $\mathfrak{a}$ be an ideal lying in an ambiguous class. Then $\mathfrak{a}^2 = (\alpha)$ with $N_K(\alpha) > 0$. But we have $\mathfrak{a}\mathfrak{a}' = (N_K(\mathfrak{a}))$. Therefore

$$\mathfrak{a}\mathfrak{a}' = (N_K(\mathfrak{a})/\alpha)(\alpha) = (N_K(\mathfrak{a})/\alpha)\mathfrak{a}^2$$

so that $\mathfrak{a} = \mathfrak{a}'$. By the previous question, $\mathfrak{a}$ can be written as

$$r\wp_1^{a_1} \cdots \wp_t^{a_t},$$

$r > 0$, $a_i = 0, 1$. In the restricted class group, these form at most 2^t ideal classes.

Exercise 8.3.10 With the notation as in the previous two questions, show that there is exactly one relation of the form

$$\wp_1^{a_1} \cdots \wp_t^{a_t} = \rho\mathcal{O}_K, \qquad N_K(\rho) > 0,$$

with $a_i = 0$ or 1, $\sum_{i=1}^t a_i > 0$.

Solution. We first show there is at least one such relation. If $d < 0$, then $\wp_1 \cdots \wp_t = (\sqrt{d})$. If $d > 0$, let ε be a generator of all the totally positive units of K (i.e., units $\varepsilon > 0$ and $\varepsilon' > 0$). Set $\mu = \sqrt{d}(1-\varepsilon)$. Then $\mu' = -\sqrt{d}(1-\varepsilon')$ so that $\varepsilon\mu' = -\sqrt{d}(\varepsilon - 1) = \mu$. Therefore $(\mu) = (\mu')$ and by the penultimate question, we have the result.

To establish uniqueness, consider first the case $d < 0$. If

$$\wp_1^{a_1} \cdots \wp_t^{a_t} = \rho\mathcal{O}_K, \qquad N_K(\rho) > 0,$$

then $(\rho) = (\rho')$ since $\wp_i' = \wp_i$. Therefore $\rho = b\sqrt{d}$ for some $b \in \mathbb{Q}$. The other two cases of $\mathbb{Q}(\sqrt{-1})$ are $\mathbb{Q}(\sqrt{-3})$ are similarly dealt with. If now $d > 0$, then $\rho = \eta\rho'$ with η a totally positive unit. We write $\eta = \varepsilon^m$ for some m. With μ defined as above, notice that $\rho/\mu^m = \rho'/(\mu')^m$ so that $\rho/\mu^m = r \in \mathbb{Q}$. Hence $(\rho) = (\mu)^m$. Since (μ) has order 2 in the (restricted) class group we are done.

Exercise 8.3.11 Let K be a quadratic field of discriminant d. Show that the number of ambiguous ideal classes is 2^{t-1} where t is the number of distinct primes dividing d. Deduce that 2^{t-1} divides the order of the class group.

Solution. This is now immediate from the previous two exercises. Since the restricted class group has index 1 or 2 in the ideal class group, the divisibility assertion follows.

Exercise 8.3.12 If K is a quadratic field of discriminant d and class number 1, show that d is prime or $d = 4$ or 8.

Solution. If d has t distinct prime divisors, then 2^{t-1} divides the class number. Thus $t \leq 1$. Since the discriminant is either squarefree or four times a squarefree number, the result is now clear.

Exercise 8.3.13 If a real quadratic field K has odd class number, show that K has a unit of norm -1.

Solution. Since K has odd class number, $H = H_0$. This means there is a unit of norm -1.

Exercise 8.3.14 Show that $15 + 4\sqrt{14}$ is the fundamental unit of $\mathbb{Q}(\sqrt{14})$.

Solution. The continued fraction development of $\sqrt{14}$ is

$$[3, \overline{1, 2, 1, 6}]$$

and the convergents are easily computed:

$$\frac{3}{1}, \frac{4}{1}, \frac{11}{3}, \frac{15}{4}, \ldots.$$

By Theorem 8.2.9, we find that $15 + 4\sqrt{14}$ is the fundamental unit.

Exercise 8.3.15 In Chapter 6 we showed that $\mathbb{Z}[\sqrt{14}]$ is a PID (principal ideal domain). Assume the following hypothesis: given $\alpha, \beta \in \mathbb{Z}[\sqrt{14}]$, such that $\gcd(\alpha, \beta) = 1$, there is a prime $\pi \equiv \alpha \pmod{\beta}$ for which the fundamental unit $\varepsilon = 15+4\sqrt{14}$ generates the coprime residue classes $\pmod{\pi}$. Show that $\mathbb{Z}[\sqrt{14}]$ is Euclidean.

Solution. We define the Euclidean algorithm inductively as follows. For $u \in \mathbb{Z}[\sqrt{14}]$, define $\varphi(u) = 1$. If ε generates the residue classes $\pmod{\pi}$, where π is prime, define $\varphi(\pi) = 2$. If β is a prime for which ε does not generate the residue classes $\pmod{\beta}$, define $\varphi(\beta) = 3$. Now for any $\gamma \in \mathbb{Z}[\sqrt{14}]$, factor

$$\gamma = \pi_1^{e_1} \cdots \pi_r^{e_r} \beta_1^{f_1} \cdots \beta_s^{f_s},$$

where $\varphi(\pi_i) = 2$, $\varphi(\beta_i) = 3$. Define

$$\varphi(\gamma) = 2(e_1 + \cdots + e_r) + 3(f_1 + \cdots + f_s).$$

An easy induction argument using the Chinese Remainder Theorem (Theorem 5.3.13) shows that φ is a Euclidean algorithm.

Exercise 8.3.16 Let $d = a^2 + 1$. Show that if $|u^2 - dv^2| \neq 0, 1$ for integers u, v, then

$$|u^2 - dv^2| > \sqrt{d}.$$

Solution. The continued fraction expansion of $\sqrt{d}$ is

$$[a, 2a, 2a, \ldots]$$

and the convergents p_k/q_k always satisfy $p_k^2 - dq_k^2 = \pm 1$. If $|u^2 - dv^2| < \sqrt{d}$, then u/v is a convergent of $\sqrt{d}$, by Exercise 8.2.5. Since $|u^2 - dv^2| \neq 0, 1$, we are done.

Exercise 8.3.17 Suppose that n is odd, $n \geq 5$, and that $n^{2g}+1 = d$ is squarefree. Show that the class group of $\mathbb{Q}(\sqrt{d})$ has an element of order $2g$.

Solution. We have $(n)^{2g} = (\sqrt{d} - 1)(\sqrt{d} + 1)$. Since n is odd, each of the ideals $(-1 + \sqrt{d})$ and $(1 + \sqrt{d})$ must be coprime. By Exercise 5.3.12, each of them must be a $2g$th power. Therefore

$$\mathfrak{a}^{2g} = (\sqrt{d} - 1)$$

and

$$(\mathfrak{a}')^{2g} = (\sqrt{d} + 1).$$

We claim that $\mathfrak{a}$ has order greater than or equal to g. Observe that $n^{2g}+1 \equiv 2 \pmod 4$ because n is odd. Therefore $1, \sqrt{d}$ is an integral basis of $\mathbb{Q}(\sqrt{d})$. If $\mathfrak{a}^m$ were principal, then

$$\mathfrak{a}^m = (u + v\sqrt{d})$$

implies that

$$n^m = |u^2 - dv^2|,$$

which by the previous exercise is either $0, 1$ or $> \sqrt{d}$. The former cannot hold since $n \geq 5$. Thus,

$$n^m > \sqrt{d} = \sqrt{n^{2g} + 1} > n^g.$$

Hence, $m > g$. Since $m \mid 2g$, we must have that $\mathfrak{a}$ has order $2g$.

It is conjectured that there are infinitely many squarefree numbers of the form $n^{2g} + 1$. Thus, this argument does not establish that there are infinitely many real quadratic fields whose class number is divisible by g. However, by a simple modification of this argument, we can derive such a result. We leave it as an exercise to the interested reader.

Chapter 9

Higher Reciprocity Laws

9.1 Cubic Reciprocity

Exercise 9.1.1 If π is a prime of $\mathbb{Z}[\rho]$, show that $N(\pi)$ is a rational prime or the square of a rational prime.

Solution. Let $N(\pi) = n > 1$. Then $\pi\overline{\pi} = n$. Now n is a product of rational prime divisors. Since π is prime, $\pi \mid p$ for some rational prime p. Write $p = \pi\gamma$. Then $N(p) = N(\pi)N(\gamma) = p^2$. Thus, either $N(\pi) = p$ or $N(\pi) = p^2$.

Exercise 9.1.2 If $\pi \in \mathbb{Z}[\rho]$ is such that $N(\pi) = p$, a rational prime, show that π is a prime of $\mathbb{Z}[\rho]$.

Solution. If π factored in $\mathbb{Z}[\rho]$, then $\pi = \alpha\beta$ and $p = N(\pi) = N(\alpha)N(\beta)$ which implies that $N(\alpha) = 1$ or $N(\beta) = 1$ so that π cannot be factored nontrivially in $\mathbb{Z}[\rho]$.

Exercise 9.1.3 If p is a rational prime congruent to 2 (mod 3), show that p is prime in $\mathbb{Z}[\rho]$. If $p \equiv 1 \pmod 3$, show that $p = \pi\overline{\pi}$ where π is prime in $\mathbb{Z}[\rho]$.

Solution. Let $p \equiv 2 \pmod 3$ be a rational prime. If $p = \pi\gamma$, with $N(\gamma), N(\pi) > 1$, then $p^2 = N(\pi)N(\gamma)$ implies that $N(\pi) = p$ and $N(\gamma) = p$.

Writing $\pi = a + b\rho$, we find $p = N(\pi) = a^2 - ab + b^2$ so that

$$4p = 4a^2 - 4ab + 4b^2 = (2a - b)^2 + 3b^2.$$

Hence $p \equiv (2a - b)^2 \pmod 3$, a contradiction since 2 is not a square mod 3.

Finally, if $p \equiv 1 \pmod 3$, then by quadratic reciprocity:

$$\left(\frac{-3}{p}\right) = \left(\frac{-1}{p}\right)\left(\frac{3}{p}\right) = \left(\frac{p}{3}\right) = \left(\frac{1}{3}\right) = 1$$

so that $x^2 \equiv -3 \pmod{p}$ has a solution. Hence $py = x^2 + 3$ for some $x, y \in \mathbb{Z}$. Therefore p divides $(x+\sqrt{-3})(x-\sqrt{-3}) = (x+1+2\rho)(x-1-2\rho)$. If p were prime in $\mathbb{Z}[\rho]$, it would divide one of these two factors, which is not the case. Thus $p = \alpha\beta$ for some $\alpha, \beta \in \mathbb{Z}[\rho]$ and $N(\alpha) > 1, N(\beta) > 1$. Hence $N(\alpha) = p$ so that $\alpha\overline{\alpha} = p$. Moreover, α is prime by Exercise 9.1.2.

Recall that in Section 2.3 we found that $3 = -\rho^2(1-\rho)^2$ and $(1-\rho)$ is irreducible, so that 3 is not a prime in $\mathbb{Z}[\rho]$.

Exercise 9.1.4 Let π be a prime of $\mathbb{Z}[\rho]$. Show that $\alpha^{N(\pi)-1} \equiv 1 \pmod{\pi}$ for all $\alpha \in \mathbb{Z}[\rho]$ which are coprime to π.

Solution. Since π is prime, the ideal (π) is prime. Hence $\mathbb{Z}[\rho]/(\pi)$ is a field, containing $N(\pi)$ elements. Its multiplicative group, consisting of classes coprime to π, has $N(\pi) - 1$ elements. Thus, by Lagrange's theorem, the result is immediate.

Exercise 9.1.5 Let π be a prime not associated to $(1-\rho)$. First show that $3 \mid N(\pi) - 1$. If $(\alpha, \pi) = 1$, show that there is a unique integer $m = 0, 1$ or 2 such that

$$\alpha^{(N(\pi)-1)/3} \equiv \rho^m \pmod{\pi}.$$

Solution. By Exercise 9.1.3, we know that $\pi \mid (\alpha^{N(\pi)-1} - 1)$. By Exercise 9.1.4, we know $N(\pi) \equiv 1 \pmod{3}$. Thus, we can write $\beta = \alpha^{(N(\pi)-1)/3}$ and observe that

$$\beta^3 - 1 = (\beta - 1)(\beta - \rho)(\beta - \rho^2).$$

Since π is prime and divides $\beta^3 - 1$, it must divide one of the three factors on the right. If π divides at least two factors, then $\pi \mid (1-\rho)$ which means π is an associate of $1-\rho$, contrary to assumption. Thus, $\beta \equiv 1, \rho$, or $\rho^2 \pmod{\pi}$ as desired.

Exercise 9.1.6 Show that:

(a) $(\alpha/\pi)_3 = 1$ if and only if $x^3 \equiv \alpha \pmod{\pi}$ is solvable in $\mathbb{Z}[\rho]$;

(b) $(\alpha\beta/\pi)_3 = (\alpha/\pi)_3(\beta/\pi)_3$; and

(c) If $\alpha \equiv \beta \pmod{\pi}$, then $(\alpha/\pi)_3 = (\beta/\pi)_3$.

Solution. Clearly if $x^3 \equiv \alpha \pmod{\pi}$ has a solution, then by Exercise 9.1.4, $\alpha^{(N(\pi)-1)/3} \equiv 1 \pmod{\pi}$ so that $(\alpha/\pi)_3 = 1$. For the converse, let g be a primitive root of $\mathbb{Z}[\rho]/(\pi)$. Then, writing $\alpha = g^r$ we find $g^{(rN(\pi)-1)/3} \equiv 1 \pmod{\pi}$ so that

$$\frac{r(N(\pi)-1)}{3} \equiv 0 \pmod{N(\pi)-1}.$$

Hence $3 \mid r$, and α is a cube mod π. That is, $x^3 \equiv \alpha \pmod{\pi}$ has a solution. This proves (a).

For (b),

$$\left(\frac{\alpha\beta}{\pi}\right)_3 \equiv (\alpha\beta)^{(N(\pi)-1)/3} \equiv \alpha^{(N(\pi)-1)/3}\beta^{(N(\pi)-1)/3} \equiv \left(\frac{\alpha}{\pi}\right)_3\left(\frac{\beta}{\pi}\right)_3 \pmod{\pi}.$$

For (c), if $\alpha \equiv \beta \pmod{\pi}$, then

$$\left(\frac{\alpha}{\pi}\right)_3 \equiv \alpha^{(N(\pi)-1)/3} \equiv \beta^{(N(\pi)-1)/3} \equiv \left(\frac{\beta}{\pi}\right)_3 \pmod{\pi}.$$

Exercise 9.1.7 Show that:
(a) $\overline{\chi_\pi(\alpha)} = \chi_\pi(\alpha)^2 = \chi_\pi(\alpha^2)$; and
(b) $\overline{\chi_\pi(\alpha)} = \chi_{\overline{\pi}}(\overline{\alpha})$.

Solution. $\chi_\pi(\alpha)$ is by definition one of $1, \rho$, or ρ^2 so that (a) is immediate.
For (b), observe that

$$\alpha^{(N(\pi)-1)/3} \equiv \chi_\pi(\alpha) \pmod{\pi}$$

implies

$$\overline{\alpha}^{(N(\pi)-1)/3} \equiv \overline{\chi_\pi(\alpha)} \pmod{\overline{\pi}}$$

on the one hand. On the other hand,

$$\overline{\alpha}^{(N(\pi)-1)/3} \equiv \chi_{\overline{\pi}}(\overline{\alpha}) \pmod{\overline{\pi}}$$

by definition. Part (b) is now immediate.

Exercise 9.1.8 If $q \equiv 2 \pmod 3$, show that $\chi_q(\overline{\alpha}) = \chi_q(\alpha^2)$ and $\chi_q(n) = 1$ if n is a rational integer coprime to q.

Solution. Since $\overline{q} = q$,

$$\chi_q(\overline{\alpha}) = \chi_{\overline{q}}(\overline{\alpha}) = \overline{\chi_q(\alpha)} = \chi_q(\alpha^2)$$

by the previous exercise. Also,

$$\chi_q(n) = \chi_q(\overline{n}) = \chi_q(n^2) = \chi_q(n)^2.$$

Since $\chi_q(n) \neq 0$, we deduce $\chi_q(n) = 1$.

Exercise 9.1.9 Let $N(\pi) = p \equiv 1 \pmod 3$. Among the associates of π, show there is a unique one which is primary.

Solution. Write $\pi = a + b\rho$. All the associates of π can be written down: $a + b\rho, -b + (a - b)\rho, (b - a) - a\rho, -a - b\rho, b + (b - a)\rho, (a - b) + a\rho$. Since $p = a^2 - ab + b^2$, not both a and b are divisible by 3.

If $b \equiv 0 \pmod 3$, then $\pi \equiv 1 \pmod 3$ implies $a \equiv 1 \pmod 3$ so that $-a \equiv 2 \pmod 3$ and hence $-a - b\rho$ is the primary. If $a \equiv 0 \pmod 3$ then one of $(b - a) - a\rho$ or $(a - b) + a\rho$ is primary. If both a and b are coprime to 3, then we must have $a \equiv b \pmod 3$, for otherwise $3 \mid p$, contrary to assumption. Thus $a \equiv b \equiv \pm 1 \pmod 3$, so that one of $b + (b - a)\rho$ or $-b + (a - b)\rho$ is primary, as desired.

Exercise 9.1.11 If $\chi_1, \dots, \chi_r$ are nontrivial and the product $\chi_1 \cdots \chi_r$ is also nontrivial prove that $g(\chi_1) \cdots g(\chi_r) = J(\chi_1, \dots, \chi_r) g(\chi_1 \cdots \chi_r)$.

Solution. Define $\psi : \mathbb{F}_p \to \mathbb{C}$ by $\psi(t) = \zeta^t$. Then $\psi(t_1 + t_2) = \psi(t_1)\psi(t_2)$ and we can write $g(\chi) = \sum_t \chi(t)\psi(t)$. This is just for notational convenience. Now

$$\begin{aligned} g(\chi_1) \cdots g(\chi_r) &= \left(\sum_{t_1} \chi_1(t_1)\psi(t_1) \right) \cdots \left(\sum_{t_r} \chi_r(t_r)\psi(t_r) \right) \\ &= \sum_s \psi(s) \left(\sum_{t_1 + \cdots + t_r = s} \chi_1(t_1) \cdots \chi_r(t_r) \right). \end{aligned}$$

If $s \neq 0$, writing $t_i = su_i$, the inner sum becomes

$$(\chi_1 \cdots \chi_r)(s) J(\chi_1, \dots, \chi_r).$$

If $s = 0$, then the inner sum is

$$\sum_{t_1 + \cdots + t_r = 0} \chi_1(t_1) \cdots \chi_r(t_r) = 0$$

since $t_1, \dots, t_{r-1}$ can be chosen arbitrarily so that $t_r = -t_1 - \cdots - t_{r-1}$, and each of the sums corresponding to $t_1, \dots, t_{r-1}$ is zero since $\chi_1, \dots, \chi_r$ are nontrivial. This completes the proof.

Exercise 9.1.12 If $\chi_1, \dots, \chi_r$ are nontrivial, and $\chi_1 \cdots \chi_r$ is trivial, show that

$$g(\chi_1) \cdots g(\chi_r) = \chi_r(-1) p J(\chi_1, \dots, \chi_{r-1}).$$

Solution. By the previous exercise,

$$g(\chi_1) \cdots g(\chi_{r-1}) = J(\chi_1, \dots, \chi_{r-1}) g(\chi_1 \cdots \chi_{r-1}).$$

Multiplying both sides of the equation by $g(\chi_r)$ leads us to evaluate

$$g(\chi_r) g(\chi_1 \cdots \chi_{r-1}).$$

However, $(\chi_1 \cdots \chi_{r-1})\chi_r = \chi_0$ means that

$$g(\chi_1 \cdots \chi_{r-1}) = g(\overline{\chi_r}) = \chi_r(-1) p$$

by the proof of Theorem 9.1.10.

Exercise 9.1.14 Show that $g(\chi)^3 = p\pi$.

Solution. By Exercise 9.1.12, $g(\chi)^3 = pJ(\chi, \chi)$ and by the previous exercise, $J(\chi, \chi) = \pi$.

Exercise 9.1.17 Let π be a prime of $\mathbb{Z}[\rho]$. Show that $x^3 \equiv 2 \pmod{\pi}$ has a solution if and only if $\pi \equiv 1 \pmod{2}$.

Solution. We first observe that $x^3 \equiv 2 \pmod{\pi}$ is solvable if and only if $x^3 \equiv 2 \pmod{\pi'}$ is solvable for any associate of π. Thus we may assume that π is primary.

If $\pi = q$ is a rational prime, then $\chi_q(2) = \chi_2(q) = \chi_2(1) = 1$ so that 2 is a cubic residue for all such primes. If $\pi = a + b\rho$ is primary, by cubic reciprocity, $\chi_\pi(2) = \chi_2(\pi)$. The norm of (2) is 4 and

$$\pi = \pi^{(4-1)/3} \equiv \chi_2(\pi) \pmod{2}.$$

Thus $\chi_\pi(2) = \chi_2(\pi) = 1$ if and only if $\pi \equiv 1 \pmod{2}$.

9.2 Eisenstein Reciprocity

Exercise 9.2.1 Show that $q \equiv 1 \pmod{m}$ and that $1, \zeta_m, \zeta_m^2, \dots, \zeta_m^{m-1}$ are distinct coset representatives mod $\wp$.

Solution. Observe that

$$\frac{x^m - 1}{x - 1} = 1 + x + \cdots + x^{m-1} = \prod_{i=1}^{m-1} (x - \zeta_m^i).$$

Putting $x = 1$ in this identity gives

$$m = \prod_{i=1}^{m-1} (1 - \zeta_m^i).$$

If $\zeta_m^i \equiv \zeta_m^j \pmod{\wp}$ (say), then $\zeta_m^{j-i} \equiv 1 \pmod{\wp}$ so that $m \equiv 0 \pmod{\wp}$, contrary to $m \notin \wp$. Thus $1, \zeta_m, \dots, \zeta_m^{m-1}$ are distinct mod $\wp$. Moreover, the cosets they represent form a multiplicative subgroup of $\mathbb{Z}[\zeta_m]/\wp$ of order m. Since $\left(\mathbb{Z}[\zeta_m]/\wp\right)^*$ has order $q - 1 = N(\wp) - 1$, we must have $m \mid q - 1$.

Exercise 9.2.2 Let $\alpha \in \mathbb{Z}[\zeta_m]$, $\alpha \notin \wp$. Show that there is a unique integer i (modulo m) such that

$$\alpha^{(q-1)/m} \equiv \zeta_m^i \pmod{\wp}.$$

Solution. Since $\left(\mathbb{Z}[\zeta_m]/\wp\right)^*$ has $q-1$ elements, we have $\alpha^{q-1} \equiv 1 \pmod{\wp}$. Thus

$$\prod_{i=1}^{m-1} (\alpha^{(q-1)/m} - \zeta_m^i) \equiv 0 \pmod{\wp}.$$

Since $\wp$ is a prime ideal, there is an integer i, $0 \leq i < m$, such that

$$\alpha^{(q-1)/m} \equiv \zeta_m^i \pmod{\wp}.$$

This i is unique by Exercise 9.2.1.

Exercise 9.2.3 Show that:

(a) $(\alpha/\wp)_m = 1$ if and only if $x^m \equiv \alpha \pmod{\wp}$ is solvable in $\mathbb{Z}[\zeta_m]$;

(b) for all $\alpha \in \mathbb{Z}[\zeta_m]$, $\alpha^{\frac{N(\wp)-1}{m}} \equiv (\alpha/\wp)_m \pmod{\wp}$;

(c) $(\alpha\beta/\wp)_m = (\alpha/\wp)_m(\beta/\wp)_m$; and

(d) if $\alpha \equiv \beta \pmod{\wp}$, then $(\alpha/\wp)_m = (\beta/\wp)_m$.

Solution. If $x^m \equiv \alpha \pmod{\wp}$ has a solution, then

$$\alpha^{(N(\wp)-1)/m} \equiv x^{N(\wp)-1} \equiv 1 \pmod{\wp}$$

by the analogue of Fermat's little Theorem. Thus, $(\alpha/\wp)_m = 1$. For the converse, we know that $\left(\mathbb{Z}[\zeta_m]/\wp\right)^*$ is cyclic, being the multiplicative group of a finite field. Let g be a generator, and set $\alpha = g^r$. If $(\alpha/\wp) = 1$, then $\alpha^{(q-1)/m} \equiv 1 \pmod{\wp}$. Hence $g^{r(q-1)/m} \equiv 1 \pmod{\wp}$. Since g has order $q-1$, we must have

$$r\frac{q-1}{m} \equiv 0 \pmod{q-1}.$$

Hence, $m \mid r$, so that α is an mth power. This proves (a).

For (b), we need only note the case $\alpha \equiv 0 \pmod{\wp}$ which is clear. Parts (c) and (d) are proved exactly as in the case of the cubic residue symbol in Exercise 9.1.6.

Exercise 9.2.4 If $\wp$ is a prime ideal of $\mathbb{Z}[\zeta_m]$ not containing m show that

$$\left(\frac{\zeta_m}{\wp}\right)_m = \zeta_m^{(N(\wp)-1)/m}.$$

Solution. By definition,

$$\left(\frac{\zeta_m}{\wp}\right)_m \equiv \zeta_m^{(N(\wp)-1)/m} \pmod{\wp}.$$

Since both $(\zeta_m/\wp)_m$ and $\zeta_m^{(N(\wp)-1)/m}$ are mth roots of unity and by Exercise 9.2.1, distinct roots represent distinct classes, we must have

$$\left(\frac{\zeta_m}{\wp}\right)_m = \zeta_m^{(N(\wp)-1)/m}.$$

Exercise 9.2.5 Suppose $\mathfrak{a}$ and $\mathfrak{b}$ are ideals coprime to (m). Show that:

(a) $(\alpha\beta/\mathfrak{a})_m = (\alpha/\mathfrak{a})_m(\beta/\mathfrak{a})_m$;

(b) $(\alpha/\mathfrak{ab})_m = (\alpha/\mathfrak{a})_m(\beta/\mathfrak{b})_m$; and

(c) if α is prime to $\mathfrak{a}$ and $x^m \equiv \alpha \pmod{\mathfrak{a}}$ is solvable in $\mathbb{Z}[\zeta_m]$, then $(\alpha/\mathfrak{a})_m = 1$.

Solution. Parts (a) and (b) are immediate from Exercise 9.2.3. For (c), we note that $x^m \equiv \alpha \pmod{\wp}$ has a solution for every prime ideal $\wp$ dividing $\mathfrak{a}$. Thus by Exercise 9.2.3 (a), $(\alpha/\wp)_m = 1$ for every prime ideal $\wp$ dividing $\mathfrak{a}$. thus, $(\alpha/\mathfrak{a})_m = 1$.

Exercise 9.2.6 Show that the converse of (c) in the previous exercise is not necessarily true.

Solution. Choose two distinct prime ideals $\wp_1, \wp_2$ coprime to (m). The map

$$\begin{array}{rcl} \mathbb{Z}[\zeta_m]/\wp_1 & \to & \mathbb{Z}[\zeta_m]/\wp_1, \\ \beta \pmod{\wp_1} & \mapsto & \beta^{(N(\wp_1)-1)/m} \pmod{\wp_1}, \end{array}$$

is a homomorphism with kernel consisting of the subgroup of mth powers by Exercise 9.2.3 (a). This subgroup has size $(N(\wp_1) - 1)/m$. Hence the image is the subgroup of mth roots of unity. Thus, we can find β, γ so that

$$\begin{array}{rcl} \beta^{(N(\wp_1)-1)/m} & \equiv & \zeta_m \pmod{\wp_1}, \\ \gamma^{(N(\wp_2)-1)/m} & \equiv & \zeta_m^{m-1} \pmod{\wp_2}. \end{array}$$

By the Chinese Remainder Theorem (Theorem 5.3.13), we can find

$$\begin{array}{rcl} \alpha & \equiv & \beta \pmod{\wp_1}, \\ \alpha & \equiv & \gamma \pmod{\wp_2}. \end{array}$$

Then $(\alpha/\wp_1)_m = (\beta/\wp_1)_m = \zeta_m, (\alpha/\wp_2)_m = (\gamma/\wp_2)_m = \zeta_m^{m-1}$, and therefore $(\alpha/\wp_1\wp_2)_m = 1$. But then $x^m \equiv \alpha \pmod{\wp_1\wp_2}$ has no solution because neither $x^m \equiv \alpha \pmod{\wp_1}$ nor $x^m \equiv \alpha \pmod{\wp_2}$ has a solution.

Exercise 9.2.7 If $\alpha \in \mathbb{Z}[\zeta_\ell]$ is coprime to ℓ, show that there is an integer $c \in \mathbb{Z}$ (unique mod ℓ) such that $\zeta_\ell^c\alpha$ is primary.

Solution. Let $\lambda = 1 - \zeta_\ell$. Since the prime ideal (λ) has degree 1, there is an $a \in \mathbb{Z}$ such that $\alpha \equiv a \pmod{\lambda}$. Hence $(\alpha - a)/\lambda \equiv b \pmod{\lambda}$. So we can write $\alpha \equiv a + b\lambda \pmod{\lambda^2}$. Since $\zeta_\ell = 1 - \lambda$, we have $\zeta_\ell^c \equiv 1 - c\lambda \pmod{\lambda^2}$. Thus,

$$\zeta_\ell^c\alpha \equiv (1 - c\lambda)(a + b\lambda) \equiv a + (b - ac)\lambda \pmod{\lambda^2}.$$

Now, $(a, \ell) = 1$ for otherwise $\lambda \mid \alpha$ contrary to assumption. Choose c so that $ac \equiv b \pmod{\lambda}$. Then $\zeta_\ell^c\alpha \equiv a \pmod{\lambda^2}$. Moreover c is clearly unique mod ℓ.

Exercise 9.2.8 With notation as above, show that $(x+\zeta^i y)$ and $(x+\zeta^j y)$ are coprime in $\mathbb{Z}[\zeta_\ell]$ whenever $i \neq j$, $0 \leq i,j < \ell$.

Solution. Suppose $\mathfrak{a}$ is an ideal of $\mathbb{Z}[\zeta_\ell]$ which contains both $(x+\zeta^i y)$ and $(x+\zeta^j y)$. Then $(\zeta^j - \zeta^i)y \in \mathfrak{a}$ and

$$(\zeta^j - \zeta^i)x = \zeta^j(x+\zeta^i y) - \zeta^i(x+\zeta^j y) \in \mathfrak{a}.$$

Since x and y are coprime, we deduce that $(\zeta^j - \zeta^i) \in \mathfrak{a}$. By Exercise 4.3.7 or by 4.5.9, we deduce that $\mathfrak{a}$ and (ℓ) are not coprime. Since (ℓ) is totally ramified and $(\ell) = (1-\zeta)^{\ell-1}$ we see that $(1-\zeta) = \lambda \in \mathfrak{a}$. Thus $(z) \in (\lambda)$ which implies $\ell \mid z$ contrary to assumption.

Exercise 9.2.9 Show that the ideals $(x+\zeta^i y)$ are perfect ℓth powers.

Solution. Since the ideals $(x+\zeta^i y)$, $1 \leq i \leq \ell - 1$, are mutually coprime, and their product is an ℓth power, each ideal must be an ℓth power (see Exercise 5.3.12).

Exercise 9.2.10 Consider the element

$$\alpha = (x+y)^{\ell-2}(x+\zeta y).$$

Show that:

(a) the ideal (α) is a perfect ℓth power.

(b) $\alpha \equiv 1 - u\lambda \pmod{\lambda^2}$ where $u = (x+y)^{\ell-2}y$.

Solution. (a) is immediate from the previous exercise. To prove (b), observe that $x + \zeta y = x + y - \lambda y$. Thus

$$\alpha = (x+y)^{\ell-1} - \lambda y(x+y)^{\ell-2} = (x+y)^{\ell-1} - \lambda u.$$

Now $x^\ell + y^\ell + z^\ell \equiv x + y + z \pmod{\ell}$, by Fermat's little Theorem. If $\ell \mid (x+y)$, then $\ell \mid z$, contrary to assumption. Therefore

$$(x+y)^{\ell-1} \equiv 1 \pmod{\ell}$$

since $\ell \nmid (x+y)$. Since $(\ell) = (\lambda)^{\ell-1}$ we find $\alpha \equiv 1 - u\lambda \pmod{\lambda^2}$ which gives (b).

Exercise 9.2.11 Show that $\zeta^{-u}\alpha$ is primary.

Solution. We have

$$\zeta^{-u}\alpha = (1-\lambda)^{-u}\alpha \equiv (1+u\lambda)(1-u\lambda) \pmod{\lambda^2}$$

so that $\zeta^{-u}\alpha \equiv 1 \pmod{\lambda^2}$. Hence, $\zeta^{-u}\alpha$ is primary.

Exercise 9.2.12 Use Eisenstein reciprocity to show that if $x^\ell + y^\ell + z^\ell = 0$ has a solution in integers, $\ell \nmid xyz$, then for any $p \mid y$, $(\zeta/p)_\ell^{-u} = 1$. (Hint: Evaluate $(p/\zeta^{-u}\alpha)_\ell$.)

Solution. By Exercise 9.2.11, $\zeta^{-u}\alpha$ is primary so by Eisenstein reciprocity,

$$\left(\frac{p}{\zeta^{-u}\alpha}\right)_\ell = \left(\frac{\zeta^{-u}\alpha}{p}\right)_\ell = \left(\frac{\zeta}{p}\right)_\ell^{-u}\left(\frac{\alpha}{p}\right)_\ell.$$

Now $(\zeta^{-u}\alpha) = (\alpha)$ is an ℓth power by Exercise 9.2.10 (a). So the left-hand side of the above equation is 1. To evaluate $(\alpha/p)_\ell$, note that

$$\alpha \equiv (x+y)^{\ell-1} \pmod{p},$$

since $p \mid y$. Thus, since p is primary,

$$\left(\frac{\alpha}{p}\right)_\ell = \left(\frac{(x+y)^{\ell-1}}{p}\right)_\ell = \left(\frac{p}{(x+y)^{\ell-1}}\right)_\ell,$$

again by Eisenstein reciprocity. By Exercise 9.2.9, $(x+y)$ is an ℓth power of an ideal so that $(\alpha/p)_\ell = 1$. Therefore $(\zeta/p)_\ell^{-u} = 1$ for every $p \mid y$.

Exercise 9.2.13 Show that if

$$x^\ell + y^\ell + z^\ell = 0$$

has a solution in integers, $l \nmid xyz$, then for any $p \mid xyz$, $(\zeta/p)_\ell^{-u} = 1$.

Solution. We proved this for $p \mid y$ in the previous exercise. Since the equation is symmetric in x, y, z the same applies for $p \mid x$ or $p \mid z$.

Exercise 9.2.14 Show that $(\zeta/p)_\ell^{-u} = 1$ implies that $p^{\ell-1} \equiv 1 \pmod{\ell^2}$.

Solution. Let us factor $(p) = \wp_1 \cdots \wp_g$ as a product of prime ideals. We know that $N(\wp_i) = p^f$, and that $gf = \ell - 1$. Thus, by Exercises 9.2.4 and 9.2.5,

$$\left(\frac{\zeta}{p}\right)_\ell = \prod_{i=1}^{g}\left(\frac{\zeta}{\wp_i}\right)_\ell = \prod_{i=1}^{g}\zeta^{(N(\wp_i)-1)/\ell} = \zeta^{g(p^f-1)/\ell}.$$

Since $(\zeta/p)_\ell^u = 1$, we have

$$ug\frac{p^f-1}{\ell} \equiv 0 \pmod{\ell}.$$

Moreover, $(g,\ell) = 1$, $(u,\ell) = 1$, so that $p^f \equiv 1 \pmod{\ell^2}$. Since $f \mid \ell - 1$, we deduce that $p^{\ell-1} \equiv 1 \pmod{\ell^2}$.

Exercise 9.2.14 is a famous result of Furtwangler proved in 1912.

Exercise 9.2.15 If ℓ is an odd prime and

$$x^\ell + y^\ell + z^\ell = 0$$

for some integers x, y, z coprime to ℓ, then show that $p^{\ell-1} \equiv 1 \pmod{\ell^2}$ for every $p \mid xyz$. Deduce that $2^{\ell-1} \equiv 1 \pmod{\ell^2}$.

Solution. By Exercise 9.2.14, the first assertion is immediate. For the second part, observe that at least one of x, y, z must be even.

9.3 Supplementary Problems

Exercise 9.3.1 Show that there are infinitely many primes p such that $(2/p) = -1$.

Solution. We know that $(2/p) = -1$ if and only if $p \equiv \pm 3 \pmod 8$. Suppose there are only finitely many such primes $q_1, \ldots, q_k$ (say) excluding 3. Consider the number $b = 8q_1q_2\cdots q_k + 3$. By construction b is not divisible by any q_i. Moreover, $b \equiv 3 \pmod 8$ so that $(2/b) = -1$. Let $b = p_1 \cdots p_m$ be the prime decomposition of b with p_i not necessarily distinct. Since

$$-1 = \left(\frac{2}{b}\right) = \prod_{i=1}^{m} \left(\frac{2}{p_i}\right)$$

we must have $(2/p_i) = -1$ for some i. Since p_i is distinct from $q_1, \ldots, q_k$, this is a contradiction.

Exercise 9.3.2 Let a be a nonsquare integer greater than 1. Show that there are infinitely many primes p such that $(a/p) = -1$.

Solution. Without loss of generality, we may suppose a is squarefree and greater than 2 by the previous exercise. Suppose there are only finitely many primes $q_1, \ldots, q_k$, (say) such that $(a/q_i) = -1$. (This set could possibly be empty.) Write $a = 2^e r_1 \cdots r_m$ for the prime factorization of a with $e = 0$ or 1 and r_i odd and distinct. By the Chinese Remainder Theorem, we can find a solution to the simultaneous congruences

$$\begin{aligned} x &\equiv 1 \pmod{q_i}, \quad 1 \leq i \leq k, \\ x &\equiv 1 \pmod 8, \\ x &\equiv 1 \pmod{r_i}, \quad 1 \leq i \leq m-1, \\ x &\equiv c \pmod{r_m}, \end{aligned}$$

with c any nonresidue mod r_m. (It is here that we are assuming a has at least one odd prime divisor.) Let b be a solution greater than 1 and write $b = p_1 \cdots p_t$ as its prime decomposition with p_i not necessarily distinct.

Since $b \equiv 1 \pmod 8$, $(2/b) = 1$. Also, by the quadratic reciprocity law for the Jacobi symbol, $(r_i/b) = (b/r_i)$. Thus

$$\left(\frac{a}{b}\right) = \left(\frac{2}{b}\right)^e \prod_{i=1}^{m} \left(\frac{r_i}{b}\right) = \left(\frac{c}{r_m}\right) = -1.$$

Also,

$$\left(\frac{a}{b}\right) = \prod_{i=1}^{t} \left(\frac{a}{p_i}\right).$$

Therefore, $(a/p_i) = -1$ for some p_i. Moreover, $p_i \mid b$ and b is coprime to $q_1, \ldots, q_k$ by construction. This is a contradiction.

Exercise 9.3.3 Suppose that $x^2 \equiv a \pmod p$ has a solution for all but finitely many primes. Show that a is a perfect square.

Solution. Write $a = b^2 c$ with c squarefree. Then $(c/p) = 1$ for all but finitely many primes. By the previous exercise, this is not possible if $c > 1$. Thus, $c = 1$.

Exercise 9.3.4 Let K be a quadratic extension of $\mathbb{Q}$. Show that there are infinitely many primes which do not split completely in K.

Solution. Let $K = \mathbb{Q}(\sqrt{D})$, with D squarefree and greater than 1. By the previous question, there are infinitely many primes p such that $x^2 \equiv D \pmod p$ has no solution. Hence $(D/p) = -1$. By the theory of the Kronecker symbol, we deduce that p does not split in K.

Exercise 9.3.5 Suppose that a is an integer coprime to the odd prime q. If $x^q \equiv a \pmod p$ has a solution for all but finitely many primes, show that a is a perfect qth power. (This generalizes the penultimate exercise.)

Solution. We must show that if a is not a qth power, then there are infinitely many primes p such that $x^q \equiv a \pmod p$ has no solution. We will work in the field $K = \mathbb{Q}(\zeta_q)$. Write

$$(a) = \wp_1^{e_1} \cdots \wp_n^{e_n},$$

where the $\wp_i$ are distinct prime ideals of $\mathcal{O}_K$. We claim that $q \nmid e_i$ for some i. To see this, let $p_i = \wp_i \cap \mathbb{Z}$. Since $(q, a) = 1$ we have $q \neq p_i$ for any i, so each p_i is unramified in $\mathcal{O}_K$. Thus

$$\operatorname{ord}_{p_i} a = \operatorname{ord}_{\wp_i}(a) = e_i.$$

If $q \mid e_i$ for all i, then a would be a qth power. So we may suppose $q \nmid e_n$. Now let $\mathfrak{q}_1, \ldots, \mathfrak{q}_k$ be a finite set of prime ideals different from $\wp_1, \ldots, \wp_n$

and $(1-\zeta_q)$. By the Chinese Remainder Theorem (Exercise 5.3.13), we can find an $x \in \mathcal{O}_K$ satisfying the simultaneous congruences

$$\begin{aligned} x &\equiv 1 \pmod{\mathfrak{q}_i}, \quad 1 \le i \le k, \\ x &\equiv 1 \pmod{q}, \\ x &\equiv 1 \pmod{\wp_j}, \quad 1 \le j \le n-1, \\ x &\equiv c \pmod{\wp_n}, \end{aligned}$$

where c is chosen so that $(c/\wp_n) = \zeta_q$. Let b be a solution greater than 1. Then $b \equiv 1 \pmod{q}$ and hence is primary. Hence

$$\left(\frac{a}{b}\right)_q = \left(\frac{b}{a}\right)_q$$

by the Eisenstein reciprocity law. Now

$$\left(\frac{b}{a}\right)_q = \prod_{i=1}^{n} \left(\frac{b}{\wp_i}\right)_q^{e_i} = \zeta_q^{e_n} \neq 1,$$

since $q \nmid e_n$. On the other hand, factoring b into a product of prime ideals and using the multiplicativity of the qth power residue symbol, we deduce that

$$\left(\frac{a}{\wp}\right)_q \neq 1$$

for some prime ideal $\wp$ dividing b. Hence $x^q \equiv a \pmod{\wp}$ has no solution. Since b is coprime to the given set $\mathfrak{q}_1, \dots, \mathfrak{q}_k$ of prime ideals (possibly empty), we can produce inductively infinitely many prime ideals $\wp$ such that $x^q \equiv a \pmod{\wp}$ has no solution. A fortiori, there are infinitely many primes p such that $x^q \equiv a \pmod{p}$ has no solution.

Exercise 9.3.6 Let $p \equiv 1 \pmod{3}$. Show that there are integers A and B such that

$$4p = A^2 + 27B^2.$$

A and B are unique up to sign.

Solution. We work in the ring of Eisenstein integers $\mathbb{Z}[\rho]$. Since $p \equiv 1 \pmod{3}$, p splits as $\pi\overline{\pi}$ in $\mathbb{Z}[\rho]$. Writing $\pi = a + b\rho$, we see that

$$p = a^2 - ab + b^2.$$

Thus,

$$4p = (2a-b)^2 + 3b^2 = (2b-a)^2 + 3a^2 = (a+b)^2 + 3(a-b)^2.$$

A simple case by case examination $\pmod{3}$ shows that one of a, b or $a-b$ is divisible by 3 because $p \equiv 1 \pmod{3}$. The uniqueness is evident from the uniqueness of $p = a^2 - ab + b^2$.

Exercise 9.3.7 Let $p \equiv 1 \pmod 3$. Show that $x^3 \equiv 2 \pmod p$ has a solution if and only if $p = C^2 + 27D^2$ for some integers C, D.

Solution. If $x^3 \equiv 2 \pmod p$ has a solution, so does $x^3 \equiv 2 \pmod \pi$ where $p = \pi\overline{\pi}$ is the factorization of p in $\mathbb{Z}[\rho]$. By Exercise 9.1.17, $\pi \equiv 1 \pmod 2$, and we write $\pi = a + b\rho$. By the previous exercise, we can assume

$$p = a^2 - ab + b^2$$

with $3 \mid b$ (without loss of generality). Thus, $\pi \equiv 1 \pmod 2$ implies that $a \equiv 1 \pmod 2$, $b \equiv 0 \pmod 2$. Writing

$$4p = A^2 + 27B^2,$$

we have $A = 2a-b$, $B = b/3$. Since B is even, so is A. Thus, $p = C^2+27D^2$. Conversely, if $p = C^2 + 27D^2$, then $4p = (2C)^2 + 27(2D)^2$. By uniqueness, $B = \pm 2D$, by the previous exercise. Thus B is even and so is b. Therefore $\pi = a + b\rho \equiv 1 \pmod 2$. By Exercise 9.1.7, $x^3 \equiv 2 \pmod \pi$ has a solution in $\mathbb{Z}[\rho]$. Since $\mathbb{Z}[\rho]/(\pi)$ has p elements, we can find an integer $y \equiv x \pmod \pi$. Thus $y^3 \equiv 2 \pmod \pi$. But then $y^3 \equiv 2 \pmod{\overline{\pi}}$ so that $y^3 \equiv 2 \pmod p$.

Exercise 9.3.8 Show that the equation

$$x^3 - 2y^3 = 23z^m$$

has no integer solutions with $\gcd(x, y, z) = 1$.

Solution. Reduce the equation mod 23 to find $x^3 \equiv 2y^3 \pmod{23}$. If $23 \nmid y$, 2 is a cubic residue $\pmod{23}$. By the previous exercise, we can write

$$23 = C^2 + 27D^2$$

which is not possible. If $23 \mid y$, then $23 \mid x$ and then $23 \mid \gcd(x, y, z)$.

Chapter 10

Analytic Methods

10.1 The Riemann and Dedekind Zeta Functions

Exercise 10.1.1 Show that for $\text{Re}(s) > 1$,

$$\zeta(s) = \prod_p \left(1 - \frac{1}{p^s}\right)^{-1},$$

where the product is over prime numbers p.

Solution. Since every natural number can be factored uniquely as a product of prime powers, it is clear that when we expand the product

$$\prod_p \left(1 + \frac{1}{p^s} + \frac{1}{p^{2s}} + \frac{1}{p^{3s}} + \cdots\right)$$

the term $1/n^s$ occurs exactly once. The assertion is now evident.

Exercise 10.1.2 Let K be an algebraic number field and $\mathcal{O}_K$ its ring of integers. The Dedekind zeta function $\zeta_K(s)$ is defined for $\text{Re}(s) > 1$ as the infinite series

$$\zeta_K(s) = \sum_{\mathfrak{a}} \frac{1}{(N\mathfrak{a})^s},$$

where the sum is over all ideals of $\mathcal{O}_K$. Show that the infinite series is absolutely convergent for $\text{Re}(s) > 1$.

Solution. For any s with $\text{Re}(s) > 1$, it suffices to show that the partial sums

$$\sum_{N\mathfrak{a} \le x} \frac{1}{(N\mathfrak{a})^s}$$

are bounded. Indeed, since any ideal can be expressed as a product of powers of prime ideals, it is evident that

$$\sum_{N\mathfrak{a}\le x} \frac{1}{(N\mathfrak{a})^\sigma} \le \prod_{N\wp\le x} \left(1 + \frac{1}{(N\wp)^\sigma} + \frac{1}{(N\wp)^{2\sigma}} + \cdots\right),$$

where $\sigma = \mathrm{Re}(s) > 1$. Hence

$$\sum_{N\mathfrak{a}\le x} \frac{1}{(N\mathfrak{a})^\sigma} \le \prod_{N\wp\le x} \left(1 - \frac{1}{(N\wp)^\sigma}\right)^{-1}.$$

For each prime ideal $\wp$, we have a unique prime number p such that $N(\wp) = p^f$ for some integer f. Moreover, there are at most $[K : \mathbb{Q}]$ prime ideals corresponding to the same prime p. In fact, they are determined from the factorization $p\mathcal{O}_K = \wp_1^{e_1} \cdots \wp_g^{e_g}$ and by Exercise 5.3.17, $\sum_{i=1}^g e_i f_i = [K : \mathbb{Q}]$ where $N\wp_i = p^{f_i}$. Since $e_i \ge 1$ and $f_i \ge 1$, we find $g \le [K : \mathbb{Q}]$. Hence

$$\sum_{N\mathfrak{a}\le x} \frac{1}{(N\mathfrak{a})^\sigma} \le \prod_{p\le x} \left(1 - \frac{1}{p^\sigma}\right)^{-[K:\mathbb{Q}]}.$$

Since the product $\prod_p (1 - p^{-\sigma})^{-1}$ converges absolutely for $\sigma > 1$, the result follows.

Exercise 10.1.3 Prove that for $\mathrm{Re}(s) > 1$,

$$\zeta_K(s) = \prod_{\wp} \left(1 - \frac{1}{(N\wp)^s}\right)^{-1}.$$

Solution. Since every ideal $\mathfrak{a}$ can be written uniquely as a product of prime ideals (see Theorem 5.3.6), we find that when the product

$$\prod_{\wp} \left(1 + \frac{1}{(N\wp)^s} + \frac{1}{(N\wp)^{2s}} + \cdots\right)$$

is expanded, $1/(N\mathfrak{a})^s$ occurs exactly once for each ideal $\mathfrak{a}$ of $\mathcal{O}_K$.

Exercise 10.1.5 Show that $(s-1)\zeta(s)$ can be extended analytically for $\mathrm{Re}(s) > 0$.

Solution. We apply Theorem 10.1.4 with $a_m = 1$. Then $A(x) = [x]$, the greatest integer less than or equal to x. Thus,

$$\sum_{n=1}^\infty \frac{1}{n^s} = s \int_1^\infty \frac{[x]\, dx}{x^{s+1}}.$$

Writing $[x] = x - \{x\}$ we obtain for $\mathrm{Re}(s) > 1$,

$$\zeta(s) = \frac{s}{s-1} - s\int_1^\infty \frac{\{x\}\,dx}{x^{s+1}}.$$

Since $\{x\}$ is bounded by 1, the latter integral converges for $\mathrm{Re}(s) > 0$. Thus, $(s-1)\zeta(s)$ is analytic for $\mathrm{Re}(s) > 0$.

Exercise 10.1.6 Evaluate

$$\lim_{s\to 1}(s-1)\zeta(s).$$

Solution. From the previous exercise we have

$$\lim_{s\to 1}(s-1)\zeta(s) = \lim_{s\to 1} s - \lim_{s\to 1} s(s-1)\int_1^\infty \frac{\{x\}\,dx}{x^{s+1}}$$

and the latter limit is zero since the integral is bounded. Hence, the desired limit is 1.

Exercise 10.1.8 For $K = \mathbb{Q}(i)$, evaluate

$$\lim_{s\to 1+}(s-1)\zeta_K(s).$$

Solution. Clearly, this limit is $\pi/4$.

Exercise 10.1.9 Show that the number of integers (a, b) with $a > 0$ satisfying $a^2 + Db^2 \leq x$ is

$$\frac{\pi x}{2\sqrt{D}} + O(\sqrt{x}).$$

Solution. Corresponding to each such (a, b) we associate $(a, \sqrt{D}b)$ which lies inside the circle $u^2 + v^2 \leq x$. We now count these "lattice" points.

We will call $(a, \sqrt{D}b)$ internal if $(a+1)^2 + D(b+1)^2 \leq x$. Otherwise, call it a boundary lattice point. Let I be the number of internal lattice points, and B the number of boundary lattice points. Each lattice point has area $\sqrt{D}$. Thus

$$\sqrt{D}I \leq \frac{\pi}{2}x \leq \sqrt{D}(I+B)$$

since in our count $a > 0$ and $b \in \mathbb{Z}$. A little reflection shows that any boundary point is contained in the annulus

$$\left(\sqrt{x} - \sqrt{D+1}\right)^2 \leq u^2 + v^2 \leq \left(\sqrt{x} + \sqrt{D+1}\right)^2$$

which has area $O(\sqrt{xD})$. Thus

$$\sqrt{D}B = O(\sqrt{xD})$$

and we get $B = O(\sqrt{x})$. Thus,

$$I = \frac{\pi x}{2\sqrt{D}} + O(\sqrt{x}).$$

Exercise 10.1.10 Suppose $K = \mathbb{Q}(\sqrt{-D})$ where $D > 0$ and $-D \not\equiv 1 \pmod 4$ and $\mathcal{O}_K$ has class number 1. Show that $(s-1)\zeta_K(s)$ extends analytically to $\mathrm{Re}(s) > \frac{1}{2}$ and find

$$\lim_{s\to 1}(s-1)\zeta_K(s).$$

(Note that there are only finitely many such fields.)

Solution. Each ideal of $\mathcal{O}_K$ is principal, of the form $(a + b\sqrt{D})$. We may choose $a > 0$. Thus

$$\zeta_K(s) = \sum_{\substack{a>0 \\ b\in\mathbb{Z}}} \frac{1}{(a^2+Db^2)^s} = \sum_{n=1}^{\infty} \frac{a_n}{n^s},$$

where a_n is the number of solutions of $a^2 + Db^2 = n$ with $a > 0, b \in \mathbb{Z}$. By Theorem 10.1.4, we have

$$\zeta_K(s) = s\int_1^{\infty} \frac{A(x)}{x^{s+1}}\,dx,$$

where $A(x) = \sum_{n\le x} a_n$. By the previous exercise,

$$A(x) = \frac{\pi x}{2\sqrt{D}} + O(\sqrt{x})$$

so that

$$\zeta_K(s) = \frac{\pi s}{2\sqrt{D}(s-1)} + s\int_1^{\infty} \frac{E(x)\,dx}{x^{s+1}},$$

where $E(x) = O(\sqrt{x})$. The latter integral converges for $\mathrm{Re}(s) > \frac{1}{2}$. This gives the desired analytic continuation. Moreover,

$$\lim_{s\to 1}(s-1)\zeta_K(s) = \frac{\pi}{2\sqrt{D}} = \frac{\pi}{\sqrt{|d_K|}}.$$

(In the next section, we will establish a similar result for any quadratic field K.)

10.2 Zeta Functions of Quadratic Fields

Exercise 10.2.1 Let $K = \mathbb{Q}(\sqrt{d})$ with d squarefree, and denote by a_n the number of ideals in $\mathcal{O}_K$ of norm n. Show that a_n is multiplicative. (That is, prove that if $(n, m) = 1$, then $a_{nm} = a_n a_m$.)

Solution. Let $\mathfrak{a}$ be an ideal of norm n and let

$$n = \prod_{i=1}^{k} p_i^{\alpha_i}$$

be the unique factorization of n into distinct prime powers. Then by unique factorization of ideals,

$$\mathfrak{a} = \prod_{j=1}^{s} \wp_j^{e_j}.$$

We see immediately that

$$a_n = \prod_{i=1}^{k} a_{p_i^{\alpha_i}}.$$

This implies that a_n is multiplicative.

Exercise 10.2.2 Show that for an odd prime p, $a_p = 1 + (d/p)$.

Solution. By Theorem 7.4.2, we see that an odd prime p splits in $\mathbb{Q}(\sqrt{d})$ if and only if $(d/p) = 1$, in which case there are two ideals of norm p. If p does not split, there are no ideals of norm p. Finally, if p ramifies, then $(d/p) = 0$ and there is only one ideal of norm p, by Exercise 7.4.3.

Exercise 10.2.3 Let d_K be the discriminant of $K = \mathbb{Q}(\sqrt{d})$. Show that for all primes p, $a_p = 1 + (d_K/p)$.

Solution. Since $d_K = d$ or $4d$, the result is clear for odd primes p from the previous exercise. We therefore need only consider $p = 2$. If $2 \mid d_K$, then by Theorem 7.4.5, 2 ramifies and there is only one ideal of norm 2. If $2 \nmid d_K$, then

$$\left(\frac{d_K}{2}\right) = \begin{cases} 1 & \text{if } d_K \equiv 1 \pmod 8, \\ -1 & \text{if } d_K \equiv 5 \pmod 8, \end{cases}$$

by the definition of the Kronecker symbol. The result is now immediate from Theorem 7.4.5.

Exercise 10.2.4 Show that for all primes p,

$$a_{p^\alpha} = \sum_{j=1}^{\alpha} \left(\frac{d_K}{p^j}\right) = \sum_{\delta \mid p^\alpha} \left(\frac{d_K}{\delta}\right).$$

Solution. The norm of any prime ideal is either p or p^2, the latter occurring if and only if $(d/p) = -1$. Thus for $\alpha = 2$, the formula is established and for $\alpha = 1$, the previous exercise applies. If $(d/p) = -1$, then clearly $a_{p^\alpha} = 0$ if α is odd and if α is even, then there is only one ideal of norm p^α. If p splits, then any ideal of norm p^α must be of the form

$$\wp^j (\wp')^{\alpha - j}$$

for some j, where $p\mathcal{O}_K = \wp\wp'$. It is now clear that $a_{p^\alpha} = \alpha + 1$ which is the sum

$$\sum_{j=0}^{\alpha} \left(\frac{d_K}{p^j}\right).$$

Finally, if p ramifies, there is only one ideal of norm p^α, namely $\wp^\alpha$ where $p\mathcal{O}_K = \wp^2$. This completes the proof.

Exercise 10.2.5 Prove that

$$a_n = \sum_{\delta | n} \left(\frac{d_K}{\delta} \right).$$

Solution. Since a_n is multiplicative,

$$a_n = \prod_{p^\alpha \| n} \left(\sum_{j=0}^{\alpha} \left(\frac{d_K}{p} \right) \right)$$

by the previous exercise. The result is now immediate upon expanding the product.

Exercise 10.2.6 Let d_K be the discriminant of the quadratic field K. Show that there is an $n > 0$ such that $(d_K/n) = -1$.

Solution. We know that $d_K \equiv 0$ or $1 \pmod 4$. If $d_K \equiv 1 \pmod 4$, then for any odd n we have

$$\left(\frac{d_K}{n} \right) = \left(\frac{n}{|d_K|} \right).$$

Let $|d_K| = pa$ where p is an odd prime. Since d_K is squarefree, $p \nmid a$. Let u be a quadratic nonresidue modulo p. We can find an odd $n \equiv u \pmod p$ and $n \equiv 1 \pmod{2a}$ by the Chinese Remainder Theorem. Then

$$\left(\frac{d_K}{n} \right) = \left(\frac{n}{p} \right) \left(\frac{n}{a} \right) = \left(\frac{u}{p} \right) \left(\frac{1}{a} \right) = -1,$$

as desired. If d_K is even, let $d_K = d_1 d_2$ where d_1 is 4 or 8 and d_2 is an odd discriminant. Then

$$\left(\frac{d_K}{n} \right) = \left(\frac{d_1}{n} \right) \left(\frac{d_2}{n} \right)$$

by definition of the Kronecker symbol. Since $d_1 = 4$ or 8, it is easy to find an a such that $(d_1/a) = -1$. Choose $n \equiv a \pmod{d_1}$, $n \equiv 1 \pmod{d_2}$. Then $(d_K/n) = -1$, as desired.

Exercise 10.2.7 Show that

$$\left| \sum_{n \le x} \left(\frac{d_K}{n} \right) \right| \le |d_K|.$$

Solution. Let

$$S = \sum_{\substack{n \bmod |d_K| \\ (n,d_K)=1}} \left(\frac{d_K}{n}\right).$$

Choose n_0 such that $(n_0, |d_K|) = 1$ and $(d_K/n_0) = -1$. (This is possible by the previous exercise.) Then

$$\left(\frac{d_K}{n_0}\right) S = \sum_{\substack{n \bmod |d_K| \\ (n,|d_K|)=1}} \left(\frac{d_K}{nn_0}\right).$$

As n ranges over residue classes mod $|d_K|$ so does nn_0. Hence

$$-S = \left(\frac{d_K}{n_0}\right) S = S$$

so that $S = 0$.

Now define v by

$$v|d_K| \leq x < (v+1)|d_K|.$$

Then,

$$\sum_{n \leq x} \left(\frac{d_K}{n}\right) = \sum_{|d_K|v \leq n < x} \left(\frac{d_K}{n}\right)$$

since

$$\sum_{n<|d_K|v} \left(\frac{d_K}{n}\right) = \sum_{j=1}^{v} \left(\sum_{(j-1)|d_K|<n<j|d_K|} \left(\frac{d_K}{n}\right)\right)$$

and the inner sum is zero because it is equal to S. Thus

$$\left|\sum_{n \leq x} \left(\frac{d_K}{n}\right)\right| \leq |d_K|.$$

Exercise 10.2.10 If K is a quadratic field, show that $(s-1)\zeta_K(s)$ extends to an analytic function for $\text{Re}(s) > \frac{1}{2}$.

Solution. By Theorem 10.1.4,

$$\zeta_K(s) = \frac{cs}{s-1} + s\int_1^\infty \frac{E(x)\,dx}{x^{s+1}},$$

where $E(x) = O(\sqrt{x})$. The integral therefore converges for $\text{Re}(s) > \frac{1}{2}$.

10.3 Dirichlet's L-Functions

Exercise 10.3.1 Show that $L(s,\chi)$ converges absolutely for $\mathrm{Re}(s) > 1$.

Solution. Since $|\chi(n)| \leq 1$, we have

$$\left|\sum_{n=1}^{\infty} \frac{\chi(n)}{n^s}\right| \leq \sum_{n=1}^{\infty} \frac{1}{n^{\sigma}},$$

where $\sigma = \mathrm{Re}(s)$. The latter series converges absolutely for $\mathrm{Re}(s) > 1$.

Exercise 10.3.2 Prove that

$$\left|\sum_{n \leq x} \chi(n)\right| \leq m.$$

Solution. If χ is nontrivial, there is an $a \pmod m$ coprime to m such that $\chi(a) \neq 1$. Then

$$\sum_{\substack{b \bmod m \\ (b,m)=1}} \chi(b) = \sum_{\substack{b \bmod m \\ (b,m)=1}} \chi(ab) = \chi(a) \sum_{\substack{b \bmod \mathrm{m} \\ (b,m)=1}} \chi(b).$$

Hence

$$\sum_{\substack{b \bmod m \\ (b,m)=1}} \chi(b) = 0.$$

Now, partition the interval $[1, x]$ into subintervals of length m and suppose that $km < x \leq (k+1)m$. Then

$$\sum_{n \leq x} \chi(n) = \sum_{n \leq km} \chi(n) + \sum_{km < n \leq x} \chi(n).$$

The first sum on the right-hand side is zero and the second sum is bounded by m .

Exercise 10.3.3 If χ is nontrivial, show that $L(s,\chi)$ extends to an analytic function for $\mathrm{Re}(s) > 0$.

Solution. By Theorem 10.1.4, this is now immediate.

Exercise 10.3.4 For $\mathrm{Re}(s) > 1$, show that

$$L(s,\chi) = \prod_{p} \left(1 - \frac{\chi(p)}{p^s}\right)^{-1}.$$

Solution. Since χ is completely multiplicative,

$$\begin{aligned} L(s,\chi) &= \prod_p \left(1+\frac{\chi(p)}{p^s}+\frac{\chi(p^2)}{p^{2s}}+\cdots\right) \\ &= \prod_p \left(1+\frac{\chi(p)}{p^s}+\frac{\chi(p)^2}{p^s}+\cdots\right) \\ &= \prod_p \left(1-\frac{\chi(p)}{p^s}\right)^{-1}. \end{aligned}$$

Exercise 10.3.5 Show that

$$\sum_{\chi \bmod m} \overline{\chi}(a)\chi(b) = \begin{cases} \varphi(m) & \text{if } a \equiv b \pmod m, \\ 0 & \text{otherwise.} \end{cases}$$

Solution. If $a \equiv b \pmod m$, the result is clear. If $a \not\equiv b \pmod m$, let ψ be a character such that $\psi(a) \neq \psi(b)$. Then

$$\begin{aligned} \sum_{\chi \bmod m} \psi(ba^{-1})\chi(ba^{-1}) &= \sum_{\chi \bmod m} (\psi\chi)(ba^{-1}) \\ &= \sum_{\chi \bmod m} \chi(ba^{-1}), \end{aligned}$$

because as χ ranges over characters mod m, so does $\psi\chi$. But

$$(1-\psi(ba^{-1})) \sum_{\chi \bmod m} \chi(ba^{-1}) = 0$$

so the result follows.

Exercise 10.3.6 For $\text{Re}(s) > 1$, show that

$$\sum_{\chi \bmod m} \log L(s,\chi) = \varphi(m) \sum_{p^n \equiv 1 \bmod m} \frac{1}{np^{ns}}.$$

Solution. By Exercises 10.3.3 and 10.3.5, we find

$$\begin{aligned} \sum_{\chi \bmod m} \log L(s,\chi) &= \sum_{n,p} \frac{1}{np^{ns}} \sum_{\chi \bmod m} \chi(p^n) \\ &= \varphi(m) \sum_{p^n \equiv 1 (\bmod m)} \frac{1}{np^{ns}}. \end{aligned}$$

Exercise 10.3.7 For $\text{Re}(s) > 1$, show that

$$\sum_{\chi \bmod m} \overline{\chi}(a) \log L(s,\chi) = \varphi(m) \sum_{p^n \equiv a \bmod m} \frac{1}{np^{ns}}.$$

Solution. By Exercise 10.3.3,

$$\sum_{\chi \bmod m} \overline{\chi}(a) \log L(s,\chi) = \sum_{n,p} \frac{1}{np^{ns}} \sum_{\chi \bmod m} \overline{\chi}(a)\chi(p^n)$$

since the series converges absolutely in $\mathrm{Re}(s) > 1$. By Exercise 10.3.5, the inner sum on the right-hand side is $\varphi(m)$ when $p^n \equiv a \pmod{m}$ and zero otherwise. The result is now immediate.

Exercise 10.3.8 Let $K = \mathbb{Q}(\zeta_m)$. Set

$$f(s) = \prod_{\chi} L(s,\chi).$$

Show that $\zeta_K(s)/f(s)$ is analytic for $\mathrm{Re}(s) > \frac{1}{2}$.

Solution. The primes that split completely in $\mathbb{Q}(\zeta_m)$ are those primes $p \equiv 1 \pmod{m}$. Thus, because there are $\phi(m)$ ideals of norm p for $p \equiv 1 \pmod{m}$,

$$\begin{aligned} \zeta_K(s) &= \prod_{\wp} \left(1 - \frac{1}{(N\wp)^s}\right)^{-1} \\ &= \prod_{p\equiv 1 \bmod m} \left(1 - \frac{1}{p^s}\right)^{-\phi(m)} g(s), \end{aligned}$$

where $g(s)$ is analytic for $\mathrm{Re}(s) > \frac{1}{2}$. By Exercise 10.3.6

$$\prod_{\chi} L(s,\chi) = \prod_{p\equiv 1 \bmod m} \left(1 - \frac{1}{p^s}\right)^{-\phi(m)} h(s),$$

where $h(s)$ is analytic and nonzero for $\mathrm{Re}(s) > \frac{1}{2}$ (Why?).

Thus, $\zeta_K(s)/f(s)$ is analytic for $\mathrm{Re}(s) > \frac{1}{2}$. This gives the analytic continuation of $\zeta_K(s)$ for $\mathrm{Re}(s) > \frac{1}{2}$. We can in fact show that

$$\zeta_K(s) = \prod_{\chi} L(s,\chi).$$

10.4 Primes in Arithmetic Progressions

Exercise 10.4.3 With the notation as in Section 10.3, write

$$f(s) = \prod_{\chi} L(s,\chi) = \sum_{n=1}^{\infty} \frac{c_n}{n^s}.$$

Show that $c_n \geq 0$.

Solution. This is immediate from Exercise 10.3.6 because

$$f(s) = \exp\left(\phi(m) \sum_{p^n \equiv 1 \bmod m} \frac{1}{np^{ns}}\right).$$

Exercise 10.4.4 With notation as in the previous exercise, show that

$$\sum_{n=1}^{\infty} \frac{c_n}{n^s}$$

diverges for $s = 1/\phi(m)$.

Solution. By Euler's theorem (Theorem 1.1.14), $p^{\phi(m)} \equiv 1 \pmod{m}$ for prime $p \nmid m$. Thus,

$$\begin{aligned} f(1/\phi(m)) &= \exp\left(\phi(m) \sum_{p^n \equiv 1 \bmod m} \frac{1}{np^{n/\phi(m)}}\right) \\ &\geq \exp\left(\sum_{p \nmid m} \frac{1}{p}\right). \end{aligned}$$

Since

$$\sum_p \frac{1}{p} = +\infty,$$

we are done (by Exercise 1.4.18).

Exercise 10.4.6 Show that

$$\sum_{p \equiv 1 \bmod m} \frac{1}{p} = +\infty.$$

Solution. By Exercise 10.3.6,

$$\sum_{p^n \equiv 1 \bmod m} \frac{1}{np^{ns}} = \frac{1}{\phi(m)} \sum_{\chi \bmod m} \log L(s, \chi).$$

As $s \to 1^+$, we see that

$$\sum_{p \equiv 1 \bmod m} \frac{1}{p} = +\infty$$

because $L(1, \chi) \neq 0$ for $\chi \neq \chi_0$ and

$$\lim_{s \to 1^+} \log L(s, \chi_0) = \lim_{s \to 1^+} \log\left(\zeta(s) \prod_{p \mid m}\left(1 - \frac{1}{p^s}\right)\right) = +\infty.$$

Note that

$$\sum_{\substack{n\geq 2\\ p}} \frac{1}{np^n} \leq \sum_p \frac{1}{p(p-1)} < \infty.$$

Exercise 10.4.7 Show that if $\gcd(a,m) = 1$, then

$$\sum_{p\equiv a \bmod m} \frac{1}{p} = +\infty.$$

Solution. By Exercise 10.3.7,

$$\sum_{p^n\equiv a \bmod m} \frac{1}{np^{ns}} = \frac{1}{\phi(m)} \sum_{\chi \bmod m} \overline{\chi}(a) \log L(s,\chi).$$

As $s \to 1^+$, we see that

$$\sum_{p\equiv a \bmod m} \frac{1}{p} = +\infty$$

again because $L(1,\chi) \neq 0$ for $\chi \neq \chi_0$ and

$$\lim_{s\to 1^+} \log L(s,\chi_0) = +\infty.$$

Hence there are infinitely many primes in any given coprime residue class mod m.

10.5 Supplementary Problems

Exercise 10.5.1 Define for each character χ (mod m) the Gauss sum

$$g(\chi) = \sum_{a \bmod m} \chi(a) e^{2\pi i a/m}.$$

If $(n,m) = 1$, show that

$$\chi(n) g(\overline{\chi}) = \sum_{b \bmod m} \overline{\chi}(b) e^{2\pi i b n/m}.$$

Solution.

$$\begin{aligned}
\chi(n) g(\overline{\chi}) &= \sum_{a \bmod m} \chi(n)\overline{\chi}(a) e^{2\pi i a/m} \\
&= \sum_{b \bmod m} \overline{\chi}(b) e^{2\pi i b n/m}
\end{aligned}$$

upon setting $a = bn$ in the first sum. Observe that as a ranges over coprime residue classes (mod m), so does bn since $(n,m) = 1$.

Exercise 10.5.2 Show that $|g(\chi)| = \sqrt{m}$.

Solution.

$$\sum_{n=1}^{m-1} |\chi(n)|^2 |g(\overline{\chi})|^2 = \sum_{n=0}^{m-1} \left| \sum_{b \bmod m} \overline{\chi}(b) e^{2\pi i b n/m} \right|^2$$
$$= \sum_{b_1, b_2} \overline{\chi}(b_1)\chi(b_2) \sum_{n=0}^{m-1} e^{2\pi i (b_1 - b_2)n/m}.$$

The last sum is the sum of a geometric progression and is 0 unless $b_1 \equiv b_2 \pmod{m}$ in which case it is m. Thus,

$$\phi(m)|g(\overline{\chi})|^2 = m\phi(m)$$

from which the result follows.

Exercise 10.5.3 Establish the Pólya–Vinogradov inequality:

$$\left| \sum_{n \le x} \chi(n) \right| \le \tfrac{1}{2} m^{1/2}(1 + \log m)$$

for any nontrivial character $\chi \pmod{m}$.

Solution. By the two previous exercises, we get

$$\sum_{n \le x} \chi(n) = g(\overline{\chi})^{-1} \sum_{n \le x} \left(\sum_{b \bmod m} \overline{\chi}(b) e^{2\pi i b n/m} \right).$$

Interchanging summation gives

$$\sum_{n \le x} e^{2\pi i b n/m} = \frac{e^{2\pi i b([x]+1)/m} - 1}{e^{2\pi i b/m} - 1},$$

provided $b \not\equiv 0 \pmod{m}$. Observe that the numerator is bounded by 2 and the denominator can be written as

$$e^{\pi i b/m}(e^{\pi i b/m} - e^{-\pi i b/m}) = 2ie^{\pi i b/m} \sin \frac{\pi b}{m}$$

so that for $b \not\equiv 0 \pmod{m}$

$$\left| \sum_{n \le x} e^{2\pi i b n/m} \right| \le \frac{1}{|\sin \frac{\pi b}{m}|}.$$

Since $|\sin \frac{\pi b}{m}| = |\sin \frac{\pi(m-b)}{m}|$, we may suppose $b \le m/2$. In that case,

$$\left| \sin \frac{\pi b}{m} \right| \ge \frac{2}{\pi} \left(\frac{\pi b}{m} \right)$$

because $\sin x \geq 2x/\pi$ for $0 \leq x \leq \pi/2$, as is seen by looking at the graph of $\sin x$. Therefore

$$\begin{aligned}\left|\sum_{n\leq x}\chi(n)\right| &\leq |g(\overline{\chi}|^{-1}\sum_{b \bmod m}\frac{m}{2b}\\ &\leq \frac{\sqrt{m}}{2}\log m.\end{aligned}$$

The last inequality follows by noting that

$$\sum_{b\leq m}\frac{1}{b}\leq 1+\int_1^m\frac{dt}{t}=1+\log m.$$

Exercise 10.5.4 Let p be prime. Let χ be a character $\bmod p$. Show that there is an $a \leq p^{1/2}(1+\log p)$ such that $\chi(a) \neq 1$.

Solution. If each $a \leq p^{1/2}(1+\log p) = u$ (say) satisfies $\chi(a) = 1$, then

$$\sum_{a\leq u}\chi(a)=u.$$

By the Pólya–Vinogradov inequality, the left-hand side is $\leq \frac{1}{2}p^{1/2}(1+\log p)$, which is a contradiction.

Exercise 10.5.5 Show that if χ is a nontrivial character $\bmod m$, then

$$L(1,\chi)=\sum_{n\leq u}\frac{\chi(n)}{n}+O\left(\frac{\sqrt{m}\log m}{u}\right).$$

Solution. By Theorem 10.1.4,

$$\sum_{n>u}\frac{\chi(n)}{n}=\int_1^\infty\frac{A(x)\,dx}{x^2},$$

where

$$A(x)=\begin{cases}0 & \text{if } x<u,\\ \sum_{u<n\leq x}\chi(n) & \text{if } x\geq u.\end{cases}$$

By Pólya–Vinogradov, $A(x) = O(\sqrt{m}\log m)$ and so

$$\sum_{n>u}\frac{\chi(n)}{n}=\int_u^\infty\frac{A(x)\,dx}{x^2}=O\left(\frac{\sqrt{m}\log m}{u}\right).$$

Therefore

$$L(1,\chi)=\sum_{n=1}^{\infty}\frac{\chi(n)}{n}=\sum_{n\leq u}\frac{\chi(n)}{n}+O\left(\frac{\sqrt{m}\log m}{u}\right).$$

Exercise 10.5.6 Let D be a bounded open set in $\mathbb{R}^2$ and let $N(x)$ denote the number of lattice points in xD. Show that

$$\lim_{x\to\infty} \frac{N(x)}{x^2} = \text{vol}(D).$$

Solution. Without loss of generality, we may translate our region by a lattice point to the first quadrant. A lattice point $(u, v) \in xD$ if and only if $(u/x, v/x) \in D$. This suggests we partition the plane into squares of length $1/x$ with sides parallel to the coordinate axes. This then partitions D into small squares each of area $1/x^2$. The number of lattice points of xD is then clearly the number of "interior squares." For this partition of the region, we write down the lower and upper Riemann sums. Let I_x denote the number of "interior" squares, and B_x the number of "boundary" squares. Then, by the definition of the Riemann integral

$$\frac{I_x}{x^2} \le \text{vol}(D) \le \frac{I_x + B_x}{x^2}$$

so that

$$\lim_{x\to\infty} \frac{I_x}{x^2} = \lim_{x\to\infty} \frac{I_x + B_x}{x^2} = \text{vol}(D).$$

On the other hand,

$$I_x \le N(x) \le I_x + B_x.$$

Thus,

$$\lim_{x\to\infty} \frac{N(x)}{x^2} = \text{vol}(D).$$

Exercise 10.5.7 Let K be an algebraic number field, and C an ideal class of K. Let $N(x, C)$ be the number of nonzero ideals of $\mathcal{O}_K$ belonging to C with norm $\le x$. Fix an integral ideal $\mathfrak{b}$ in C^{-1}. Show that $N(x, C)$ is the number of nonzero principal ideals (α) with $\alpha \in \mathfrak{b}$ with $|N_K(\alpha)| \le xN(\mathfrak{b})$.

Solution. For any $\mathfrak{a} \in C$, $\mathfrak{ab} = (\alpha)$ so that $(\alpha) \subseteq \mathfrak{b}$. Moreover $N_K(\alpha) = N(\mathfrak{a})N(\mathfrak{b})$. Conversely, if $\alpha \in \mathfrak{b}$, let $\mathfrak{a} = \mathfrak{b}^{-1}(\alpha)$ which is an integral ideal of norm $\le x$.

Exercise 10.5.8 Let K be an imaginary quadratic field, C an ideal class of $\mathcal{O}_K$, and d_K the discriminant of K. Prove that

$$\lim_{x\to\infty} \frac{N(x, C)}{x} = \frac{2\pi}{w\sqrt{|d_K|}},$$

where w is the number of roots of unity in K.

Solution. By the previous exercise, $wN(x, C)$ is the number of integers $\alpha \in \mathfrak{b}$ with $0 < |N_K(\alpha)| < x|N(\mathfrak{b})|$ where $\mathfrak{b} \in C^{-1}$ is fixed. Let β_1, β_2 be an integral basis of $\mathfrak{b}$, and let β_1', β_2' be the conjugates of β_1, β_2, respectively. Define

$$D = \{(u, v) \in \mathbb{R}^2 : 0 < |u\beta_1 + v\beta_2|^2 < 1\}.$$

Clearly D is bounded (as is easily verified). Then $wN(x,C)$ is the number of lattice points in $\sqrt{x}D$ so that (by the penultimate exercise)

$$\lim_{x\to\infty}\frac{wN(x,C)}{x} = \mathrm{vol}(D).$$

Set $u_1 = \mathrm{Re}(u\beta_1 + v\beta_2)$, $u_2 = \mathrm{Im}(u\beta_1 + v\beta_2)$ so that

$$\begin{aligned}\mathrm{vol}(D) &= \iint_{|u\beta_1+v\beta_2|^2<1} du\,dv \\ &= \frac{2}{N\mathfrak{b}\sqrt{|d_K|}}\cdot \iint_{u_1^2+u_2^2<1} du_1\,du_2 \\ &= \frac{2\pi}{N\mathfrak{b}\sqrt{|d_K|}}.\end{aligned}$$

Exercise 10.5.9 Let K be a real quadratic field with discriminant d_K, and fundamental unit ε. Let C be an ideal class of $\mathcal{O}_K$. Show that

$$\lim_{x\to\infty}\frac{N(x,C)}{x} = \frac{2\log\varepsilon}{\sqrt{d_K}},$$

where $N(x,C)$ denotes the number of integral ideals of norm $\leq x$ lying in the class C.

Solution. As in the previous two exercises, we fix an ideal $\mathfrak{b}\in C^{-1}$ and we are reduced to counting the number (α), with $(N_K(\alpha)) \leq xN\mathfrak{b}$. Therefore, fix an integral basis β_1, β_2 of $\mathfrak{b}$, β_1', β_2' denote the conjugates, respectively. Notice that we have infinitely many choices of α since $(\alpha) = (\varepsilon^m\alpha)$ for any integer m. Our first step is to isolate only one generator. Since α/α' is a unit of norm ±1, there is an integer m so that

$$-2m\log\varepsilon \leq \log\left|\frac{\alpha}{\alpha'}\right| < (-2m+2)\log\varepsilon.$$

Thus, setting $\omega = \varepsilon^m\alpha$, we find

$$0 \leq \log\left|\frac{\omega}{|N_K(\omega)^{1/2}|}\right| < \log\varepsilon.$$

If ω_1, ω_2 are associated elements of $\mathfrak{b}$ satisfying the same inequality, then the fact that $\omega_1 = \eta\omega_2$ for some unit η gives $1 \leq |\eta| < \varepsilon$. Thus, $\eta = \pm1$ because ε is a fundamental unit. Therefore $2N(x,C)$ is the number of $\omega\in\mathfrak{b}$ such that $0 < |N_K(\omega)| < (N\mathfrak{b})x$,

$$0 \leq \log\left|\frac{\omega}{|N_K(\omega)|^{1/2}}\right| < \log\varepsilon.$$

Now consider

$$D = \left\{ (u,v) \in \mathbb{R}^2 : \begin{array}{l} 0 < |u\beta_1 + v\beta_2||u\beta_1' + v\beta_2'| < 1, \\ 0 < \log \left| \frac{u\beta_1 + v\beta_2}{|u\beta_1 + v\beta_2|^{1/2}|u\beta_1' + v\beta_2'|^{1/2}} \right| < \log \varepsilon. \end{array} \right\}$$

Proceeding as in the previous question gives

$$\lim_{x\to\infty} \frac{2N(x,C)}{x} = \frac{4\log\varepsilon}{\sqrt{|d_K|}}.$$

Exercise 10.5.10 Let K be an imaginary quadratic field. Let $N(x;K)$ denote the number of integral ideals of norm $\leq x$. Show that

$$\lim_{x\to\infty} \frac{N(x;K)}{x} = \frac{2\pi h}{w\sqrt{|d_K|}},$$

where h denotes the class number of K.

Solution. Clearly,

$$N(x;K) = \sum_C N(x,C),$$

where the (finite) sum is over the ideal classes of K. Thus,

$$\lim_{x\to\infty} \frac{N(x;K)}{x} = \sum_C \lim_{x\to\infty} \frac{N(x,C)}{x}$$

and by the penultimate question,

$$\lim_{x\to\infty} \frac{N(x,C)}{x} = \frac{2\pi}{w\sqrt{|d_K|}}.$$

Exercise 10.5.11 Let K be a real quadratic field. Let $N(x,K)$ denote the number of integral ideals of norm $\leq x$. Show that

$$\lim_{x\to\infty} \frac{N(x;K)}{x} = \frac{2h\log\varepsilon}{\sqrt{|d_K|}},$$

where h is the class number of K.

Solution. This follows exactly as in the previous question except that we invoke the corresponding limit of $N(x,C)$ for the real quadratic case.

Exercise 10.5.12 (Dirichlet's Class Number Formula) Suppose that K is a quadratic field with discriminant d_K. Show that

$$\sum_{n=1}^{\infty} \left(\frac{d_K}{n}\right)\frac{1}{n} = \begin{cases} \frac{2\pi h}{w\sqrt{|d_K|}} & \text{if } d_K < 0, \\ \frac{2h\log\varepsilon}{\sqrt{|d_K|}} & \text{if } d_K > 0, \end{cases}$$

where h denotes the class number of K.

Solution. By Example 10.2.9, the number of integral ideals of norm $\leq x$ is $cx + O(\sqrt{x})$ where

$$c = \sum_{n=1}^{\infty} \left(\frac{d_K}{n}\right)\frac{1}{n}.$$

Thus,

$$\lim_{x\to\infty} \frac{N(x,K)}{x} = \sum_{n=1}^{\infty} \left(\frac{d_K}{n}\right)\frac{1}{n}.$$

Comparing this limit with the previous two questions gives the desired result.

Exercise 10.5.13 Let d be squarefree and positive. Using Dirichlet's class number formula, prove that the class number of $\mathbb{Q}(\sqrt{-d})$ is $O(\sqrt{d}\log d)$.

Solution. Let D be the discriminant of $\mathbb{Q}(\sqrt{-d})$. By Dirichlet's class number formula,

$$\frac{2\pi h}{w\sqrt{|D|}} = \sum_{n=1}^{\infty} \left(\frac{D}{n}\right)\frac{1}{n}.$$

Since $|D| = |d|$ or $4|d|$, it suffices to prove that

$$\sum_{n=1}^{\infty} \left(\frac{D}{n}\right)\frac{1}{n} = O(\log d).$$

By a previous exercise (Exercise 10.5.5), we have for any $u \geq 1$,

$$\sum_{n=1}^{\infty} \left(\frac{D}{n}\right)\frac{1}{n} = \sum_{n\leq u} \left(\frac{D}{n}\right)\frac{1}{n} + O\left(\frac{\sqrt{d}\log d}{u}\right),$$

which was derived using the Pólya-Vinogradov inequality. Choosing $u = \sqrt{d}$, and noting that

$$\sum_{n\leq\sqrt{d}} \left(\frac{D}{n}\right)\frac{1}{n} = O(\log d),$$

we obtain the result.

Exercise 10.5.14 Let d be squarefree and positive. Using Dirichlet's class number formula, prove that the class number h of $\mathbb{Q}(\sqrt{d})$ is $O(\sqrt{d})$.

Solution. Let D be the discriminant of $\mathbb{Q}(\sqrt{d})$. By Dirichlet's class number formula

$$\frac{2h\log\varepsilon}{\sqrt{|D|}} = \sum_{n=1}^{\infty} \left(\frac{D}{n}\right)\frac{1}{n},$$

where ε is the fundamental unit of $\mathbb{Q}(\sqrt{d})$. Since

$$\varepsilon = a + b\sqrt{d} \quad \text{or} \quad \frac{a + b\sqrt{d}}{2}$$

for integers a, b and $a^2 - db^2 = \pm 1$ or ± 4 we deduce that

$$\log \varepsilon \gg \log d.$$

Estimating the infinite series as in the previous question, the result is now immediate.

Exercise 10.5.15 With $\psi(x)$ defined (as in Chapter 1) by

$$\psi(x) = \sum_{p^\alpha \leq x} \log p,$$

prove that for $\mathrm{Re}(s) > 1$,

$$-\frac{\zeta'}{\zeta}(s) = s \int_1^\infty \frac{\psi(x)}{x^{s+1}} dx.$$

Solution. Taking logs of the identity

$$\zeta(s) = \prod_p \left(1 - \frac{1}{p^s}\right)^{-1}$$

we differentiate to obtain

$$-\frac{\zeta'}{\zeta}(s) = \sum_{n=1}^\infty \frac{\Lambda(n)}{n^s},$$

where $\Lambda(n) = \log p$ if $n = p^\alpha$ and 0 otherwise. It is clear that

$$\psi(x) = \sum_{n \leq x} \Lambda(n)$$

and so the result now follows by Theorem 10.1.4.

Exercise 10.5.16 If for any $\varepsilon > 0$,

$$\psi(x) = x + O(x^{1/2+\varepsilon}),$$

show that $\zeta(s) \neq 0$ for $\mathrm{Re}(s) > \frac{1}{2}$.

Solution. If the given estimate holds, we obtain an analytic continuation of

$$-\frac{\zeta'}{\zeta}(s)$$

for $\mathrm{Re}(s) > \frac{1}{2}$, apart from a simple pole at $s = 1$. Thus $\zeta(s)$ has no zeros in $\mathrm{Re}(s) > \frac{1}{2}$.

Chapter 11

Density Theorems

11.1 Counting Ideals in a Fixed Ideal Class

Exercise 11.1.1 Show that B_x is a bounded region in $\mathbb{R}^n$.

Solution. Since the integral basis $\beta_1, ..., \beta_n$ is linearly independent over $\mathbb{Q}$,

$$d_{K/\mathbb{Q}}(\beta_1, ..., \beta_n) = [\det(\beta_i^{(j)})]^2 \neq 0.$$

Thus the linear map

$$\phi(x_1, ..., x_n) = (\alpha^{(1)}, ..., \alpha^{(n)})$$

is invertible. Let M be the largest of the values $|\log|\epsilon_j^{(i)}||$ for $1 \leq i, j \leq r$. Then for each element of B_x, we have from the second relation defining B_x,

$$|\alpha^{(i)}| \leq e^{rM}(N(\mathfrak{b})x)^{1/n},$$

holds for $1 \leq i \leq n$. Therefore the image of ϕ is a bounded set and thus the inverse image is bounded.

Exercise 11.1.2 Show that $tB_1 = B_{t^n}$ for any $t > 0$.

Solution. There are two conditions defining B_x. The second one involving units is invariant under the homogenous change of variables. The first inequality gets multiplied by t^n.

Exercise 11.1.3 Show that $N(x, C) = O(x)$. Deduce that $N(x; K) = O(x)$.

Solution. We may write the x_i's in terms of the $\alpha^{(i)}$'s by inverting the transformation matrix, as noted in Exercise 11.1.1. As the $\alpha^{(i)}$'s are $O(x^{1/n})$, we deduce that the x_i's are $O(x^{1/n})$. The result is now immediate.

Exercise 11.1.6 Prove that $\zeta(s, C)$ extends to the region $\Re(s) > 1 - \frac{1}{n}$ except for a simple pole at $s = 1$ with residue

$$\frac{2^{r_1}(2\pi)^{r_2} R_K}{w\sqrt{|d_K|}}.$$

Deduce that $\zeta_K(s)$ extends to $\Re(s) > 1 - \frac{1}{n}$ except for a simple pole at $s = 1$ with residue

$$\rho_K := \frac{2^{r_1}(2\pi)^{r_2} h_K R_K}{w\sqrt{|d_K|}},$$

where h_K denotes the class number of K.

Solution. If we let $\alpha_K = \rho_K / h_K$, and consider the Dirichlet series

$$f(s) = \sum_{m=1}^{\infty} \frac{a_m}{m^s} := \zeta(s, C) - \alpha_K \zeta(s),$$

then by Theorem 11.1.5, we have

$$\sum_{m \leq x} a_m = O(x^{1-\frac{1}{n}}).$$

By Theorem 10.1.4, $f(s)$ converges for $\Re(s) > 1 - \frac{1}{n}$. As $\zeta(s)$ has a simple pole at $s = 1$ with residue 1, the latter assertions are immediate.

Exercise 11.1.7 Prove that there are infinitely many prime ideals in $\mathcal{O}_K$ which are of degree 1.

Solution. We have

$$\log \zeta_K(s) = \sum_{\wp} \frac{1}{N(\wp)^s} + \sum_{n \geq 2, \wp} \frac{1}{nN(\wp)^{ns}}.$$

The second sum is easily seen to converge for $\Re(s) > 1/2$. The first sum can be separated into two parts, one over primes of first degree and the other over primes of degree ≥ 2. Again, the second sum converges for $\Re(s) > 1/2$. If the first sum consisted of only finitely many terms, the right hand side would tend to a finite limit as $s \to 1^+$, which is not the case as the Dedekind zeta function has a simple pole at $s = 1$.

Exercise 11.1.8 Prove that the number of prime ideals $\wp$ of degree ≥ 2 and with norm $\leq x$ is $O(x^{1/2} \log x)$.

Solution. If $\wp$ is a prime ideal of degree $r \geq 2$, then $N(\wp) = p^r \leq x$ implies $r \leq (\log x)/\log 2$ and $p \leq x^{1/2}$. For each prime, there are a bounded number of prime ideals in K above p. Thus, the final estimate is obtained by counting the number of possible pairs (p, r) and this is $O(x^{1/2} \log x)$.

Exercise 11.1.9 Let μ be defined on integral ideals $\mathfrak{a}$ of $\mathcal{O}_K$ as follows. $\mu(\mathcal{O}_K) = 1$, and if $\mathfrak{a}$ is divisible by the square of a prime ideal, we set $\mu(\mathfrak{a}) = 0$. Otherwise, we let $\mu(\mathfrak{a}) = (-1)^k$ when $\mathfrak{a}$ is the product of k distinct prime ideals. Show that

$$\sum_{\mathfrak{b}|\mathfrak{a}} \mu(\mathfrak{b}) = 0$$

unless $\mathfrak{a} = \mathcal{O}_K$.

Solution. Clearly, the function

$$f(\mathfrak{a}) = \sum_{\mathfrak{b}|\mathfrak{a}} \mu(\mathfrak{b})$$

is multiplicative and so, it suffices to evaluate it on prime ideals. But this is clearly zero.

Exercise 11.1.10 Prove that the number of ideals of $\mathcal{O}_K$ of odd norm $\leq x$ is

$$\rho_K x \prod_{\wp|2}\left(1 - \frac{1}{N(\wp)}\right) + O(x^{1-\frac{1}{n}}),$$

where the product is over prime ideals $\wp$ of $\mathcal{O}_K$ dividing $2\mathcal{O}_K$.

Solution. By the previous exercise, the number of such ideals is

$$\sum_{N(\mathfrak{a})\leq x} \sum_{\mathfrak{b}|(\mathfrak{a},2)} \mu(\mathfrak{b})$$

since an ideal has odd norm if and only if it has no prime ideal divisor above 2. Interchanging summation and using Theorem 11.1.5, we obtain the result.

Exercise 11.1.11 Let $A(x)$ be the number of ideals of $\mathcal{O}_K$ of even norm $\leq x$ and $B(x)$ of odd norm $\leq x$. Show that

$$\lim_{x\to\infty} \frac{A(x)}{B(x)} = 1$$

if and only if $K = \mathbb{Q}$ or K is a quadratic field in which 2 ramifies.

Solution. From the previous exercise, we see that

$$B(x) = \prod_{\wp|2}\left(1 - \frac{1}{N(\wp)}\right)\rho_K x + O(x^{1-\frac{1}{n}}).$$

We see that $A(x) \sim B(x)$ if and only if $A(x) + B(x) \sim 2B(x)$. That is, if and only if

$$\frac{1}{2} = \prod_{\wp|2}\left(1 - \frac{1}{N(\wp)}\right).$$

Let $\wp_1, ..., \wp_t$ be the prime ideals above 2 with norms $2^{m_1}, ..., 2^{m_t}$ respectively. The above equation implies

$$\frac{1}{2} = \prod_{i=1}^{t} \left(\frac{2^{m_i} - 1}{2^{m_i}} \right).$$

In other words, we have

$$2^{m_1 + \cdots + m_t - 1} = \prod_{i=1}^{t} (2^{m_i} - 1).$$

The right hand side is a product of odd numbers and so the only way the left hand side can be odd is if $m_1 + \cdots + m_t = 1$ which means that there is only one prime ideal above 2 and it has norm 2. This can only happen if $K = \mathbb{Q}$ or if K is a quadratic field in which 2 ramifies.

Exercise 11.1.12 With notation as in the discussion preceding Theorem 11.1.4, let V_x denote the set of n-tuples $(x_1, ..., x_n)$ satisfying

$$|\alpha^{(1)} \cdots \alpha^{(n)}| \leq xN(\mathfrak{b}).$$

Let $t = x^{1/n}$. Show that there is a $\delta > 0$ (independent of x) such that for each lattice point P contained in $V_{(t-\delta)^n}$, all the points contained in the translate of the standard unit cube by P belong to V_x.

Solution. We fix $\delta > 0$ and choose it appropriately later. We may write the norm form $\alpha^{(1)} \cdots \alpha^{(n)}$ as

$$\sum_{i_1, ..., i_n} a_{i_1, ..., i_n} x_1^{i_1} \cdots x_n^{i_n}$$

where the summation is over all positive integers $i_1, ..., i_n$ such that $i_1 + \cdots + i_n = n$ and the $a_{i_1, ..., i_n}$'s are rational integers. If $P = (u_1, ..., u_n)$, then any point contained in the translate of the standard unit cube by P is of the form $(u_1 + t_1, ..., u_n + t_n)$ with t_i's bounded by 1. Thus, by the solution of Exercise 11.1.1, we deduce that the norm of any such point is

$$\sum_{i_1, ..., i_n} a_{i_1, ..., i_n} u_1^{i_1} \cdots u_n^{i_n} + O(x^{\frac{n-1}{n}}).$$

This has absolute norm

$$\leq (t - \delta)^n + O(x^{\frac{n-1}{n}}) \leq x - n\delta x^{\frac{n-1}{n}} + O(x^{\frac{n-1}{n}}) \leq x$$

if we choose δ sufficiently large so that the negative sign dominates. (This result is important to make the intuitive argument preceding Theorem 11.1.4 rigorous. Indeed, a similar argument shows that there is a $\delta > 0$ so that for any lattice point P contained in V_x, the entire translate of the standard unit cube by P is also contained in $V_{(t+\delta)^n}$. If U is the group generated by the fundamental units, there is a natural action of U on $\mathbb{R}^n$ which preserves the absolute value of the norm form. Then B_x can be described as the set of orbits under this action. That is, $B_x = V_x/U$ so that from the containments established above, we deduce the result stated before Theorem 11.1.4.)

11.2 Distribution of Prime Ideals

Exercise 11.2.1 Show that $L(s,\chi)$ converges absolutely for $\Re(s) > 1$ and that

$$L(s,\chi) = \prod_{\wp}\left(1 - \frac{\chi(\wp)}{N(\wp)^s}\right)^{-1},$$

in this region. Deduce that $L(s,\chi) \neq 0$ for $\Re(s) > 1$.

Solution. We have by multiplicativity of χ,

$$L(s,\chi) = \prod_{\wp}\left(1 - \frac{\chi(\wp)}{N(\wp)^s}\right)^{-1}$$

and the product converges absolutely for $\Re(s) > 1$ if and only if

$$\sum_{\wp} \frac{1}{N(\wp)^s}$$

converges in this region, which is certainly the case as there are only a bounded number of prime ideals above a given prime p. The non-vanishing is also clear.

Exercise 11.2.2 If χ is not the trivial character, show that

$$\sum_{C} \chi(C) = 0$$

where the summation is over the ideal classes C of $\mathcal{H}$.

Solution. If χ is not the trivial character, there is a C_0 such that $\chi(C_0) \neq 1$. Thus,

$$\sum_{C} \chi(CC_0) = \chi(C_0) \sum_{C} \chi(C),$$

where we have used the fact that as C runs over elements of the ideal class group, so does CC_0. The result is now immediate.

Exercise 11.2.3 If C_1 and C_2 are distinct ideal classes, show that

$$\sum_{\chi} \overline{\chi(C_1)}\chi(C_2) = 0.$$

If $C_1 = C_2$, show that the sum is h_K. (This is analogous to Exercise 10.3.5.)

Solution. If $C_1 = C_2$, the result is clear. We consider the sum

$$\sum_{\chi} \chi(A)$$

for $A \neq 1$. We can then take a non-trivial character ψ of the subgroup generated by A and extend this character to the full ideal class group in the usual way. Then,

$$\sum_{\chi} (\chi\psi)(A) = \psi(A) \sum_{\chi} \chi(A)$$

and as before, the result is now evident.

Exercise 11.2.5 Let C be an ideal class of $\mathcal{O}_K$. For $\Re(s) > 1$, show that

$$\sum_{\chi} \overline{\chi}(C) \log L(s,\chi) = h_K \sum_{\wp^m \in C} \frac{1}{mN(\wp)^{ms}}$$

where the first summation is over the characters of the ideal class group and the second summation is over all prime ideals $\wp$ of $\mathcal{O}_K$ and natural numbers m such that $\wp^m \in C$.

Solution. In the left hand side, we insert the series for $\log L(s,\chi)$. By interchanging the summation and using the orthogonality relations established in the Exercise 11.2.3, we obtain the desired result.

Exercise 11.2.6 Show that

$$\sum_{n \geq 2, \wp^m \in C} \frac{1}{mN(\wp)^{ms}}$$

converges for $\Re(s) > 1/2$.

Solution. As noted earlier, the number of prime ideals above a fixed prime p is at most the degree of the number field. Thus, the result is clear from the fact that

$$\sum_{m \geq 2, p} \frac{1}{mp^{ms}}$$

converges for $\Re(s) > 1/2$.

Exercise 11.2.7 If $\chi^2 \neq \chi_0$ show that $L(1,\chi) \neq 0$.

Solution. We use the classical inequality

$$3 + 4\cos\theta + \cos 2\theta \geq 0,$$

as follows. We write

$$\chi(\wp) = e^{i\theta_p}$$

so that for real $\sigma > 1$,

$$\Re\left(3\log\zeta_K(\sigma) + 4\log L(\sigma,\chi) + \log L(\sigma,\chi^2)\right)$$

$$= \sum_{m,\wp} \frac{1}{mN(\wp)^{\sigma m}}\left(3 + 4\cos\theta_p + \cos 2\theta_p\right) \geq 0.$$

Hence,

$$|\zeta_K(\sigma)^3 L(\sigma,\chi)^4 L(\sigma,\chi^2)| \geq 1.$$

If $L(1,\chi) = 0$, the left hand side of this inequality tends to zero as $\sigma \to 1^+$, which is a contradiction.

Exercise 11.2.8 Let C be a fixed ideal class in $\mathcal{O}_K$. Show that the set of prime ideals $\wp \in C$ has Dirichlet density $1/h_K$.

Solution. We have by the orthogonality relation,

$$\sum_{\chi} \overline{\chi(C)} \log L(s,\chi) = h_K \sum_{\wp\in C} \frac{1}{N(\wp)^s} + O(1)$$

as $s \to 1^+$. Since $L(1,\chi) \neq 0$, we may take limits of the left hand side as $s \to 1^+$ and obtain a bounded quantity from the non-trivial characters. Since $L(s,\chi_0) = \zeta_K(s)$, we deduce immediately that

$$\lim_{s\to 1^+} \frac{\sum_{\wp\in C} 1/N(\wp)^s}{\log\zeta_K(s)} = \frac{1}{h_K},$$

as desired.

Exercise 11.2.9 Let m be a natural number and $(a,m) = 1$. Show that the set of primes $p \equiv a (\operatorname{mod} m)$ has Dirichlet density $1/\phi(m)$.

Solution. This is immediate from Exercise 10.3.7.

Exercise 11.2.10 Show that the set of primes p which can be written as $a^2 + 5b^2$ has Dirichlet density $1/4$.

Solution. We have already seen that the class number of $\mathbb{Q}(\sqrt{-5})$ is 2. The set of prime ideals lying in the principal class are of the form $(a + b\sqrt{-5})$ and have norm $a^2 + 5b^2$. By Hecke's theorem, the Dirichlet density of these prime ideals is $1/2$ and taking into account that there are two ideals of norm p in the principal class gives us the final density of $1/4$.

Exercise 11.2.11 Show that if $K = \mathbb{Q}$, the principal ray class group mod m is isomorphic to $(\mathbb{Z}/m\mathbb{Z})^*$.

Solution. The elements of the principal ray class group are the ideals (α) modulo m with a totally positive generator and $(\alpha, m) = 1$. The result is now clear.

11.3 The Chebotarev density theorem

Exercise 11.3.1 Show that action of the Galois group on the set of prime ideals lying above a fixed prime $\mathfrak{p}$ of k is a transitive action.

Solution. Suppose not. Take a prime ideal $\wp$ which is not in the Galois orbit of $\wp_i$ (say) lying above the prime ideal $\mathfrak{p}$. By the Chinese remainder theorem (Theorem 5.3.13), we may find an element $x \in \wp$ and $x-1 \in \sigma(\wp_i)$ for all σ in the Galois group. But then, $N_{K/k}(x)$ is an integer of $\mathcal{O}_k$ which on one hand is divisible by $\wp$ and on the other coprime to $\mathfrak{p}$, a contradiction.

Exercise 11.3.4 By taking $k = \mathbb{Q}$ and $K = \mathbb{Q}(\zeta_m)$, deduce from Chebotarev's theorem the infinitude of primes in a given arithmetic progression $a \pmod{m}$ with $(a, m) = 1$.

Solution. The Galois group consists of automorphisms τ_a satisfying

$$\tau_a(\zeta_m) = \zeta_m^a.$$

Comparing this with the action of the Frobenius automorphism of p, we see that $\sigma_p = \tau_a$ where $p \equiv a \pmod{p}$. By Chebotarev, the Dirichlet density of primes p for which $\sigma_p = \tau_a$ is $1/\phi(m)$.

Exercise 11.3.5 If $k = \mathbb{Q}$ and $K = \mathbb{Q}(\sqrt{D})$, deduce from Chebotarev's theorem that the set of primes p with Legendre symbol $(D/p) = 1$ is $1/2$.

Solution. By Theorem 7.4.2, we see that these are precisely the set of primes which split completely in K and by Chebotarev, the density of such primes is $1/2$.

Exercise 11.3.6 If $f(x) \in \mathbb{Z}[x]$ is an irreducible normal polynomial of degree n (that is, its splitting field has degree n over $\mathbb{Q}$), then show that the set of primes p for which $f(x) \equiv 0 \pmod{p}$ has a solution is of Dirichlet density $1/n$.

Solution. By Theorem 5.5.1 and Exercise 5.5.2, we see that the set of primes p for which $f(x) \equiv 0 \pmod{p}$ has a solution coincides with the set of primes p which split completely in the field obtained by adjoining a root of f. By our assumption, this is a Galois extension of degree n and to say p splits completely is equivalent to saying that $\sigma_p = 1$. By Chebotarev, the Dirichlet density of such primes is $1/n$.

Exercise 11.3.7 If $f(x) \in \mathbb{Z}[x]$ is an irreducible polynomial of degree $n > 1$, show that the set of primes p for which $f(x) \equiv 0 \pmod{p}$ has a solution has Dirichlet density < 1.

Solution. Let K be the splitting field of f over $\mathbb{Q}$ with Galois group G. Let H be the subgroup corresponding to the field obtained by adjoining a root of f to $\mathbb{Q}$. It is not difficult to see that $f(x) \equiv 0 \pmod{p}$ has a solution if and only if the Artin symbol σ_p lies in some conjugate of H. This is a set stable under conjugation. If we take into account that the identity element is common to all the conjugate subgroups of H, we obtain

$$|\cup_{g\in G} gHg^{-1}| \leq [G:H](|H|-1)+1 = |G|+1-[G:H] < |G|$$

if $[G:H] = n > 1$, which is the case.

Exercise 11.3.8 Let q be prime. Show that the set of primes p for which $p \equiv 1 \pmod{q}$ and

$$2^{\frac{p-1}{q}} \equiv 1 (\operatorname{mod} p),$$

has Dirichlet density $1/q(q-1)$.

Solution. The second condition happens if and only if $x^q \equiv 2 \pmod{p}$ has a solution and together with $p \equiv 1 \pmod{p}$, the conditions are equivalent to saying p splits completely in the field $\mathbb{Q}(\zeta_q, 2^{1/q})$. As this field has degree $q(q-1)$, the result now follows from Chebotarev's theorem.

Exercise 11.3.9 If a natural number n is a square mod p for a set of primes p which has Dirichlet density 1, show that n must be a square.

Solution. If n is not a square, the field $\mathbb{Q}(\sqrt{n})$ is quadratic over $\mathbb{Q}$ and by Chebotarev, the density of primes for which n is not a square is $1/2$. (This shows that we can assert the conclusion of the theorem if the set of primes p for which n is a square has density $> 1/2$.)

11.4 Supplementary Problems

Exercise 11.4.1 Let G be a finite group and for each subgroup H of G and each irreducible character ψ of H, define $a_H(\psi, \chi)$ by

$$\operatorname{Ind}_H^G \psi = \sum_{\chi} a_H(\psi, \chi)\chi$$

where the summation is over irreducible characters χ of G. For each χ, let A_χ be the vector $(a_H(\psi, \chi))$ as H varies over all cyclic subgroups of G and ψ varies over all irreducible characters of H. Show that the A_χ's are linearly independent over $\mathbb{Q}$.

Solution. If

$$\sum_{\chi} c_{\chi} A_{\chi} = 0$$

for some integers c_{χ}, then by Frobenius reciprocity, the character

$$\phi = \sum_{\chi} c_{\chi} \chi$$

restricts to the zero character on every cyclic subgroup. By the linear independence of characters (or equivalently, by the orthogonality relations), each c_{χ} is equal to zero.

Exercise 11.4.2 Let G be a finite group with t irreducible characters. By the previous exercise, choose a set of cyclic subgroups H_i and characters ψ_i of H_i so that the $t \times t$ matrix $(a_{H_i}(\psi_i, \chi))$ is non-singular. By inverting this matrix, show that any character χ of G can be written as a rational linear combination of characters of the form $\operatorname{Ind}_{H_i}^G \psi_i$, with H_i cyclic and ψ_i one-dimensional. (This result is usually called *Artin's character theorem* and is weaker than Brauer's induction theorem.)

Solution. Since the row rank of a matrix is equal to the column rank, it is clear that we can choose a set of such H_i's and ψ_i's. Thus,

$$\operatorname{Ind}_{H_i}^G \psi_i = \sum_{\chi} a_{H_i}(\psi_i, \chi)\chi.$$

Moreover,

$$a_H(\psi, \chi) = (\operatorname{Ind}_H^G \psi, \chi)$$

are all non-negative integers. Thus, the inverse matrix consists of rational entries.

Exercise 11.4.3 Deduce from the previous exercise that some positive integer power of the Artin L-function $L(s, \chi; K/k)$ attached to an irreducible character χ admits a meromorphic continuation to $\Re(s) = 1$.

Solution. By the previous exercise, we may write $L(s, \chi; K/k)$ as a product of functions of the form $L(s, \psi_i, K/K^{H_i})^{m_i}$ with m_i's rational numbers. By the Artin reciprocity law, each $L(s, \psi_i, K/K^{H_i})$ coincides with $L(s, \chi_i)$ with χ_i a Hecke character of K^{H_i}. As $L(s, \chi_i)$ has a meromorphic continuation to $\Re(s) = 1$, the result follows.

Exercise 11.4.4 If K/k is a finite Galois extension of algebraic number fields with group G, show that

$$\zeta_K(s) = \prod_{\chi} L(s, \chi; K/k)^{\chi(1)},$$

where the product is over all irreducible characters χ of G.

Solution. Since the right hand side represents the L-function attached to the regular representation, which is $\mathrm{Ind}_1^G 1$, we have that it is equal to

$$L(s, 1, K/K) = \zeta_K(s),$$

by the invariance property under induction of Artin L-series.

Exercise 11.4.5 Fix a complex number $s_0 \in \mathbb{C}$ with $\Re(s_0) \geq 1$ and any finite Galois extension K/k with Galois group G. For each subgroup H of G define the *Heilbronn character* θ_H by

$$\theta_H(g) = \sum_\chi n(H, \chi)\chi(g)$$

where the summation is over all irreducible characters χ of H and $n(H, \chi)$ is the order of $L(s, \chi; K/K^H)$ at $s = s_0$. (By Exercise 11.4.3, the order is a rational number.) Show that $\theta_G|H = \theta_H$.

Solution. We have

$$\theta_G|_H = \sum_\chi n(G, \chi)\chi|_H.$$

But

$$\chi|_H = \sum_\psi (\chi|_H, \psi)\psi$$

where the sum is over irreducible characters ψ of H. Thus,

$$\theta_G|_H = \sum_\psi \left(\sum_\chi n(G, \chi)(\chi|_H, \psi) \right) \psi.$$

By Frobenius reciprocity,

$$(\chi|_H, \psi) = (\chi, \mathrm{Ind}_H^G \psi)$$

so that the inner sum is

$$\sum_\chi n(G, \chi)(\chi, \mathrm{Ind}_H^G \psi).$$

This is equal to $n(H, \psi)$ since

$$L(s, \psi, K/K^H) = L(s, \mathrm{Ind}_H^G \psi, K/k) = \prod_\chi L(s, \chi, K/k)^{(\chi, \mathrm{Ind}_H^G \psi)}.$$

Thus,

$$\theta_G|_H = \sum_\psi n(H, \psi)\psi = \theta_H.$$

Exercise 11.4.6 Show that $\theta_G(1)$ equals the order at $s = s_0$ of the Dedekind zeta function $\zeta_K(s)$.

Solution. We have

$$\theta_G(1) = \sum_\chi n(G,\chi)\chi(1)$$

which by Exercise 11.4.4 is the order of $\zeta_K(s)$ at $s = s_0$.

Exercise 11.4.7 Show that

$$\sum_\chi n(G,\chi)^2 \leq (\mathrm{ord}_{s=s_0}\, \zeta_K(s))^2 .$$

Solution. We compute (θ_G, θ_G) using the orthogonality relations to obtain

$$\sum_\chi n(G,\chi)^2.$$

On the other hand,

$$(\theta_G, \theta_G) = \frac{1}{|G|} \sum_{g \in G} |\theta_G(g)|^2.$$

By Exercise 11.4.5, $\theta_G(g) = \theta_{\langle g \rangle}(g)$. But if H is abelian, $n(H,\psi) \geq 0$ by Artin's reciprocity law and so for $h \in H$,

$$|\theta_H(h)| \leq \sum_\psi n(H,\psi)|\psi(h)|.$$

Thus,

$$|\theta_H(h)| \leq \sum_\psi n(H,\psi) = \theta_H(1),$$

which by Exercise 11.4.6 is the order of the Dedekind zeta function $\zeta_K(s)$ at $s = s_0$. The result is now immediate.

Exercise 11.4.8 For any irreducible non-trivial character χ, deduce that

$$L(s,\chi;K/k)$$

admits an analytic continuation to $s = 1$ and that $L(1,\chi;K/k) \neq 0$.

Solution. Since the Dedekind zeta function has a simple pole at $s = 1$, we see that the previous exercise applied to the point $s_0 = 1$ implies

$$\sum_{\chi \neq 1} n(G,\chi)^2 \leq 0$$

because $n(G,1) = 1$. Hence, $n(G,\chi) = 0$ for any $\chi \neq 1$. By Exercise 11.4.3 we are done.

Exercise 11.4.9 Fix a conjugacy class C in $G = \text{Gal}(K/k)$ and choose $g_C \in C$. Show that

$$\sum_{m, \mathfrak{p}, \sigma_\mathfrak{p}^m \in C} \frac{1}{N(\mathfrak{p})^{ms}} = \frac{|C|}{|G|} \sum_{\chi} \overline{\chi(g_C)} \log L(s, \chi; K/k).$$

Solution. This is an immediate consequence of the orthogonality relations.

Exercise 11.4.10 Show that

$$\lim_{s \to 1+} \frac{\sum_{\mathfrak{p}, \sigma_\mathfrak{p} \in C} 1/N(\mathfrak{p})^s}{\log \zeta_k(s)} = \frac{|C|}{|G|}$$

which is Chebotarev's theorem.

Solution. We take the limit as $s \to 1^+$ in both sides of the equation of the previous exercise. Observe that by Exercise 11.4.8, the limit as $s \to 1^+$ of $\log L(s, \chi; K/k)$ for $\chi \neq 1$ is finite. Since $L(s, 1; K/k) = \zeta_k(s)$, the result is now immediate.

Exercise 11.4.11 Show that $\zeta_K(s)/\zeta_k(s)$ is entire. (This is called the *Brauer-Aramata theorem.*)

Solution. By Exercise 11.4.7, we have

$$n(G, 1)^2 \leq (\text{ord}_{s=s_0} \zeta_K(s))^2.$$

But $n(G, 1) = \text{ord}_{s=s_0} \zeta_k(s)$ and as both $\zeta_K(s)$ and $\zeta_k(s)$ are regular everywhere except at $s = 1$, we deduce that

$$\text{ord}_{s=s_0} \zeta_k(s) \leq \text{ord}_{s=s_0} \zeta_K(s),$$

for $s_0 \neq 1$. But for $s_0 = 1$, this inequality is also true. The result is now immediate. (It is a famous conjecture of Dedekind that if K is an arbitrary extension of k (not necessarily Galois), then $\zeta_K(s)/\zeta_k(s)$ is always entire. This exercise shows that the conjecture is true in the Galois case. The result is also known if K is contained in a solvable extension of k.)

Exercise 11.4.12 (Stark) Let K/k be a finite Galois extension of algebraic number fields. If $\zeta_K(s)$ has a simple zero at $s = s_0$, then $L(s, \chi; K/k)$ is analytic at $s = s_0$ for every irreducible character χ of $\text{Gal}(K/k)$.

Solution. If $\zeta_K(s)$ has a simple zero at $s = s_0$, then

$$\sum_{\chi} n(G, \chi)^2 \leq 1.$$

By the meromorphy of Artin L-series, we have that each $n(G, \chi)$ is an integer. The inequality implies that for at most one χ, we have $|n(G, \chi)| = 1$. If χ is non-abelian, then the factorization of $\zeta_K(s)$ as in Exercise 11.4.4 gives a contradiction for the corresponding L-function introduces a pole or zero of order greater than 1. Hence χ is abelian, but in this case the result is known by Artin reciprocity and Hecke's theorem.

Exercise 11.4.13 (Foote- K. Murty) For any irreducible character χ of $\mathrm{Gal}(K/k)$, show that

$$L(s,\chi,K/k)\zeta_K(s)$$

is analytic for $s \neq 1$.

Solution. This is immediate from the inequality

$$|n(G,\chi)| \leq \mathrm{ord}_{s=s_0}\zeta_K(s)$$

for $s_0 \neq 1$.

Exercise 11.4.14 If K/k is solvable, show that

$$\sum_{\chi\neq 1} n(G,\chi)^2 \leq (\mathrm{ord}_{s=s_0}\zeta_K(s)/\zeta_k(s))^2.$$

Solution. Let $f = \theta_G - n(G,1)1$ and note that

$$(f,f) = \sum_{\chi\neq 1} n(G,\chi)^2.$$

Now by Exercise 11.4.5

$$f(g) = \theta_{\langle g\rangle}(g) - n(G,1) = n(\langle g\rangle,1) - n(G,1) + \sum_{\psi\neq 1} n(\langle g\rangle,\psi)\psi(g).$$

The subfield fixed by $\langle g\rangle$ is a subfield of K and we know Dedekind's conjecture for this extension. Thus,

$$n(\langle g\rangle,1) - n(G,1) \geq 0.$$

Therefore,

$$|f(g)| \leq n(\langle g\rangle,1) - n(G,1) + \sum_{\psi\neq 1} n(\langle g\rangle,\psi) = \mathrm{ord}_{s=s_0}(\zeta_K(s)/\zeta_k(s)),$$

from which the inequality follows.

Bibliography

[B] Bell, E.T. *Men of Mathematics.* Simon and Schuster, New York, 1937.

[BL] Bateman, P. and Low, M.E. Prime numbers in arithmetic progression with difference 24. *Amer. Math. Monthly* **72** (1965), 139–143.

[Er] Erdös, P. Problems and results on consecutive integers. *Eureka* **38** (1975/6), 3–8.

[FM] R. Foote and V. Kumar Murty, Zeros and Poles of Artin L-series, *Math. Proc. Camb. Phil. Soc.,* **105** (1989), 5-11.

[Ga] Gauss, C.F. *Disquisitiones Arithmeticae.* Springer-Verlag, New York, 1986. (Translated by Arthur A. Clarke, revised by William C. Waterhouse.)

[Ha] Harper, M., $\mathbb{Z}[\sqrt{14}]$ is Euclidean, *Canadian Journal of Mathematics,* **56**(1), (2004), 55–70.

[He] Hecke, E., *Lectures on the Theory of Algebraic Numbers,* GTM 77, Springer-Verlag, 1981.

[HM] Harper, M., and Murty, R., Euclidean rings of algebraic integers, *Canadian Journal of Mathematics,* **56**(1), (2004), 71-76.

[IR2] Ireland, K. and Rosen, M. *A Classical Introduction to Modern Number Theory,* 2nd Ed. Springer-Verlag, New York, 1990.

[La] Lang, S., *Algebraic Number Theory,* Springer-Verlag, 1986.

[L-K] Loo-Keng, Hua, *Introduction to Number Theory.* Springer-Verlag, Berlin, 1982.

[Ma] Marcus, D. *Number Fields.* Springer-Verlag, New York, 1977.

[Mat] Matsumura, H., *Commutative Ring Theory,* Cambridge University Press, 1989.

[Mu] Murty, R. Primes in certain arithmetic progressions. *J. Madras University*, Section B **51** (1988), 161–169.

[Mu2] Murty, R., *Problems in Analytic Number Theory*, GTM/RIM 206, Springer-Verlag, 2001.

[Na] Narasimhan, R. et al. *Algebraic Number Theory.* Commercial Printing Press, Bombay.

[N] Narkiewicz, W., *Elementary and analytic theory of algebraic numbers*, Second edition, Springer-Verlag, Berlin, PWN - Polish Scientific Publishers, Warsaw, 1990.

[Ro] Rosen, K. *Elementary Number Theory and its Applications.* Addison-Wesley, Murray Hill, 1988.

[S] Schur, I. Über die Existenz unendlich vieler Primzahlen in einigen speziellen arithmetischen Progressionen. *S-B Berlin Math. Ges.*, **11** (1912), 40–50, appendix to *Archiv der Math. und Phys.* (3), vol. 20 (1912–1913). (See also his *Collected Papers*, Springer-Verlag, New York, 1973.)

[Sc] Schoof, R. and Corrales-Rodrigáñez, C. The support problem and its elliptic analogue. *J. Number Theory* **64** (1997), no. 2, 276–290.

[Se] J.-P. Serre, Linear Representations of Finite Groups, GTM 42, Springer-Verlag, 1977.

[Sil] Silverman, J. Wieferich's criterion and the *ABC* conjecture. *J. Number Theory* **30** (1988), no. 2, 226–237.

[St] Stewart, I. and Tall, D. *Algebraic Number Theory.* Chapman and Hall, London, 1979.

Index

(continued from page ii)

64 EDWARDS. Fourier Series. Vol. I. 2nd ed.
65 WELLS. Differential Analysis on Complex Manifolds. 2nd ed.
66 WATERHOUSE. Introduction to Affine Group Schemes.
67 SERRE. Local Fields.
68 WEIDMANN. Linear Operators in Hilbert Spaces.
69 LANG. Cyclotomic Fields II.
70 MASSEY. Singular Homology Theory.
71 FARKAS/KRA. Riemann Surfaces. 2nd ed.
72 STILLWELL. Classical Topology and Combinatorial Group Theory. 2nd ed.
73 HUNGERFORD. Algebra.
74 DAVENPORT. Multiplicative Number Theory. 3rd ed.
75 HOCHSCHILD. Basic Theory of Algebraic Groups and Lie Algebras.
76 IITAKA. Algebraic Geometry.
77 HECKE. Lectures on the Theory of Algebraic Numbers.
78 BURRIS/SANKAPPANAVAR. A Course in Universal Algebra.
79 WALTERS. An Introduction to Ergodic Theory.
80 ROBINSON. A Course in the Theory of Groups. 2nd ed.
81 FORSTER. Lectures on Riemann Surfaces.
82 BOTT/TU. Differential Forms in Algebraic Topology.
83 WASHINGTON. Introduction to Cyclotomic Fields. 2nd ed.
84 IRELAND/ROSEN. A Classical Introduction to Modern Number Theory. 2nd ed.
85 EDWARDS. Fourier Series. Vol. II. 2nd ed.
86 VAN LINT. Introduction to Coding Theory. 2nd ed.
87 BROWN. Cohomology of Groups.
88 PIERCE. Associative Algebras.
89 LANG. Introduction to Algebraic and Abelian Functions. 2nd ed.
90 BRØNDSTED. An Introduction to Convex Polytopes.
91 BEARDON. On the Geometry of Discrete Groups.
92 DIESTEL. Sequences and Series in Banach Spaces.
93 DUBROVIN/FOMENKO/NOVIKOV. Modern Geometry—Methods and Applications. Part I. 2nd ed.
94 WARNER. Foundations of Differentiable Manifolds and Lie Groups.
95 SHIRYAEV. Probability. 2nd ed.
96 CONWAY. A Course in Functional Analysis. 2nd ed.
97 KOBLITZ. Introduction to Elliptic Curves and Modular Forms. 2nd ed.
98 BRÖCKER/TOM DIECK. Representations of Compact Lie Groups.
99 GROVE/BENSON. Finite Reflection Groups 2nd ed.
100 BERG/CHRISTENSEN/RESSEL. Harmonic Analysis on Semigroups: Theory of Positive Definite and Related Functions.
101 EDWARDS. Galois Theory.
102 VARADARAJAN. Lie Groups, Lie Algebras and Their Representations.
103 LANG. Complex Analysis. 3rd ed.
104 DUBROVIN/FOMENKO/NOVIKOV. Modern Geometry—Methods and Applications. Part II.
105 LANG. $SL_2(\mathbf{R})$.
106 SILVERMAN. The Arithmetic of Elliptic Curves.
107 OLVER. Applications of Lie Groups to Differential Equations. 2nd ed.
108 RANGE. Holomorphic Functions and Integral Representations in Several Complex Variables.
109 LEHTO. Univalent Functions and Teichmüller Spaces.
110 LANG. Algebraic Number Theory.
111 HUSEMÖLLER. Elliptic Curves. 2nd ed.
112 LANG. Elliptic Functions.
113 KARATZAS/SHREVE. Brownian Motion an Stochastic Calculus. 2nd ed.
114 KOBLITZ. A Course in Number Theory an Cryptography. 2nd ed.
115 BERGER/GOSTIAUX. Differential Geometr Manifolds, Curves, and Surfaces.
116 KELLEY/SRINIVASAN. Measure and Integral. Vol. I.
117 J.-P. SERRE. Algebraic Groups and Class Fields.
118 PEDERSEN. Analysis Now.
119 ROTMAN. An Introduction to Algebraic Topology.
120 ZIEMER. Weakly Differentiable Functions Sobolev Spaces and Functions of Bound Variation.
121 LANG. Cyclotomic Fields I and II. Combined 2nd ed.
122 REMMERT. Theory of Complex Functions *Readings in Mathematics*
123 EBBINGHAUS/HERMES et al. Numbers. *Readings in Mathematics*

124 DUBROVIN/FOMENKO/NOVIKOV. Modern Geometry—Methods and Applications. Part III
125 BERENSTEIN/GAY. Complex Variables: An Introduction.
126 BOREL. Linear Algebraic Groups. 2nd ed.
127 MASSEY. A Basic Course in Algebraic Topology.
128 RAUCH. Partial Differential Equations.
129 FULTON/HARRIS. Representation Theory: A First Course.
Readings in Mathematics
130 DODSON/POSTON. Tensor Geometry.
131 LAM. A First Course in Noncommutative Rings. 2nd ed.
132 BEARDON. Iteration of Rational Functions.
133 HARRIS. Algebraic Geometry: A First Course.
134 ROMAN. Coding and Information Theory.
135 ROMAN. Advanced Linear Algebra.
136 ADKINS/WEINTRAUB. Algebra: An Approach via Module Theory.
137 AXLER/BOURDON/RAMEY. Harmonic Function Theory. 2nd ed.
138 COHEN. A Course in Computational Algebraic Number Theory.
139 BREDON. Topology and Geometry.
140 AUBIN. Optima and Equilibria. An Introduction to Nonlinear Analysis.
141 BECKER/WEISPFENNING/KREDEL. Gröbner Bases. A Computational Approach to Commutative Algebra.
142 LANG. Real and Functional Analysis. 3rd ed.
143 DOOB. Measure Theory.
144 DENNIS/FARB. Noncommutative Algebra.
145 VICK. Homology Theory. An Introduction to Algebraic Topology. 2nd ed.
146 BRIDGES. Computability: A Mathematical Sketchbook.
147 ROSENBERG. Algebraic K-Theory and Its Applications.
148 ROTMAN. An Introduction to the Theory of Groups. 4th ed.
149 RATCLIFFE. Foundations of Hyperbolic Manifolds.
150 EISENBUD. Commutative Algebra with a View Toward Algebraic Geometry.
151 SILVERMAN. Advanced Topics in the Arithmetic of Elliptic Curves.
152 ZIEGLER. Lectures on Polytopes.
153 FULTON. Algebraic Topology: A First Course.
154 BROWN/PEARCY. An Introduction to Analysis.
155 KASSEL. Quantum Groups.
156 KECHRIS. Classical Descriptive Set Theory.
157 MALLIAVIN. Integration and Probability.
158 ROMAN. Field Theory.
159 CONWAY. Functions of One Complex Variable II.
160 LANG. Differential and Riemannian Manifolds.
161 BORWEIN/ERDÉLYI. Polynomials and Polynomial Inequalities.
162 ALPERIN/BELL. Groups and Representations.
163 DIXON/MORTIMER. Permutation Groups.
164 NATHANSON. Additive Number Theory: The Classical Bases.
165 NATHANSON. Additive Number Theory: Inverse Problems and the Geometry of Sumsets.
166 SHARPE. Differential Geometry: Cartan's Generalization of Klein's Erlangen Program.
167 MORANDI. Field and Galois Theory.
168 EWALD. Combinatorial Convexity and Algebraic Geometry.
169 BHATIA. Matrix Analysis.
170 BREDON. Sheaf Theory. 2nd ed.
171 PETERSEN. Riemannian Geometry.
172 REMMERT. Classical Topics in Complex Function Theory.
173 DIESTEL. Graph Theory. 2nd ed.
174 BRIDGES. Foundations of Real and Abstract Analysis.
175 LICKORISH. An Introduction to Knot Theory.
176 LEE. Riemannian Manifolds.
177 NEWMAN. Analytic Number Theory.
178 CLARKE/LEDYAEV/STERN/WOLENSKI. Nonsmooth Analysis and Control Theory.
179 DOUGLAS. Banach Algebra Techniques in Operator Theory. 2nd ed.
180 SRIVASTAVA. A Course on Borel Sets.
181 KRESS. Numerical Analysis.
182 WALTER. Ordinary Differential Equations.

183 MEGGINSON. An Introduction to Banach Space Theory.
184 BOLLOBAS. Modern Graph Theory.
185 COX/LITTLE/O'SHEA. Using Algebraic Geometry.
186 RAMAKRISHNAN/VALENZA. Fourier Analysis on Number Fields.
187 HARRIS/MORRISON. Moduli of Curves.
188 GOLDBLATT. Lectures on the Hyperreals: An Introduction to Nonstandard Analysis.
189 LAM. Lectures on Modules and Rings.
190 ESMONDE/MURTY. Problems in Algebraic Number Theory. 2nd ed.
191 LANG. Fundamentals of Differential Geometry.
192 HIRSCH/LACOMBE. Elements of Functional Analysis.
193 COHEN. Advanced Topics in Computational Number Theory.
194 ENGEL/NAGEL. One-Parameter Semigroups for Linear Evolution Equations.
195 NATHANSON. Elementary Methods in Number Theory.
196 OSBORNE. Basic Homological Algebra.
197 EISENBUD/HARRIS. The Geometry of Schemes.
198 ROBERT. A Course in p-adic Analysis.
199 HEDENMALM/KORENBLUM/ZHU. Theory of Bergman Spaces.
200 BAO/CHERN/SHEN. An Introduction to Riemann–Finsler Geometry.
201 HINDRY/SILVERMAN. Diophantine Geometry: An Introduction.
202 LEE. Introduction to Topological Manifolds.
203 SAGAN. The Symmetric Group: Representations, Combinatorial Algorithms, and Symmetric Functions.
204 ESCOFIER. Galois Theory.
205 FÉLIX/HALPERIN/THOMAS. Rational Homotopy Theory. 2nd ed.
206 MURTY. Problems in Analytic Number Theory.
Readings in Mathematics
207 GODSIL/ROYLE. Algebraic Graph Theory.
208 CHENEY. Analysis for Applied Mathematics.
209 ARVESON. A Short Course on Spectral Theory.
210 ROSEN. Number Theory in Function Fields.
211 LANG. Algebra. Revised 3rd ed.
212 MATOUŠEK. Lectures on Discrete Geometry.
213 FRITZSCHE/GRAUERT. From Holomorphic Functions to Complex Manifolds.
214 JOST. Partial Differential Equations.
215 GOLDSCHMIDT. Algebraic Functions and Projective Curves.
216 D. SERRE. Matrices: Theory and Applications.
217 MARKER. Model Theory: An Introduction.
218 LEE. Introduction to Smooth Manifolds.
219 MACLACHLAN/REID. The Arithmetic of Hyperbolic 3-Manifolds.
220 NESTRUEV. Smooth Manifolds and Observables.
221 GRÜNBAUM. Convex Polytopes. 2nd ed.
222 HALL. Lie Groups, Lie Algebras, and Representations: An Elementary Introduction.
223 VRETBLAD. Fourier Analysis and Its Applications.
224 WALSCHAP. Metric Structures in Differential Geometry.
225 BUMP: Lie Groups

9 780387 221823